GLOBAL ENVIRONMENTAL POLICY

Concepts, Principles, and Practice

GLOBAL ENVIRONMENTAL POLICY

Concepts, Principles, and Practice

Charles H. Eccleston
Frederic March

CRC Press
Taylor & Francis Group
Boca Raton London New York

CRC Press is an imprint of the
Taylor & Francis Group, an **informa** business

CRC Press
Taylor & Francis Group
6000 Broken Sound Parkway NW, Suite 300
Boca Raton, FL 33487-2742

First issued in paperback 2017

© 2011 by Taylor and Francis Group, LLC
CRC Press is an imprint of Taylor & Francis Group, an Informa business

No claim to original U.S. Government works

ISBN-13: 978-1-4398-4766-4 (hbk)
ISBN-13: 978-1-138-11625-2 (pbk)

Library of Congress Cataloging-in-Publication Data

Eccleston, Charles H.
  Global environmental policy : concepts, principles, and practice / Charles H. Eccleston and Frederic March.
       p. cm.
    Includes bibliographical references and index.
    ISBN 978-1-4398-4766-4 (alk. paper)
    1. Environmental policy--International cooperation. I. Title.

GE170.E24 2010
333.72--dc22   2010026021

Visit the Taylor & Francis Web site at
http://www.taylorandfrancis.com

and the CRC Press Web site at
http://www.crcpress.com

Dedication

*This book is dedicated to Professor Lynton K. Caldwell,
the father of modern environmental policy.*

(Photograph courtesy of Wendy Read.)

Contents

Section three: Environmental ethics and economics

Preface

It's hard for the modern generation to understand
Thoreau, who lived beside a pond but didn't own
water skis or a snorkel.

Loudon Wainwright

Environmental policy is often practiced reactively—each crisis is addressed
as an isolated event. Random destructive events of nature, technology, war,
and terrorism are common occurrences, and emergencies must indeed be
dealt with as they occur. Hence, this book includes a focus on policies that
would minimize their probability and the damage done when they occur.
Our principal policy concerns are the cumulative environmental impacts
of over 6 billion people on the globe, people whose very lifestyles have
come to threaten the well-being of humanity as a whole. As explained in
this book, the issues we are confronted with today are very complex and
potentially catastrophic, ranging from threats as diverse as water scarcity, topsoil erosion, depleted fisheries, loss of biodiversity, deforestation,
nanotechnology, genetically modified organisms, exhaustion of natural
resources, and climate change.

The book aims to offer the reader a balanced skill set of concepts,
principles, and practices for developing and implementing environmental
policy solutions. We describe the global system of environmental policy.
We introduce the principal theories employed in analyzing, planning,
and implementing programs and projects. Several case studies and many
examples are provided that illustrate both the power and the limitations
of theoretical approaches. After the Introduction, we have divided the
book into four parts to organize the large variety of concepts, principles,
policies, practices, and issues in a logical, yet connected, manner that is
reader friendly.

The Introduction defines the nature and scope of the environmental
policy problem, outlines its origins and evolution, and introduces the policy
"frameworks" of the United Nations, the European Union, and the United
States. It also presents an in-depth discussion of the concept of the

"integrated interdisciplinary approach," which characterizes environmental policy as a process and defines the environmental policy framework.

Section one, "Introduction to Global Environmental Concepts and Principles," describes the historical role of the environmental movement and the unifying principles and concepts of environmental policy, and defines the nature and scope of the environmental policy problem. It also develops the concept of "sustainability" in depth and describes the role of international treaties in the United Nations framework for global environmental policy.

Section two, "Sustainability, Environmental Impact Assessment, and Decision Making," examines the environmental impact assessment (EIA) process and its importance in managing environmental policy. It reviews decision-making approaches including the roles of human subjectivity and bias along with objective analysis and models. Further, it discusses the role that environmental management systems play in implementing environmental policies.

Section three, "Environmental Ethics and Economics," discusses the underlying environmental, ethical, and economic issues that govern virtually every aspect of environmental policy formulation and implementation. Ethics and economics are closely related because they are both essential to understanding who benefits from the policy and who pays the price (monetary, property, health, and environmental) for its adverse effects. The issue of costs external to the investment decision maker is a key element of both economic and ethical assessments.

Section four, "Critical Global Environmental Issues," discusses how the world as a whole is rapidly approaching or is already experiencing critical environmental stresses in the following areas: water resources, food supply, fossil energy, global climate, and human population growth. Some aspects of these issues have already reached crisis levels in a number of regions, and left unchecked could ultimately threaten the quality of human life on a global scale. Each chapter critically examines these issues, describes how policy organizations are currently dealing with them, discusses the limitations of current policies, and identifies the need for new and improved global-scale policies to reverse the adverse trends.

Each chapter begins with a case study and ends with a problem set. Most of the questions are designed to elicit critical thinking. The book ends with two capstone problems. These problems involve consideration of nearly every major topic and aspect presented in this book. Upon completion, students should be able to take a realistic policy problem, evaluate it in terms of the concepts, principles, and tools described throughout the book, and develop a pragmatic policy solution to that problem. To this end, the capstone problems are designed to test a student's competency in developing an integrated, systematic, and comprehensive policy approach to complex environmental problems at both a local or regional level (Epilogue, capstone problem 1) as well as the national/international context (Epilogue, capstone problem 2).

Audience

It is our sincere hope that this foundation will facilitate lifelong learning from the experience of working in the many aspects of environmental policy. The expertise we seek to develop is that of the integrator, who can relate his or her own special professional training to the entire environmental policy context. Curricular areas of relevance include the following:

Environmental curricula: Environmental science, engineering, planning, and policy students who require an interdisciplinary perspective on society's environmental problems and their solutions

Other science and engineering curricula: Biology, ecology, premed, chemistry, meteorology, oceanography, and various engineering disciplines whose students are likely to encounter environmental issues and problems in their professional careers

Social science curricula: Geography, economics, political science, law, and behavioral sciences (psychology, sociology, and anthropology) students who may become involved in various aspects of environmental policy

Liberal arts curricula: Especially literature, history, and philosophy students who may become writers, journalists, and filmmakers seeking to understand people's lives and experiences, and the environmental conditions that shape character, attitudes, and society

Professional persons currently involved in environmental issues: People with any of the above academic backgrounds who are involved with environmental issues as public- and private-sector leaders and policymakers, as managers of programs and projects, as university and private (think tank) organization researchers, and as environmental policy advocates in a variety of organizations

Private citizens: People who wish to advocate on behalf of environmental issues or become involved as private- or public-sector volunteers, or simply to become more informed citizens*

Acknowledgments

We are indebted to the following reviewers and greatly appreciate their many constructive comments to our book as a work in progress:

---

* The views expressed by the authors are solely their own and do not represent those of any professional organization or employer, past or present.

Dr. Roberta Sonnino, Lecturer in Environmental Policy and Planning at Cardiff University in Wales. We are especially indebted to Dr. Sonnino for her comprehensive review and generous support of our effort.

Timothy Cohen, Senior Environmental Economist at Sandia National Laboratories.

M. Ron Deverman, President of the National Association of Environmental Professionals and Principal Environmental Planning Manager for HNTB, a multidisciplinary engineering company.

David Keys, National Environmental Policy Act Coordinator for the U.S. National Oceanographic and Atmospheric Administration Fisheries Service – Southeast Region.

Andrew J. March, BS Environmental Engineering, University of Colorado, who was a student intern at the National Center for Atmospheric Research, and is currently a post-graduate student in energy systems at the University of Orleans, France.

Authors

Charles H. Eccleston is an author, trainer, project manager, and environmental policy consultant. He is recognized in *Marquis' Who's Who in America* and *Who's Who in the World* for his innovative environmental policy, planning, and National Environmental Policy Act (NEPA) achievements. With nearly 20 years of experience, he has managed and prepared numerous environmental and energy analyses and policy studies. Eccleston is the author of 60 professional papers and four other books.

He is an elected director to the National Association of Environmental Professional's (NAEP) Board of Directors. Recently, he received a national award from NAEP for Outstanding Environmental Leadership.

Currently, he serves as an elected representative to the International Organization for Standardization's 242 working group, responsible for developing an ISO 50001 Energy Management System (EnMS) standard that will be used worldwide to manage generation/use of energy.

He has served on two White House–sponsored environmental policy taskforces, including a taskforce chartered with improving the U.S. NEPA process. As part of this effort, he proposed, developed, and published the original concept for integrating NEPA with an ISO 14001 environmental management system, which has now been adopted by a number of agencies around the world. He later generalized this approach to incorporate any international environmental impact assessment process. Still later, he generalized this integrated process to incorporate the sustainable development.

Eccleston is fluent on a wide range of environmental and energy policy issues such as sustainability, climate change, water and food scarcity, radioactive/hazardous waste, peak oil, population, environmental economics, energy generation. His energy-related experience includes investigating nuclear, gas-fired, and coal-fired plants and renewable systems.

Eccleston has traveled extensively throughout Europe, the Middle East, and particularly Asia. Many of the concepts described in this global environmental policy book are an outgrowth of research he conducted on his travels. His graduate degree is in environmental geology. He can be reached at ecclestonc@msn.com or env_planning@msn.com.

Frederic March has managed projects involving environmental impact assessments (EIA) and policy studies in Latin America, Africa, and the Middle East, commissioned by the World Bank, the Inter-American Development Bank, and the United Nations. In his work, he has actively promoted sustainability development in economic policy development.

In Colombia, he managed the North Coast Tourism Infrastructure Study, comparing environmental impacts and policy implications in three candidate locations. He analyzed environmental policy options in Mexico under Plan Acapulco. In Egypt, he developed an industrial energy pilot program involving renewable and conservation technologies. In Brazil, he was co-manager for developing the Taquari River Basin Infrastructure and Policy Development Plan encompassing, which used a cost–benefit environmental analysis to evaluate water resources, transportation, energy, and public health options. He has also worked on projects in Lesotho, Somalia, India, Turkey, Barbados, Venezuela, and Guatemala.

March has held senior staff positions with the National Academy of Sciences and Sandia National Laboratories. He served on a White House team for national energy independence as leader of its water resources impact team. He has managed numerous U.S. EISs and EIAs. March is a former chairman of the NEPA Working Group of the National Association of Environmental Professionals and authored two books on the U.S. NEPA process published numerous papers. He graduated from MIT with graduate degrees in engineering systems analysis and water resources.

Introduction

Ignore the environment. It'll go away.

Seen on a bumper sticker

Case study

A sign erected in New Orleans in May of 2010 read, "The good news—mermaids are real. The bad news—they're now extinct." On April 20, 2010, a blowout caused a catastrophic explosion on the Deepwater Horizon offshore oil drilling platform, located 64 kilometers southeast of the Louisiana coast. British Petroleum (BP) was the operator and principal developer of the project. The explosion killed 11 platform workers and injured 17 others. Survivors described the incident as a sudden explosion that gave them less than 5 minutes to escape when the alarm went off. Some survivors have testified that a bubble of methane gas escaped from the well and shot up the drill column, expanding quickly as it burst through several seals and barriers before exploding. Recent reports suggest that BP engineers were aware of variations in pressure earlier that day, which were clues that gas pressure could be rising—a precursor to an explosion. When technicians on the rig tried to activate the well's blowout preventer, it failed.

The deepwater oil gusher is more than 1500 meters below the surface of the ocean. According to a *CBS News* report on July 15, 2010, the oil leak had finally been capped after 87 days, but additional measures remained to ensure that leakage would not resume. According to a *New York Times* report on August 2, 2010, about 5 million barrels have been released to the Gulf of Mexico since the leak started on April 20, 2010. This is the largest single release to the environment in the history of oil drilling.

According to an *Associated Press* report on May 1, 2010, about 3850 square miles (9972 square kilometers) had been covered by the resulting oil slick. Investigators have also discovered immense underwater plumes of oil invisible from the surface. Scientists fear that factors such as oil toxicity and oxygen depletion could result in one of the largest environmental disasters ever recorded in the United States. The Deepwater Horizon oil spill (also called the BP oil spill) has already seriously damaged the Gulf of Mexico fishing industry, tourism industry, and the habitat of hundreds of bird and other species. As of August 15, 2010, there were reports that much of the surface oil had dispersed and degraded and that some fishing had resumed. According to a *New York Times* report on July 7, 2010, there was relatively little detectable surface oil. However, there is considerable uncertainty about the long-term environmental threats from the vast plumes of oil beneath the surface.

In applying for a drilling permit, oil giant BP filed a 52-page exploration and environmental impact plan in 2009 with the former U.S. Minerals Management Service (MMS). The plan stated that it was "unlikely that an accidental surface or subsurface oil spill would occur from the proposed activities." This plan assured MMS that any spill would not seriously hurt marine wildlife and that "due to the distance to shore (48 miles) and the response capabilities that would be implemented, no significant adverse impacts are expected."

Pursuant to the National Environmental Policy Act (NEPA), the MMS issued two environmental impact statements (EIS) for regional lease sales in 2007. In 2009, the Obama administration granted BP a special exemption, known as a "categorical exclusion," from a requirement to prepare an environmental impact statement (EIS) to drill the actual exploratory well. A categorical exclusion can be applied only to activities that clearly have no significant impact. Even under the best of circumstances, such an exemption would be uncalled for. A properly prepared EIS would have considered specific mitigation measures like spill control and shut-off technology. Such an analysis might have made all the difference in the world. A memo sent by Jane Lubchenco, administrator of the National Oceanic and Atmospheric Administration (NOAA), to the Department of the Interior in October 2009 sharply criticized the very studies that had been used to approve the Deepwater Horizon site. Shortly before the accident, the President's Council on Environmental Quality (CEQ), which oversees NEPA, believed that categorical exclusions for offshore oil drilling were granted too readily. In February 2010, prior to the oil spill, the CEQ informed agencies that "they need to review how we're issuing categorical exclusions."

Oil drilling regulators in many other parts of the world often require acoustically activated triggers on offshore platforms to shut down the system in case of an emergency. While the wellhead on the Deepwater Horizon had been fitted with a blowout preventer, it was not fitted with remote control or acoustically activated triggers. MMS considered requiring such remote devices, but a report commissioned by the agency questioned its cost and effectiveness. In 2003, the agency determined that the device would not be required because the drilling rig had other backup systems to shut off the well.

There may have been other reasons that drilling the actual well was exempted from preparation of an EIS and that adequate mitigation equipment was not mandated to be installed. An inspector general's report finalized in 2008 revealed that a few MMS regulators had for years accepted gifts and money from the same oil industry executives they were supposedly tasked with regulating. This report found that a cozy relationship existed between the regulators and the oil industry, which had led to an "ethical failure." A letter written by acting Inspector General Mary Kendall in 2010 stated that "we discovered that the individuals involved in the fraternizing and gift exchange…have often known one another since childhood."

A report by the Center for Responsive Politics states that BP gave more campaign donations to the Obama campaign in the 2008 presidential election than to any other politician. Representative Vern Buchanan called on the House Committee on Government Oversight and Reform to investigate why the administration gave BP a categorical exclusion from preparing an EIS.

One policymaker stated that the responsibility for this catastrophe ran all the way from the bottom to the top-ranking levels. Weeks after the disaster, as oil continued to pour out, President Obama, frustrated by the criticism his administration had received for failing to stop the leak, reportedly ordered his aides to find a way to "plug the damn hole." Obama then moved to appoint an unaccountable commission to study the problem. Critics in New Orleans countered that such a commission merely shifted responsibility from the regulators and elected officials who, at a time when leadership is needed most, should be in charge of spearheading a solution, to unelected bureaucrats. On May 19, 2010, the Department of Interior issued Secretary Order 3299 to establish the Bureau of Ocean Energy Management to replace the Minerals Management Service. On June 29, 2010, the new organization was established as the Bureau of Ocean Energy Management, Regulation and Enforcement (BOMRE). Its Web site contains a site called "Questions and Answers: The Next Five-Year OCS Oil and Gas Leasing Program (2012–2017)" that describes its intent to ensure full compliance with the National Environmental Policy Act, which in turn requires that all other applicable federal and state environmental and safety laws and regulations be addressed. Information regarding federal investigations of MMS wrongdoing was not available on the site at the time of access. A *Wall Street Journal* report of June 10, 2010, briefly describes a Department of Interior Inspector General's report to a congressional panel.

Problem: After reading the chapter, return to this problem. Consider this case study in light of what you have read concerning the U.S. National Environmental Policy Act. In terms of environmental policy, what can this accident teach us? Had you been in charge of granting the permit for the Deepwater Horizon project, what measures would you have prescribed to prevent or mitigate

the impacts of such an accident? If you were in an important policymaking forum, what national or international policy measures
would you recommend for reducing the long-term risks from offshore drilling? As you will learn in Chapter 8, environmental ethics
is a critical attribute affecting environmental policymaking. What
changes would you require in terms of ethical conduct by regulators? To what degree may politics have clouded judgment and influenced decisions that were made?

Learning objectives

After you have read this, you should understand the following:

- The origins and evolution of threats to the global environment
- The nature of the problem for today's policy community
- The interdisciplinary nature of environmental policy
- The importance of the U.S. National Environmental Policy Act
  (NEPA) in creating a systematic, interdisciplinary planning process
  for mitigating the impact of human activities that threaten envi-
  ronmental quality—especially conducting environmental impact
  assessments
- The environmental policy frameworks of the United States, the
  European Union (EU), and the United Nations (UN)
- Your potential personal role as a future or current professional in the
  system of environmental policy functions

Pharaohs, pyramids, and policy

The authors have had the opportunity to visit and work in Egypt and to discuss our impressions of it, especially regarding the country's environmental
challenges. Egypt is a land of marked contrasts. It is a society endowed with
a rich and fascinating culture. But it is also an ancient society in the midst of
a chaotic transition—a blend of the ancient with the modern. The country
now boasts a population of over 80 million. With a population of 20 million,
Cairo is a congested, tangled mess of a city. In a single view, one can see
donkeys pulling carts, herders riding camels, jet planes landing at the international airport, and a maze of cars driving bumper-to-bumper. Although
the city boasts a population of 20 million, from an engineering standpoint, it
is designed to support perhaps a fifth of that population. Each year, 300,000
cars are added to the road system. The city is slowly grinding to a halt.

As one of the world's most polluted cities, the air pollution is so thick
that on some clear nights we were hard-pressed to make out the outline
of the pyramids, although our hotel was only a few kilometers away. The
traffic is so congested that it sometimes moves at the rate of 10 kilometers

per hour down a modern freeway. The substance of the nation, the Nile River, is a floating sewer. Horns honk incessantly as the traffic stalls to a near standstill. Clearly, this is a city in deep trouble and one that is headed for major environmental problems. Dilapidated, untuned cars belch out thick funnels of smoke. Virtually no one obeys the traffic controls, be they police, lights, or signs, as cars snake through the streets like a jumbled blob of thick spaghetti. Piles of trash are heaped on neighborhood streets, as discarded garbage rots under the hot sun. In some places, raw sewage drains into open canals. The regional water system is highly polluted.

Although the government has the Ministry of the Environment, there is little evidence of its effectiveness in Cairo. From local taxi drivers, to the police guarding the pyramids, to government officials, virtually everyone a visitor meets is on the take. Egypt, along with countless other developing nations, is on a collision course between industrialization and its urban environment. It is an important lesson for those who seek to instill environmental change. This situation is hardly unique to Egypt, but with wide variations Egypt characterizes much of the developing world, and to a lesser extent the developed world.

Cairo, along with countless other cities around the world, is on a collision course with industrialization and its urban environment. The UN's population and environmental programs that you will read about in this book are designed to guide developing and developed nations alike in improving the quality of life for their people. In the authors' opinions, Egypt is less a problem than a potential opportunity. There are vast opportunities to instill environmental ethos among the world's populace. But significant progress will not be made until pragmatic environmental polices are established. The principles and concepts laid out in the following chapters provide tools for forging realistic policies that will lead nations through this environmental transition.

Origins and scope of the problem

Human beings, as we know them, have lived on Earth for a mere speck of geologic time, perhaps less than 100,000 years or so. Over the first nine-tenths of their existence during the Old Stone Age (Paleolithic period), they were nomads living in small hunter–gatherer communities. We might envision them as a somewhat less debonair version of Fred Flintstone. They lived in crude shelters, gathered berries and roots, fished, and occasionally hunted (or were chased by) large game, including wild bison and mammoths. Their environmental footprint was so small that it took the concerted effort of trained archaeologists decades to uncover even rudimentary evidence of their encampments.

About 10,000 years ago, in the Middle Stone Age (Mesolithic period), an agricultural revolution marked a transition from nomadic life to agricultural communities. Forests were cleared, and a newly discovered technology, slash-and-burn clearance, opened large tracts of land for cultivation. Life was clearly easier than it was for their nomadic ancestors, and food became more abundant. While small in comparison to today's massive environmental footprint, our ancestors' impact on their environment was already becoming noticeable and was destined to accelerate. New evidence suggests that by 8000 years ago, ancestral forest clearing caused levels of greenhouse gas emissions to increase significantly and contributed to a warming of the Earth's climate.[1] The resulting buildup of carbon dioxide (CO_2) may have prevented another glacial period.

More recently, the world began transitioning into an industrial revolution. The middle of the eighteenth century ushered in a shift from a manual and animal workforce and wood energy for heating to an era of machine-driven mass production. England and Ireland nearly denuded their nation's forests to heat their homes and run their newly invented machinery. But a magnificent discovery would save both their forests and the Industrial Revolution—coal. This soft black rock provided the perfect substitute for wood. It was the only thing that saved what remained of their forests. However, coal had its price. Urban coal burning created a dangerous envelope of smog that descended upon London and other large cities of the day. In many instances, air pollution became so foul and concentrated that it was more harmful than the pollution found in many large cities today. And, of course, it dramatically increased greenhouse gas emissions.

The Industrial Revolution led to other equally undesirable effects. Less labor was now required for agriculture, but such increase in efficiency led to rural unemployment. Rural villagers were driven into cities and to a life of drudgery working in urban sweatshops. The ranks of cities ballooned in a manner that had never been witnessed. They became overcrowded and filthy. Men were worked to exhaustion and women were forced into prostitution, while children started lifelong careers working 12-hour shifts in filthy factories. The novels of Charles Dickens are drawn from this stage of England's industrial development.

The social and technological progress that we have witnessed over the last century has been nothing short of miraculous. But as we have seen, this progress has come at a cost. Today, we are confronted with even more difficult challenges, ranging from issues as diverse as climate change, water resources shortages and contamination, topsoil erosion, depleted fisheries, loss of biodiversity, deforestation, failures of critical nanotechnology components in control systems, and genetically modified organisms, all of which pose potentially catastrophic effects on a global scale. How do we continue to provide economic prosperity and well-being without destroying the very intricate living systems of our planet, and without depleting natural resources to unsustainable levels?

Over the last several decades, countries around the world have enacted national policies similar to the 1969 U.S. NEPA to protect their polluted air, water, and lands through a wide variety of laws and educational programs. Prior to the NEPA, these efforts lacked a common vision of the environment as an integrated system, whose many functions are highly interrelated.

Nature of the problem

Since the dawn of the Industrial Revolution, humans have been depleting resources at an astonishing and increasing rate, faster than most of these resources could possibly be replenished. Many environmentalists warn we are headed for disaster. On the other side of the aisle stands a school of economists who argue that the economic law of supply and demand will solve our problems. They point to the fact that resource prices have tended to decrease over the last several generations, while the standard of living has increased. But as an ancient proverb states, "the past does not ensure the future."

Increasingly, environmental issues such as dwindling renewable resources and peak oil (see Chapter 11) make newspaper headlines and attract Internet blogs; even so, analysis only occasionally ventures beyond the direct national-security implications to investigate the underlying causes or long-term implications. While stories of carnage receive attention-grabbing headlines and sound bites, complex stories about long-term environmental degradation command only nominal attention. They are simply not entertaining; they are sometimes even depressing. Over the long run, society may be significantly more at risk from an environmental calamity than an economic one.

Environmental complexity and uncertainties

The laws of physics tend to be surprisingly simple and elegant. Physicists even refer to them as "beautiful." Consider Newton's second law of motion, the basis of nearly all of classical physics. It can be summed up by the deceptively simple formula, $F = ma$. This simple equation provides the basis for deriving most of the mechanical laws governing the universe. While many people find physics to be difficult, describing and quantifying environmental systems can be much more complicated. This greatly increases the problem of understanding and attempting to predict the outcomes of environmental systems under stress. Such stress does not necessarily cause slow, linear, predictable changes. Consider a chemical reaction; the reaction may be incrementally slow until a certain activation level is reached, a "tipping point," after which things can change rapidly. Environmental systems can act in the same way. A critical point can be reached in which such systems experience abrupt, unexpected, and potentially catastrophic changes, such as a species population suddenly crashing instead of gradually declining. For instance, global warming could actually shut down the Gulf Stream, causing the planet to rapidly drop into an ice age climate (see Chapter 13).

The good, the bad, and the downright ugly

Like the Chinese concept of yin and yang, technology can yield global benefits as well as negative and unforeseen consequences. Take just two examples: Chlorofluorohydrocarbons were extensively marketed as great solutions to common problems such as non-stick applications in food processing and polychlorinated biphenyls, and polychlorinated biphenyls (PCBs) were widely used in electric fluids in transformers and capacitors, as pesticide extenders, adhesives, dedusting agents, cutting oils, flame retardants, heat transfer fluids, hydraulic lubricants, sealants, and paints. Their risks came to light only much later. Similarly, the adverse socioeconomic and environmental implications of a national ethanol fuels program were barely considered prior to its introduction. Some scientists had actually assessed their potentially adverse impacts before they were first marketed, but such assessments were ignored.

Consider a truly alarming example. A drug named thalidomide caused thousands of severe birth defects. However, from another perspective, it actually had a hidden blessing. The birth defects were quickly observed by an astute medical community. Imagine if it produced equally severe abnormalities such as mental retardation that would not become apparent for many years after delivery. Had this been the case, millions, rather than thousands, of children might have been irrevocably harmed. The concern today is that the ever-growing mix of new technologies and synthetic substances may have no immediate, obvious impacts.

Critics respond that revolutionary technologies such as fusion energy, nanotechnology, and biotechnology have the potential to offer revolutionary benefits akin to integrated circuits, antibiotics, and computers in the last century. Others counter that thousands of new compounds and technologies will entail subtle, long-term, unintended consequences we cannot even envision presently. It may well turn out that emerging technologies go both ways.

Policymakers in the Third World argue that if the West can have it all, then it is only fair that they also be given the opportunity to achieve a standard of living equivalent to that enjoyed in the West. Who can blame them? Yet, the stark reality is that if the remaining 90% of the developing world even began to approach the standard enjoyed in the West, the global carrying capacity would almost certainly be exceeded, with consequences that could include the breakdown of entire ecosystems followed by catastrophic socioeconomic repercussions. Whether it is fair or not, the bleak truth is that the Earth cannot sustainably support 6 billion, let alone

9 billion, people living at the standard of living enjoyed by the West. This poses a particularly vexing problem for policymakers. Can we develop environmental policies that allow for equitable development without jeopardizing the very planet we all share? We must believe that sound global environmental policies can indeed make a difference by significantly reducing the risks of current trends. The intent of this book is to provide a toolbox of principles and concepts that will enable the student to both develop and critically assess potentially beneficial policies, strategies, and programs.

Environmental dilemma

The environmental dilemma is the imbalance between economies whose technologies efficiently exploit the Earth's resources to serve the needs of a global civilization with 6 billion people and growing, and Earth's limitations yielding those resources (air, water, soils, and minerals, as well as living plant and animal resources for food, fiber, drugs, and many other products) in the quantity and quality needed to sustain the population. An important consequence of this dilemma is the growing gap between the minority of people living well and the larger number who live in poverty—sometimes extreme poverty. This situation generates social, political, and ultimately armed conflicts. Environmental professionals have a term for this situation—"unsustainable." The overriding goal of environmental and economic policy is to ensure a "sustainable" and peaceful global civilization by restoring the balance between our economies and the Earth's resources. This goal poses extraordinarily complex challenges.

Unfortunately, the term "sustainable" is often misused as a feel-good buzzword to promote products and to assuage fears about the future. For example, world-famous entertainers may fly personal jets to media interviews and fundraising concerts to save the world, politicians may appear at televised environmental festivals, and multinational corporations may advertise "sustainable products." The public may be motivated to choose paper sacks over plastic ones and reusable diapers over disposable ones. They may even scale down from large sport utility vehicles to smaller ones. But such changes, while of laudable intent, are highly unlikely to even remotely offset the scale of environmental destruction in even one region, such as the Amazon, where rainforests are being demolished at the rate of about 20,000 square miles per year (55 square miles per day).[2]

Policy remedies must address the issues that are far more likely to enable sustainable use of the world's forests than wealthy consumer choice. Now, consider similar rates of forest and other basic resource losses across other regions of the Earth in a single day. You can just begin to appreciate the enormity of the problems we face. The human consequences of such high rates of basic resource depletion include more people breathing polluted air, drinking contaminated water, eating tainted foods, suffering health effects leading to shortened lives, and living in areas prone to natural and technology hazards. Also, consider that elephants and iguanas will always be around, if only in zoos, but that ugly little toad or plant that went extinct in the Amazon yesterday morning might have held a universal cure to cancer that will have been lost forever.

National and global environmental policies

Environmental policies that aim for sustainability are by nature complex and difficult to conceive, defend, and implement. Yet experience shows that progress can be made. A common high-level goal such as sustainability can drive solutions to many environmental problems at all levels, from the local to the global. Without such goals, one is likely to end up with a hodgepodge of competing and conflicting objectives, laws, and regulations. Fortunately, there are precedents for national, regional, and global policy frameworks available. Our task as professionals is to work within and improve such frameworks to the extent possible, in order to help achieve their respective goals. Three such frameworks are briefly described as follows:

- NEPA serves to anchor the entire edifice of environmental law and regulation that has evolved in the United States since 1967.
- The EU's regional environmental policy framework, like the American system, encompasses specific laws and regulations with some accommodation to the needs and capabilities of its individual states.
- The UN global environmental policy framework largely depends on treaties and voluntary commitments by the world's nations to finding ways to contribute to global goals established by international consensus. In addition, the UN publishes many studies of environmental performance and provides extensive technical support to nations that need it.

United States environmental policy framework

U.S. environmental policy began as an environmental movement with roots dating back to nineteenth-century mining, industry, and agriculture that visibly impacted environmental quality (see Chapter 1). The United States was the first nation to promulgate a comprehensive national environmental policy. By enacting NEPA in 1969, the U.S. Congress established the following purposes:

> The purposes of this Act are: *To declare a national policy*, which will encourage productive and enjoyable

harmony between man and his environment; to promote efforts which will prevent or eliminate damage to the environment and biosphere and stimulate the health and welfare of man; to enrich the understanding of the ecological systems and natural resources important to the Nation; and to establish a Council on Environmental Quality.

NEPA Section 2

NEPA laid out the basic elements essential to virtually any national environmental policy. The strategic goals spelled out in NEPA are to (NEPA Section 101):

1. Fulfill the responsibilities of each generation as trustee of the environment for succeeding generations.
2. Assure for all Americans safe, healthful, productive, and esthetically and culturally pleasing surroundings.
3. Attain the widest range of beneficial uses of the environment without degradation, risk to health or safety, or other undesirable and unintended consequences.
4. Preserve important historic, cultural, and natural aspects of our national heritage, and maintain, wherever possible, an environment which supports diversity and variety of individual choice.
5. Achieve a balance between population and resource use which will permit high standards of living and a wide sharing of life's amenities.
6. Enhance the quality of renewable resources and approach the maximum attainable recycling of depletable resources.

NEPA also instructs the following:

Utilize a *systematic, interdisciplinary approach* which will insure the integrated use of the natural and social sciences and the environmental design arts in planning and in decision-making which may have an impact on man's environment.

NEPA Section 102 2A

Section 102 of NEPA requires federal agencies to assess the effects of and investigate alternatives to major federal actions that may significantly impact environment quality. Although NEPA announces national environmental policy goals, its actual effects on decision making have been largely limited to federal actions, including private actions that involve some form of federal enablement such as licenses, permits, or federal funding. From this perspective, NEPA's significance is profound and historic. NEPA has served as a stimulus and reference for a host of individual U.S. State Environmental Policy Acts that followed in its wake. NEPA set the stage for a host of prescriptive environmental laws and regulations that aim to protect environmental media as diverse as land, water, and air, as well as the human environment in homes, communities, and workplaces.

NEPA has also inspired a host of similar policy acts around the world and has influenced the practices of international organizations involved with the funding, development, and implementation of major projects and programs. More recently, corporations and organizations have promulgated environmental policies that reflect many of the values established in NEPA. Few other national acts have been emulated by so many nations around the globe. For a more thorough discussion of the NEPA process, the reader is directed to the texts *NEPA and Environmental Planning*[3] and *NEPA Effectiveness*[4].

NEPA by itself is not an enforcement mechanism for environmental regulation, most of which is implemented through the U.S. Environmental Protection Agency (EPA). The EPA operates extensive regulatory programs that specifically implement the laws listed in Table I.1.[5] NEPA has significantly influenced the development of EU and UN global policies that follow.

European Union environmental policy framework

Table I.2 displays the framework as it appears on the EU Web site.[6] This is a highly progressive environmental policy framework that aims to be comprehensive in scope. The program includes preparation of environmental assessments.[7]

United Nations global policy framework

The UN environmental policy framework is implemented by the UN Environmental Program (UNEP) through six "priorities" plus 22 "thematic areas," as shown in Table I.3.[8]

Chapter 4 describes the extensive environmental policy system that the UN has evolved since the end of World War II, and additional elements appear in other chapters. UNEP maintains an extensive database of environmental assessments.[9]

Table I.1 Framework of U.S. Environmental Protection Policies

Atomic Energy Act (AEA)	Federal Insecticide, Fungicide, and Rodenticide Act (FIFRA)
Chemical Safety Information, Site Security and Fuels Regulatory Relief Act	Marine Protection, Research, and Sanctuaries Act (MPRSA; also known as the Ocean Dumping Act)
Clean Air Act (CAA)	
Clean Water Act (CWA)	National Environmental Policy Act (NEPA)
Comprehensive Environmental Response, Compensation and Liability Act (CERCLA or Superfund)	National Technology Transfer and Advancement Act (NTTAA)
	Noise Control Act
Emergency Planning and Community Right-to-Know Act (EPCRA)	Nuclear Waste Policy Act (NWPA)
	Occupational Safety and Health (OSHA)
Endangered Species Act (ESA)	Ocean Dumping Act (see Marine Protection, Research, and Sanctuaries Act)
Energy Independence and Security Act (EISA)	
Energy Policy Act	Oil Pollution Act (OPA)
EO 12898: Federal actions to address environmental justice in minority populations and low-income populations	Pollution Prevention Act (PPA)
	Resource Conservation and Recovery Act (RCRA)
EO 13045: Protection of children from environmental health risks and safety risks	Response, Compensation and Liability Act
	Safe Drinking Water Act (SDWA)
	Shore Protection Act (SPA)
EO 13211: Actions concerning regulations that significantly affect energy supply, distribution, or use	Superfund (see Comprehensive Environmental Response, Compensation and Liability Act)
Federal Food, Drug, and Cosmetic Act (FFDCA)	Superfund Amendments and Reauthorization Act (SARA)
	Toxic Substances Control Act (TSCA)

Source: United States Environmental Protection Agency, "Laws and Regulations," http://www.epa.gov/lawsregs/(accessed July 7, 2010).

Applying the integrated disciplinary approach

This book has been inspired in part by NEPA's concept of a *systematic interdisciplinary* approach to environmental policy and management. The concepts "systematic" and "interdisciplinary" are not simply esoteric terms designed to impress the reader. The term *systematic* means a scientifically based step-by-step process for assessing environmental impacts and developing alternatives and mitigation measures for reducing such impacts. The term *interdisciplinary* denotes a process that incorporates

Table I.2 European Union Environmental Policy Framework

Air
Biotechnology
Chemicals
Climate change
    Air emissions
    Emissions trading
    European climate change program
    Fluorinated greenhouse gases
    Greenhouse gas emissions in the
      community
Enlargement and neighboring
  countries
    Danube and Black Sea
    Enlargement
    Financing
    Mediterranean partners
    Russia and other newly
      independent states
    Southeastern Europe
Environmental economics
    Database on environmental taxes
    Published studies
Health
Industry and technology
    Eco-label
    Eco-management and audit scheme
      (EMAS)
    Environmental technologies action
      plan (ETAP)
    Integrated product policy
    Pollution from industrial
      installations
    Retail forum
    Standardization
    The greening of public
      procurement
International issues
    Green diplomacy
    International relations

Multilateral environmental
  agreements
World Summit on Sustainable
  Development (WSSD)
Land use
    Environmental impact assessment
    Geographic information system
    Urban environment
Nature and biodiversity
Noise
    Environmental impact assessment
    Integrated coastal zone management
    Urban environment
Ozone layer protection
Soil
Sustainable development
    Sustainable cities
    Sustainable use of natural resources
    WSSD
Waste
    Batteries
    Biodegradable waste
    End-of-life vehicles
    Landfill of waste
    Mining
    Packaging
    Polychlorinated biphenyls/
      polychlorinated terphenyls
    Polyvinyl chloride
    Reporting on implementation of
      waste legislation
    Sewage sludge
    Ship dismantling
    Shipment of waste
    Waste from electrical and electronic
      equipment
    Waste incineration
    Waste management plans
    Waste oils

Table I.2 European Union Environmental Policy Framework (*Continued*)

Water and marine	Priority substances
Bathing water directive	Groundwater
Drinking water directive	Regional sea conventions
Floods directive	Urban wastewater treatment
Integrated coastal zone management	directive
Marine strategy framework	Water framework directive
directive	Water scarcity and droughts

Source: European Commission for the Environment, "Environment Policies," http://ec.europa.eu/environment/policy_en.htm (accessed July 7, 2010).

Table I.3 United Nations Environmental Policy Framework

Climate change

Disasters and conflicts

Environmental governance

Harmful substances

Resource efficiency

Other thematic areas: Art for the environment, biodiversity, biosafety, business and industry, children (Tunza programme), major groups and stakeholders, education and training, energy, environmental assessment, freshwater, gender, indigenous knowledge, land, ozone, poverty and environment, regional seas, scientists, sports and environment, urban issues, youth (Tunza programme)

Source: United Nations Environmental Program, http://www.unep.org/ (accessed July 7, 2010).

the talents and experience of a diverse array of specialists representing a range of scientific and technical disciplines; but it also refers to a process in which these specialists communicate and interact with one another rather than working in isolation. This approach is essential to working within any coherent framework of environmental policy, law, and regulation.

Our objective is to provide a foundation of knowledge and insight about the environment that will enable lifelong environmental learning in one's professional career by

1. Introducing the fundamental theoretical concepts and practical applications that underlie environmental policy
2. Introducing the tools for more effectively crafting policies, programs, and processes
3. Demonstrating how these environmental policy formulation tools can be applied to a variety of policy contexts with illustrative examples
4. Recognizing that environmental problem solving requires careful attention to the details of local as well as global environments

5. Avoiding stovepipe approaches that may cause more problems than they resolve
6. Helping students establish a foundation for whatever direction they may later pursue

An interdisciplinary approach that integrates the natural and social sciences enables students and professionals to understand how the many elements of the global environmental system function together and can be properly managed. With this in mind, we begin each chapter with learning objectives and a case study that draws on a wealth of theoretical and applied concepts. The expertise we seek to develop is that of an integrator who relates his or her own professional training to the entire policy context.

What does integrating the natural and social sciences mean in the real world of global environmental policy? To bring this seemingly abstract concept down to Earth, consider the case of global warming discussed in Chapter 13. To understand what is going on at the policy level, we need to apply the expertise of persons in a number of fields in order to help answer questions such as those depicted in Table I.4.

Informed and effective policy in every area of environmental concern must somehow deal with these issues. Upon being presented with problems of such extensive scope, complexity, and uncertainty, our first reaction may be feelings of awe and helplessness. To help overcome such feelings, this book demonstrates how a systematic interdisciplinary approach can be applied to help you navigate the complex world of environmental policy in an orderly manner. By doing so, you will learn to scope its problems by identifying the needed disciplines and developing integrated solutions.

You will discover that the practice of global environmental policy is a game that humanity is just beginning to learn how to play. It has many elements, some of which cooperate to promote an integrated policy, while others, depending on their political and socioeconomic interests, oppose a systematic approach. Although there are no magic solutions, the potential rewards of a well-crafted policy are significant; just bear in mind that no single policy is ever perfect or permanent. Every policy is a socioeconomic experiment that may or may not work as originally envisioned. Thus, a sound policy process requires the best information available, assessing the potential risks and benefits, and monitoring the policy for effectiveness.

Environmental policy as a process

With respect to the environment, the term *policy* encompasses a set of basic principles and associated guidelines, formulated and enforced by the governing body of an organization (governmental or private) to direct or limit its actions in pursuit of long-term goals.[10] Table I.5 identifies the

Table I.4 Interdisciplinary Factors That Can Affect Formulation of
Environmental Policies—The Case of Global Warming

Physical and biological sciences

How do we know that the Earth is warming? What is the scientific evidence
for and against human-induced global warming?

To what extent is global warming the result of anthropogenic (i.e., human)
factors as opposed to historical natural factors?

How do the air, water, and land environments respond to changes in mean
temperature? How do the Earth's ecological systems respond?

Given the vastness of the globe and the limitations to our ability to measure
and model the effects of climate change, how certain can we be of science-based
predictions as a basis for policy response?

To what extent can the regional and local impacts of any of the
aforementioned factors be reliably predicted?

Technology

What technology can be developed to reduce the rate of global warming?

To what extent can technology help society and its infrastructure mitigate the
potential effects of global warming?

Governance

Can we create a global governance infrastructure through international
cooperation that will promulgate and coordinate policies that achieve the
necessary transformation of the infrastructure and economic systems?

Management

How do we educate and motivate the vast numbers of people in all of the
aforementioned categories to create and manage the activities needed in
virtually every sector of global infrastructure and economy—from the top down
to the level of the urban neighborhood and the smallest rural community?

Social

What will be the effects on human habitation, food, water, and health?

If we simply acquiesce and do nothing, what will happen to human
populations around the globe if the Earth continues to warm up?

Economic

What are the economic impacts of introducing new technology to reduce
global warming on society and infrastructure worldwide?

What economic scenarios, if any, can be implemented and sustained
indefinitely for the future?

To what extent can market forces be mobilized to create technology solutions?

What role should government play in managing the economic response?

Political

Can greenhouse emissions be effectively reduced if some nations refuse to
cooperate?

Could global warming and its effects on environmental resources lead to
confrontations or regional conflicts?

Table I.5 Environmental Policy System Framework

Environmental policy goals

Maintain quality of life

Manage natural resources and promote sustainable resource and energy supplies

Maintain environmental resources and amenities

Ensure health and safety

Promote a livable and aesthetically pleasing environment

Balance the benefits of development against environmental, social, and economic costs

Environmental policy management objectives

Manage demand factors and other human activities that depend on and impact environmental systems

Manage the infrastructure systems to minimize adverse effects and otherwise improve the productivity of environmental systems

Manage the environmental systems to ensure sustainability of materials, goods, and services essential to satisfy demand

Policy scope and demand factors

Policy scope	Demand factors	Infrastructure systems	Environmental systems
International	Population	Extraction	Land
National	Economy	Production	Water
Local	Culture	Transportation	Atmosphere
	Government	Waste management	Biota
		Health and safety	Human communities—their culture and lifestyle.

Environmental policy institutions and methods

Legislative	Executive	Judicial/enforcement	Business and industry	Political	Economic/fiscal mechanisms	Management systems and methods
Laws	Analyses	Investigations	Investments	Elections	Loans/Interest rates	ISO standards
Regulations	Research and development	Litigations	Operations	Referenda	Subsidies	Hazards management
Incentives	Programs	Judgments	Marketing	Lobbying	Offsets	Waste management
Disincentives	Plans			Advocacy	Cap and trade	Ecological management
Oversight	Projects				Taxation	Infrastructure management
	Assessments					

following major activities and institutions essential for environmental policy creation and implementation:

- Environmental policy goals
- Management objectives
- Scope and demand factors
- Systems and subsystems to be managed
- Environmental policy institutions and methods

Environmental policy, as a process based on principles and societal goals, is designed to protect and enhance the natural and man-made environments for the benefits of humanity. This process is not simply a set of methods and techniques. It embodies a philosophy for coping with challenges to human well-being, lifestyles, and even survival. Environmental problems ranging from global to village and urban enclaves are inherently complex. Only rarely are there simple solutions acceptable to all stakeholders. Broadly speaking, the policy process has three stages.

Stage 1: Awareness—Virtually all future policy is foreshadowed by a period of public awareness and a political process that calls for government and private sector attention. This was the case in the 1960s when the American public responded to an alarming deterioration in environmental quality. Even today, with environmental control organizations at all levels of government and industry, new policy is continuously driven by recent findings of socially and politically unacceptable conditions.

Stage 2: Policy focus—Senior government decision makers formulate policies in response to political and public calls for action, reinforced by various sources of evidence and perceptions of what can be accomplished. The policy focus is expressed as a set of high-level goals and objectives.

Stage 3: Policy implementation—This flows from the upper echelons of government and the private sector to the most local levels, as can be inferred from Table I.5.

Policies such as NEPA and similar international policies are typically implemented through a *top-down* approach as follows:

1. A broad overarching *environmental policy* is published as public law.
2. An agency of the executive branch promulgates *regulations* that all agencies must follow in implementing the policy.
3. Each agency issues its own *agency-specific regulations* for implementing the policy.
4. Various other executive agencies are assigned roles. These typically involve *planning, analysis,* and *research and development,* through a wide range of programs and projects.

5. Enforcement agencies conduct *investigations*, pursue *litigation*, and render *judgments* to enforce compliance according to how the laws and regulations apply to the various *infrastructure* and *environmental systems*.

With respect to developing a systematic interdisciplinary process, students should note that environmental policy institutions may employ scientists, engineers, economists, political scientists, communication experts, and others. The rest of this book will illustrate and reinforce the integrated application of the environmental policy system framework. The capstone problems given at the end of the book (Epilogue) are designed to help the student review and successfully apply an understanding of the interrelated nature of the various systems and other elements depicted in Table I.5.

Notes

1. W. Ruddiman, "How did humans first alter the global climate," *Scientific American*, March 2005, 46–53.
2. Raintree Web site "Forest Facts," http://www.rain-tree.com/facts.htm (accessed July 7, 2010).
3. C. H. Eccleston, *NEPA and Environmental Planning: Tools, Techniques, and Approaches for Practitioners* (Boca Raton, FL: CRC Press, 2001).
4. F. March, *NEPA Effectiveness: Mastering the Process* (Rockville, MD: Government Institute, 1998).
5. United States Environmental Protection Agency, "Laws and Regulations." http://www.epa.gov/lawsregs/ (accessed July 7, 2010).
6. European Commission for the Environment, "Environment Policies." http://ec.europa.eu/environment/policy_en.htm (accessed July 7, 2010).
7. European Environment Agency, "Environmental Terminology." http://glossary.eea.europa.eu/terminology/sitesearch?term=Environmental+assessment (accessed July 7, 2010).
8. United Nations Environmental Program, http://www.unep.org/ (accessed July 7, 2010).
9. United Nations Environment Program, "UNEP Activities in Environmental Development," http://www.unep.org/themes/assessment/ (accessed July 7, 2010).
10. Business Dictionary, "Environmental Policy Definition," http://www.businessdictionary.com/definition/environmental-policy.html (accessed July 7, 2010).

section one

Introduction to global environmental concepts and principles

chapter one

Historical context of the environmental movement

George Washington is the only president who didn't
blame the previous administration for his troubles.

Author unknown

Case study

The quiet city of Donora, Pennsylvania, was in for one horrific Halloween.
On Halloween night, 1948, the "Donora death fog" descended like a phantom upon this sleepy steel-company town. When it cleared 6 days later,
20 people were dead, another 50 were dying, and hundreds would live out
their days with permanently damaged lungs. An investigation indicated
that one-third of the town's 14,000 residents were affected. A local physician testified that had the smog continued another evening, "the casualty
list would have been 1000 instead of 20." A decade later, Donora's mortality rate remained significantly higher than that of the neighboring areas.

The Donora event was merely a prelude to what was in store for
London. What happened a few years later sounds like a scene out of a
Stephen King horror novel. Thursday, December 4, 1952, began normally
enough, with gray skies and the temperature in the upper 30s. The cold
fog caused Londoners to burn more coal than normal. By the next morning, a foul concoction of contaminates and sooty black coal smoke began
to descend upon the city. Initially, there was no great panic. The smog
gradually increased until a lack of visibility made it almost impossible to
drive. In 4 days, it killed thousands of Londoners. A study showed that
this death fog killed some 4000 people by December 9. A total of 8000
people died in the weeks and months that followed.[1] It remains the deadliest environmental episode in recorded history.

Class project: The Donora death fog illustrates what can happen
if stringent measures are not taken to protect and preserve environmental quality. Develop a legislative plan for a hypothetical large city
that has thermal inversion characteristics similar to that of Donora.

What type of legislation, measures, and controls would you enact to prevent a recurrence of the death fog? How would you develop such measures so that they do not hamper future development of economic growth? How would you seek and gain public support for this plan?

Learning objectives

Environmental policy has evolved over time. The United States passed the National Environmental Policy Act (NEPA) in 1969, the world's first national environmental policy. NEPA has profoundly influenced environmental policy around the world. Today, well over 100 countries have passed policies similar to that of NEPA. After you read this chapter, you should understand the following:

- Factors that have driven the development of modern environmental policy
- The historical evolution of environmental policy
- The importance of NEPA in shaping environmental policy around the world

1.1   Origins of the environmental movement

The unremitting human attack on the environment has been compared to geologic cataclysms in the Earth's distant past. The resulting rate of global extinction of living species, for example, is estimated to be at least a thousand times greater than that of natural incidence. If unchecked, this assault might someday carry humanity itself to the brink of extinction. Policymakers all over the world are slowly coming to realize that continuous unchecked growth must eventually lead to a more rational use of the Earth's precious environmental resources.

Louis P. Pojman paints a graphic picture of how pervasive pollution has become. For many years, his family vacationed off the Spanish Mediterranean coast. While swimming in the ocean, they noticed a brown substance sticking to their arms and legs. This brownish substance turned out to be half-dissolved feces! He later learned, to his chagrin, that much of sewage along parts of the Mediterranean is discharged into the sea, untreated. The city's sewage system was discharging its effluent 1000 feet from the public beach. By the next day, they were all sick.[2]

The Santa Monica Bay was the filming site for the ever-popular TV show *Baywatch*. Filmed by G.G. Miller, the show portrays a beautiful and scenic Californian beach. However, Miller knew what his viewers did not: that this water was so polluted that the actors were paid extra for entering the water and were chemically cleaned afterward.[3]

Concerns such as these have profoundly shaped modern environmental policy. To fully appreciate its development, we must first understand the evolution of the environmental movement. Modern environmental policies are a postindustrial phenomenon led mainly by the United States and Europe. The environmental movement in the United States began to take root well over a century before U.S. Congress passed NEPA.

In the United States, the roots of the environmental movement can be traced to the early conservation movement and the establishment of Hot Springs National Park in 1832. The environmental movement gained momentum in the second half of the nineteenth century, largely as an effort to save the nation's wildlife heritage. The movement was at first motivated by a concern for bison, which had been hunted to near extinction.

Henry David Thoreau was concerned about the wildlife in Massachusetts; as he studied the wildlife from his cabin, he wrote *Walden; or, Life in the Woods*. In 1864, he published *The Maine Woods*, a plea for the establishment of "national preserves" of virgin forests. George Marsh, who advocated resource conservation, published *Man and Nature*, arguably the first book to describe environmental degradation and promote the recovery of damaged environments. The word "ecology" was coined by the German biologist Ernst Haeckel in 1866.

In 1872, U.S. Congress enacted a law establishing Yellowstone National Park, the world's first national park, thereby setting a precedent for the preservation of scenic federal lands. In 1873, the American Association for the Advancement of Science petitioned Congress to halt the destructive exploitation of natural resources. In the following years, Congress continued to lay the foundation for federal protection of lands by expanding the National Park System and by establishing national forests and the U.S. Soil Survey. In 1891, John Muir founded the Sierra Club, which has since become one of the world's most influential environmental organizations.

1.1.1 Early twentieth century

Giffort Pinchot was influenced by George Marsh's work, as well as by his own love of the outdoors. Appointed by President Theodore Roosevelt, Giffort Pinchot became the first chief of the U.S. Forest Service. Roosevelt set aside 125 million acres of federal lands for protection during his term of office (1901–1909). John Muir claimed that Roosevelt's policies served to stimulate economic uses of lands that should be protected in their original state. This criticism stimulated a debate over environmental management that continues to this day.

To prevent the vandalism of prehistoric Indian sites in the Southwest, Congress passed the Antiquities Act in 1906, authorizing the president to establish "national monuments"[4] on federal lands. Early federal actions on behalf of the environment, however, were not free of controversy. These policies and statutes, as important as they were, proved inadequate in the face of events to come.

During the 1930s, large parts of the Southwest suffered drought and a series of environmental calamities that created economic havoc and induced large-scale migration from Oklahoma and other locations westward. In response to these events, President Franklin D. Roosevelt created the Soil Conservation Service and the Agricultural Stabilization and Conservation Administration to promote a version of what we might today call "sustainable agriculture" through soil conservation and other beneficial land management practices.

At this time, the environmental movement had yet to move beyond land management and species conservation to deal with the larger issues of pollution and the environmental impacts of an industrial society. The impetus for a comprehensive environmental policy essentially began in the 1960s.

1.1.2   Environmental movement takes off in the 1960s

To fully appreciate the forces that led to the enactment of the world's first national environmental policy, one must appreciate the context in which NEPA was enacted. The American public and Congress alike were becoming increasingly concerned that the environment was deteriorating at an alarming rate. Satellite photographs of the Earth brought into focus just how small the planet was and its place in the universe.

The 1960s witnessed environmental disasters such as the Santa Barbara oil spill and the Love Canal incident. There were visibly polluted waterways and blighted urban landscapes in many parts of the United States and other countries. This proliferation of environmental despoliation stimulated an environmental activist movement. Lake Erie was pronounced "dead," and smog alerts were issued in major cities across the nation. To top it off, the U.S. Bureau of Reclamation proposed building a dam on the Colorado River that would flood the Grand Canyon. The fledging environment movement began to fight back.

The Sierra Club published full-page newspaper notices in the *New York Times* and *Washington Post* arguing against damming the Colorado River. The notice in question simply read, "This time it's the Grand Canyon they want to flood. *The Grand Canyon.*" The Grand Canyon dam proposal was withdrawn in 1968. The Club's action boosted its prestige and membership.

The American public was jolted by emerging scientific studies that drew attention to the existing and hypothetical threats to the environment and humanity. Among them was Paul R. Ehrlich's 1968 book *The Population Bomb*, which brought attention to the impact of exponential population growth on the environment. Noted biologist Barry Commoner generated further debate concerning growth and affluence. Arguably, no other event captured the public's imagination more than the National Broadcasting Company's (NBC's) nightly news, broadcasting scenes of the Cuyahoga River in Cleveland, Ohio. The river was so fouled with industrial waste that it caught fire and burned in June 1969.

In 1972, an association of scientists and policymakers known as the Club of Rome published their report, *The Limits to Growth*, which drew attention to the growing pressure on natural resources. Rachel Carson's book *Silent Spring* warned of the dangers of chemicals released into the environment. Carson, a mild-mannered government scientist, documented how the widespread use of the pesticide dichlorodiphenyltrichloroethane (DDT) was jeopardizing bird species.

The United Nations Conference on the Human Environment was held in Stockholm, Sweden in 1972. For the first time, representatives of multiple governments were united in a common discussion relating to the state of the global environment. This conference led directly to the creation of government environmental agencies and the United Nations Environment Program.

Events such as these led to the enactment of new laws, including NEPA. While the first version of the Clean Air Act was enacted in 1955, the larger body of environmental law encompassing protection of land and water media, as well as far more comprehensive protection of the air, remained an ideal for the future, to be realized only after the enactment of NEPA.

By the 1960s, the environmental movement was targeting specific chemicals, industrial processes, and disposal and treatment practices. The resolution to such problems required the creation of strict new policies and regulations. These requirements were principally directed at manufacturers rather than consumers. It was relatively inexpensive and easy to implement, and necessitated little in terms of the public's lifestyle or level of affluence. Today, we are faced with an array of much more complex and problematic issues, ranging from climate change to water resources, topsoil erosion, depleted fisheries, loss of biodiversity, deforestation, nanotechnology, and genetically modified organisms, all of which pose potentially catastrophic effects on a global scale.

1.2   Historical development of National Environmental Policy Act

As early as 1959, Senator James Murray had proposed the establishment of a council overseeing environmental quality. Support for a national environmental policy evolved slowly over a period of more than 10 years. Prior to NEPA's enactment, there had actually been a precedent for preparing an environmental impact assessment on proposed projects. In the early 1960s, the Atomic Energy Commission was required by Congress to prepare an environmental report for a proposal to use nuclear explosives to blast a harbor along the Alaskan coastline. This project has since been criticized as potentially one of the most environmentally catastrophic projects ever proposed, and ultimately did not go forward, in large measure because of the results of this study, which has been called the first de facto Environmental Impact Statement (EIS).[5]

By the late 1960s, Congress was hearing testimony from the scientific community warning of pollution, environmental degradation, and the potential for disaster. Mounting public concern over the looming environmental quagmire put political pressure on Congress to take action. Congress could have dealt with the impending environmental crisis by amending the laws authorizing federal programs, one statute at a time. However, it became increasingly apparent that the United States first needed an environmental policy that would set the direction for more prescriptive laws that would follow. To this end, Congress adopted NEPA—a single policy statute that set the course for laws that would follow in its wake. In doing so, Congress gave priority to addressing environmental issues in the planning phase before they evolved into larger problems. For a more thorough discussion of the NEPA process, see the texts *NEPA and Environmental Planning*[6] (Eccleston) and *NEPA Effectiveness*[7] (March).

1.2.1   Origin of National Environmental Policy Act

NEPA's policy concept was originally expressed in a proposed policy statement largely crafted by Professor Lynton Caldwell (to whom this book is dedicated), special assistant to the Senate.[8] Accordingly, Caldwell has been called the father of NEPA. The Senate Committee on Interior and Insular Affairs reported:

> that in spite of the growing public reorganization of the urgency of many environmental problems and the need to reorder national goals and priorities to deal with these problems, there is still no comprehensive national policy on environmental management. There are limited policies directed to some areas where specific problems are recognized to exist, but we do not have a considered statement of overall national goals and purposes.

Among the key environmental movement sympathizers was Senator Henry Jackson of Washington, whose special concerns included timber cutting in his home state's forests and spills from oil tankers entering Puget Sound. Following Jackson's contentious committee hearing on a Bureau of Reclamation proposal to dam the Colorado River above the Grand Canyon, he recognized that a mechanism was needed to force federal agencies to instill environmental considerations into their decision-making process. Senator Jackson's leadership provided the impetus for developing a national environmental policy. Both the U.S. House of Representatives and the Senate prepared individual drafts of the Act. The Senate's draft version of the bill was more comprehensive, but lacked a clear vision of a national environmental policy.

NEPA was largely modeled after another landmark statute, the National Employment Act of 1946, which established a Council of Economic Advisers to assist and advise the president on economic matters. Instead of creating another new and expansive bureaucracy or reengineer the existing federal agency apparatus, Congress wisely chose to craft the national policy by supplementing the existing statutory charter of federal agencies. From this point on, agencies would be expected to balance the goal of preserving the environment with other competing policies and considerations such as economic growth.

1.2.2 Environmental Impact Statement debate

Much of the debate over NEPA centered on the bill's *action-forcing mechanisms*, which appeared in the Senate version, but not in the House version. Some congressional legislators argued that an environmental policy statute alone would not be enforceable and proposed drafting an environmental amendment to the U.S. Constitution. Others, who recognized that powerful business interests would oppose environmental restrictions on the private sector, urged a statute that would overcome industrial opposition by focusing exclusively on the actions of the federal government. They argued, correctly as it turned out, that passage of such a policy bill would demonstrate the seriousness with which Congress viewed environmental protection and set a precedent for subsequent legislation that would regulate both the public and private sectors. The federal government, being the single largest organizational entity in the United States and having a vast scope and nature of its actions, accounted for a disproportionately larger share of the nation's environmental degradation. If for no other reason, such an act (even if limited to federal actions) would clearly have profound implications that would reduce future environmental degradation. Senator Jackson was adamant that such a provision be incorporated to ensure that the act not be merely a paper tiger. Respecting the need for an action-forcing mechanism, a compromise was eventually struck in a conference committee that included the following key provision[9]:

> All agencies of the Federal Government shall . . .
> include in every recommendation or report on
> proposals for legislation and other major Federal

> actions significantly affecting the quality of the human environment, a detailed statement by the responsible official.

This action-forcing mechanism would later become known as the EIS. As the debate ensued, it was argued that agencies should not only be required to compare the potential environmental impacts of alternatives to a proposed action, but should actually be required to select an environmentally benign alternative. Such a provision, however, was considered unduly restrictive and was resisted. Instead, it was successfully argued that NEPA's effectiveness would result from heightened awareness, which would ultimately lead to better environmental decisions.

Thus, Congress imposed no *substantive* requirement on federal agencies to select an environmentally benign or least damaging alternative.[10] Instead, NEPA became a balancing act. It was designed to compel federal agencies to openly consider the environmental effects of alternatives, in reaching a final decision, but not to actually tie the hands of agencies in implementing their missions. As part of its decision-making process, NEPA requires that the EIS be published and made available for review and comment by the public. For an in-depth discussion of the Environmental Impact Statements process, see the companion text *Environmental Impact Statements*[11] (Eccleston 2008).

1.2.3 Passage of National Environmental Policy Act

The final bill received the unanimous vote of the Senate Interior Committee and enjoyed widespread support among members of Congress. The significance of NEPA was reinforced when President Nixon chose to sign NEPA into law on New Year's Day 1970, proclaiming this as "my first act of the decade." Consequently, NEPA has the unique distinction of being the first law enacted during the new decade of the 1970s. Upon its enactment, few congressional members foresaw the broad ramifications that NEPA would later have for federal decision making, or that it would be a model copied by nations around the world.

Nixon authorized the Environmental Protection Agency in 1970. He later signed a flurry of landmark environmental laws, including the Clean Air Act, the Clean Water Act, and the Endangered Species Act. Representing perhaps the apogee of the early environmental movement, U.S. Congress announced the first Earth Day, which was celebrated on April 22, 1970.

1.2.4 Recent trends

A principal driver of improved NEPA compliance has undoubtedly been the accumulation of NEPA litigation that clearly sends the message that

NEPA is to be taken seriously. Perhaps a more important driver of change has been a generally improved environmental ethic within most federal agencies, and a top-down commitment to NEPA compliance. Many federal agencies have designated environmental compliance offices and have recruited a new generation of trained environmental specialists to deal with NEPA and other environmental requirements.

Most federal agencies are now making a good faith effort to incorporate NEPA's intent into their decision-making process. As a result, environmental considerations in decision making are much better integrated with economic and technical factors. Although there have been many problems with the NEPA process, including an abundance of litigation, NEPA's regulations have proven remarkably resilient. Agencies have become more adept at identifying adverse impacts and modifying their proposed actions in the early planning process to avoid such impacts. Former U.S. Secretary of Energy, James Watkins, testified thus before Congress[12]:

> As Secretary of Energy I quickly learned that the NEPA process was not being used to provide complete and unbiased information that top-level managers needed to make the best decisions. Therefore, I established new policies to enhance and reinvigorate the [U.S. Department of Energy] DOE's NEPA process.

In the wake of events such as the Exxon Valdez oil spill and the Chernobyl disaster, new issues of an increasingly global nature are coming to the forefront: acid rain, global warming, loss of biodiversity, increasing chlorofluorocarbons, and ozone depletion, to name just a few. These problems appear more menacing and intractable than those preceding them. It is in meeting such challenges that NEPA provides planners and policymakers with a powerful tool for comparing various courses of action and forging future decisions.

1.2.5   National Environmental Policy Act's influence

NEPA has been called the equivalent of the world's environmental Magna Carta.[13] Arguably, no other environmental statute has contributed more to the long-term preservation of environmental quality. As we have seen, NEPA's importance is derived not from any substantive requirement to protect the environment, but from its *procedural* (EIS) requirement to look before one leaps. This difference makes NEPA a proactive environmental initiative rather than a reactive one. In opening the early federal planning process to public review and debate, federal agencies are now compelled to take environmental considerations into account in the "light of day."

Although the process for determining the threshold level of "significance" (the primary criterion that triggers preparation of an environmental impact statement) is far from perfect, a system of environmental checks and balances has slowly emerged. Within the public forum, compromises are struck as competing environmental and nonenvironmental interests vie for influence. Slowly, this democratic process appears to be evolving toward a middle ground where environmental interests are balanced against society's need to develop and prosper, and a large domain of nonfederal actions is likewise influenced by NEPA's mandate. Privately sponsored proposals that extract a heavy environment toll often require some form of federal approval, which itself is subject to NEPA. In no small measure due to NEPA's precedent, individual American states have also enacted their own state environmental policy acts.

Beyond American shores, NEPA has set a global precedent that has been emulated by scores of other nations. The influence of NEPA has permeated virtually every corner of the globe and has the distinction of being one of the most emulated acts in the world. The Organization for Economic Cooperation commended the United States for its "exemplary practices" in environmental impact analysis and public participation. The European Economic Community now requires its members to comply with an environmental process similar to that of NEPA. Moreover, the World Bank requires funding recipients to prepare a NEPA-like analysis to evaluate environmental consequences of their projects. Chapter 5 describes international processes similar to NEPA.

1.2.6   Crisis or an opportunity

It has been said that the Chinese character for "crisis" is actually composed of two symbols, which delineate the words "danger" and "opportunity."

While some Chinese linguists debate the actual interpretation of the symbol, it nevertheless illustrates an important metaphor for the global environmental challenges we are faced with. This Introduction has described both the positive (opportunity) as well as the negative (danger) environmental forces that are in play. It is quite conceivable that the Earth is nearing a tipping point and that if its biological carrying capacity continues to decline, it will propagate a series of crisis that threaten the viability of our global civilization. Over the millennia, humankind has thus far found ways to survive many crisis of its own making. We have the col-

lective power to avert global environmental calamity. Unfortunately, we may already have reached a point where painful sacrifices will have to be made to restabilize Mother Earth. Our future will be fraught with both crises as well as opportunities to prevent them or at least minimize their harm.

Discussions, problems, and exercises

1. Explain why NEPA has been referred to as the world's Magna Carta.
2. Describe NEPA's action-forcing mechanism.
3. What is Earth Day? When was the first Earth Day?
4. Describe some of the important events that led to the passage of NEPA.
5. Describe some of the significant acts that preceded the passage of NEPA.

Notes

1. BBC News, *Historic smog death toll rises.* 2002. http://news.bbc.co.uk/2/hi/health/2545747.stm (accessed August 22, 2010).
2. L. P. Pojman, *Global Environmental Ethics*, 266. Mountain View CA: Mayfield Publishing Company, 2000.
3. G. T. Miller, *Living in the Environment*, 10th ed. (Belmont, CA: Wadsworth, 1998), 525.
4. National Monuments are essentially National Parks with a lower level of federal services to the public.
5. D. O'Neil, "Project Chariot: How Alaska Escaped Nuclear Excavation," *The Bulletin of the Atomic Scientists* Vol. 45, No. 10 (December 1989).
6. C. H. Eccleston, *NEPA and Environmental Planning: Tools, Techniques, and Approaches for Practitioners* (Boca Raton, FL: CRC Press, 2001).
7. F. March, *NEPA Effectiveness: Mastering the Process* (Government Institute, 1998).
8. L. K. Caldwell, "A Constitutional Law for the Environment - 20 Years with NEPA Indicates the Need," *Environment*, 31. December 1989.
9. Section 102(2)(C) of NEPA.
10. Senate Debate on the Conference Report to S. 1075, *To Establish a National Policy for the Environment*, December 20, 1969, *Congressional Record*, p. S 17451.
11. C. H. Eccleston, *Environmental Impact Statements: A Comprehensive Guide to Project and Strategic Planning* (New York: John Wiley & Sons Inc, 2001).
12. J. Watkins, 1992, former secretary, U.S. Department of Energy (Admiral, U.S. Navy [ret.]), testimony before the House Armed Services Committee.
13. D. R. Mandelker, *NEPA Law and Litigation.* Eagan MN: Westlaw, 2008-2010.

chapter two

Concepts and principles underlying environmental policy

It isn't pollution that's harming the environment. It's the impurities in our air and water that are doing it.

Dan Quayle
former U.S. vice president

Case study

Easter Island, which lies on a dusty speck of rock some 2000 miles off the west coast of South America, is one of the most isolated inhabited places on Earth. The first European to discover it was Admiral Roggeveen, who landed there in 1722. He discovered a primitive society of about 3000 destitute individuals living in caves and reed huts. Instead of a lush tropical paradise, Roggeveen found a nearly treeless island virtually devoid of vegetation. But even more perplexing were the 600 mysterious stone statues, averaging 20 feet in height, which lay sprawled across the landscape. The statues were a testament to the island's once thriving and relatively advanced society in which human ingenuity had enabled the inhabitants to prosper for centuries. By the time of Roggeveen's arrival, it was clear that the once harmonious relationship between the islanders and their natural environment had collapsed.

To many archaeologists, the evidence suggests that the island was initially settled perhaps as early as the fourth or fifth century AD by a small group of Polynesians who became lost at sea. These original settlers probably arrived in simple canoes and may have numbered less than 50 individuals. When they first arrived, they would have found a pristine natural environment endowed with lush forests dominated by palm trees. These early settlers began to flourish. Related families formed clans that developed their own cultural activities, which included the construction of the huge stone statues. While their real purpose remains a mystery, one thing is clear: the statues would provide a chilling testament to the islanders' downfall.

Immense amounts of human labor and environmental resources were needed to construct and move these statues. These massive figures were transported over long distances to sites across the island. The islanders' engineering solution to the transport problem provides an important clue to the demise of their society. Since they lacked beasts of burden, they performed the heavy work themselves, dragging the statues using tree trunks as rollers. Competition between opposing clans for the available timber intensified as, in their attempts to secure greater prestige and status, an increasing number of statues were erected.

At its peak in the sixteenth century, the island's population exceeded 7000 inhabitants; unfortunately, by this time almost the entire inventory of trees on the island had been cut down for fuel, for housing, and to provide rollers to transport the mysterious stone monuments. The fragile environment began to break down, but the islanders were unable to escape. Without wood to build new canoes, they were now prisoners, trapped on the land they had ruined and completely isolated from the rest of the world.

Today, archeological teams continue to seek the causes, links, and timeline of the human and environmental collapse of the island. Despite the fact that some questions remain unanswered, most (but not all) archaeologists agree that the evidence indicates that once the island had been completely deforested, chaos ensued. When wood was no longer available for building dwellings, many inhabitants were forced to live in caves. Fishing became increasingly difficult as nets, previously manufactured from tree bark and vines, rapidly dwindled. Further, as deforestation led to soil erosion and the subsequent leaching of vital nutrients, crop yields plummeted. At this point, the society began sliding toward a steep decline. The slavery and poverty that seem to have followed were exacerbated by nearly continuous warfare caused by conflicts over diminishing resources. As food supplies dwindled, the human population even appears to have resorted to cannibalism. By the eighteenth century, the population had dropped to between one-quarter and one-tenth of its former peak.

For as long as a thousand years, the islanders' way of life enabled them not only to survive but to flourish. The society eventually collapsed because the islanders failed to realize that their very existence depended on the limited natural resources of their small island. What does the story of Easter Island teach the modern world about mounting environmental problems such as dwindling petroleum and water supplies, or global warming? The history of Easter Island is a vivid reminder of the consequences that human populations may face when vital environmental resources are irreversibly damaged. The real lesson of Easter Island is that even rational societies can commit environmental suicide.

Problem: After reading this chapter, return to this problem and consider this case study in light of what you have read. This problem may be pursued either as an individual assignment or as a class project. What can modern society learn from the lesson of Easter Island? To what extent do you believe that the lessons of Easter Island are applicable to modern society? If you had time-traveled to Easter Island while the population was still thriving, how would you have tried to help them avert the disaster? Which of the policy principles would you apply and how?

Learning objectives

After reading the chapter, you should understand the following:

- The fundamental concepts that underlie global environmental policy
- The general processes that govern the development of environmental policy and law
- The problems that limit the effectiveness of environmental policy and law

2.1   Introduction

There are many principles and concepts that underlie the creation and design of environmental policy at the highest political levels, as well as policy implementation through innumerable programs and projects in the public and private sectors. The following chapters introduce the fundamental concepts and principles that apply to specific practices and disciplines essential to environmental policy:

Chapter three: Sustainability and environmental policy
Chapter four: Environmental policy treaties and their implementation
Chapter five: Environmental impact assessment
Chapter six: Environmental decision-making theory and practice
Chapter seven: Environmental management systems
Chapter eight: Environmental ethics

For this chapter, we have selected a set of concepts, principles, and practices that crosscut the entire field of environmental policy. As outlined here, this chapter is divided into three logical parts, each of which describes a set of related concepts, principles, and practices.

Part 1: Challenges to the global environment
    2.2 Tragedy of the commons
    2.3 Limits to growth
    2.4 Gaia principle

Part 1 describes the historical principles that have influenced global policy and continue to play a role in policies designed to address issues on a global scale. Part 2 discusses the two central challenges to modern environmental policy faced by virtually every nation and lower governance levels down to the village and urban neighborhood. Part 3 focuses on practices and processes applicable to creating and managing environmental policy at all governance levels.

Part 1:   Challenges to the global environment

2.2   Tragedy of the commons

The case study is a classic illustration of the *tragedy of the commons*, in which human exploitation of resources essential to survival overwhelms resource sustainability. An environmental resource is anything one uses from the environment to meet his or her needs or desires. Some resources are renewable, at least to a limited degree, such as solar energy, water, and crops. Other resources are nonrenewable, at least in terms of human experience; nonrenewable resources include species, oil, minerals, and rainforests. Still other resources are partly renewable, such as soils, groundwater aquifers, and clean air.

The term "carrying capacity" can be understood in terms of the maximum population of a given species that can be supported indefinitely by its environment in a constrained habitat without permanently impairing the productivity of that resource or habitat.

The term "commons" evolved from old English customs or common law. Until the era of the Enclosure Acts, when a long series of acts of the parliament enabled powerful landowners to fence off their properties, turning them into privately held estates, many English villages included a commons or public area of land that could be freely used by any community member to graze their domestic livestock.

In the mid-nineteenth century, William Lloyd was the first to document what is now referred to as the "tragedy of the commons."[1] During the 1960s, Garrett Hardin applied this concept to global environmental policy.[2] As he explains in his 1968 essay, when a village commons is managed judiciously, all users benefit from it. But unchecked, this prosperity inevitably leads to a dilemma—the desire to maximize individual wealth causes overgrazing, which eventually leads to the demise of the entire commons. This principle applies to the depletion of many environmental resources, way beyond that of simply grazing in a village commons, as the following examples illustrate.

Consider a village commons on which 10 villagers graze their cattle. The commons ensures increased prosperity to anyone who is able to utilize its resources. Assume that the maximum sustainable carrying capacity is sufficient to support 100 cattle. This means that each additional cow (up to 100) increases a farmer's wealth without harming anyone else. This continues until each farmer has 10 cows grazing on the commons, each producing 1 unit of utility for a total of 100 units. The carrying capacity has now been reached. From a macroperspective, it is no longer in the interest of the village to increase the number of cattle. In fact, the addition of each additional cow will now reduce the total number of units that can be produced, and the underlying soil will become increasingly compacted or be eroded away, decreasing the overall grass yield.

Now, consider this scenario from a microperspective. It is still in each farmer's interest to add additional cows. Farmer Wantamore, for example, sees a short-term gain from adding a cow but fails to appreciate the long-term adverse implications to the community. Wantamore reasons, "I stand to gain wealth by adding one cow over the carrying capacity, yet pay only a small fraction of the total negative consequences to the commons."

Wantamore gains 1 unit while causing the commons a loss of 1 unit so that total land utility is now 99 units. Each farmer, including Wantamore, loses 0.1 units. But Wantamore's total gain is now up to 10.9 units, while the nine others are down to 9.9 units. As other members catch on, they too start adding cows until the commons fails. Garrett Hardin defined the problem in the following way[3]:

> The tragedy of the commons develops in this way. Picture a pasture open to all. It is to be expected that each herdsman will try to keep as many cattle as possible on the commons. Such an arrangement may work reasonably satisfactorily for centuries because tribal wars, poaching, and disease keep the numbers of both man and beast well below the carrying capacity of the land. Finally, however, comes the day of reckoning. At this point, the inherent logic of the commons remorselessly generates tragedy.

Hardin's principle can be applied to our modern world as well. In less than 300 years, we have moved from creating environmental problems that once wrought disasters in isolated villages to problems that are now wreaking environmental havoc on a global scale. Perhaps the most practical way of managing global commons is to adopt international policies that are sustainable and enforceable (see Chapter 9).

2.2.1   Historical perspective

Humanity has always been at the mercy of nature's destructive forces; storms, hurricanes, tornadoes, volcanoes, earthquakes, floods, droughts, epidemics, plagues, and many other natural and other threats continue to unleash havoc. However, we can now predict, prevent, and to a large measure reduce the damages from such threats, in part because of the revolutionary progress in science, technology, and industry over the last century. But ironically, such progress has likewise created its own threats, and has depleted and ruined many environmental resources. The cumulative effects of these developments have come to pose a threat to the environmental resources of the global commons—its air, water, and land environments.

For some time, visionary thinkers have understood that the global community as a whole is facing environmental resource threats analogous to those of the Easter Islanders. For example, Thomas Robert Malthus (1766–1834) argued that any attempts to ameliorate the condition of the lower classes by increasing their incomes or improving agricultural productivity would be fruitless, as the extra means of subsistence would eventually be completely absorbed by their increasing population. But such visionaries had little influence because global environmental policy mechanisms had not yet been envisioned, let alone established.

How did the very idea of global environmental threats evolve? In 1907, the typical speed of a horse carriage wheeling through the streets of Manhattan was about 12 miles per hour. Given modern technological wonders, carriages are now curious exhibits in museums. These antiques have been replaced by modern high-performance, fuel-injected, turbo-charged automobiles, which can top out at 140 miles per hour—a great improvement over the old horse and buggy. Or, is it? You might be surprised to learn that our marvelous shiny automobiles with chrome wheels now jet a person through the streets of Manhattan at the lightning speed of 5 miles per hour.[4] Not only has such modern technology sometimes gone backward in terms of effectiveness, but we also pay a large environmental price: increased deaths from vehicle accidents while depleting a precious finite resource, petroleum, whose maximum production rate is close to peaking (see Chapter 11).

Consider the case of a public commons called "air." More than 50% of American air pollution is attributed to motor vehicle exhaust. This is particularly disturbing in light of recent research that links ambient air pollution exposure before birth to significantly lower IQ scores in childhood—mainly the result of general exposure to inner-city air pollutants from cars, buses, and trucks.[5] On the positive side, stringent U.S. air pollution regulations have significantly reduced vehicular air pollution and improved air quality according to the Environmental Protection Agency (EPA), as shown in Table 2.1.[6] While this is good news for Americans, air quality and related environmental quality in many other parts of the world are degrading at an alarming rate.

Some toxicologists have testified (from a cost–benefit perspective) that banning lead from gasoline was the most effective health-based measure ever enacted by the United States. Yet, air pollutant and lead pollutant exposure in many cities around the world, particularly in Third World nations, significantly exceed U.S. levels. The penalty in terms of human health in these cities may be incalculable.

Even given its stringent environmental regulations, the United States is still responsible for a sizeable amount of global environmental degradation. An average American's environmental impact is 20–50 times that of a person living in the Third World. With less than 5% of the world's population, the United States consumes 25% of the world's nonrenewable energy, nearly 35% of its mineral resources, and produces about one-third of the world's pollution. The footprints of Europe and other Western nations are not much far behind the United States.[7]

Perhaps most alarming, the developing nations with the largest populations, China and India, are rapidly expanding economically. At their current growth rates, the global environmental impact of China and India will someday greatly exceed that of Europe or the United States. In 2008, China displaced the United States as the world's leading generator of greenhouse emissions. China has shown little regard for environmental

Table 2.1 U.S. Environmental Protection Agency Report: Percent Change in U.S. Air Quality

Pollutant	1980 versus 2008	1990 versus 2008
Carbon monoxide (CO)	−79	−68
Ozone (O_3; 8 h)	−25	−14
Lead (Pb)	−92	−78
Nitrogen dioxide (NO_2)	−46	−35
PM10 (24 h)	−	−31
PM2.5 (annual)	−	−19
PM2.5 (24 h)	−	−20
Sulfur dioxide (SO_2)	−71	−59

Source: EPA, "Air Quality Trends," http://www.epa.gov/airtrends/aqtrends.html.

Note: "−" = Trend data not available; negative numbers indicate improvements in air quality; PM2.5 air quality is based on data since 2000.

quality and appears ready to trade environmental quality for economic growth.

There are two distinct and opposing camps that advocate differing paradigms for controlling pollution at all levels, from global to village. *Technophiles* believe that new technology created by economic demand will essentially solve the world's ills. They reason that if we can land a man on the moon, then with the right mix of policy, science, and investment we can produce technology that will solve any of our earthbound problems. All we have to do is unleash our technological prowess, and no problem is too complicated to solve.

Technophobes have an altogether different view of the world. They see such faith in technology as nothing less than naïve. In fact, they point to countless examples in which modern problems are the direct results of technological development. Those on the far end of the technophile scales believe that modern society has become a slave to its own devices. Consider the following excerpt, which was written by a prominent technophobe who had a particular distrust for modern technology:

> The Industrial Revolution and its consequences have been a disaster for the human race. … The continued development of technology will worsen the situation. … This is not a POLITICAL revolution. Its object will be to overthrow not governments but the economic and technological basis of the present society.[8]

Ted Kaczynski (the Unabomber) is an extreme technophobe who included these words in his manifesto. His booby-trapped bombs were designed to punish people for their role in promoting technology. In all, 16 bombs, which injured 23 people and killed 3, were attributed to him. Kaczynski is part of a very small minority of militant environmentalists who resort to violence and other extreme measures to express their views. Such views have sometimes invoked the Gaia principle, which views the Earth as a living sentient creature that civilization is harming, and that if left unchecked will ultimately destroy us. While Kaczynski is perhaps the most egregious and deranged example of technophobe behavior, the reader should be aware that public perceptions about the effects of pollution are an inherent part of the environmental policy process in a democratic society.

2.2.2   Change does not come easily

Someone once observed, "Everyone talks about the environment, but no one does anything about it." Mark Sagoff refers to it as "environmental schizophrenia." Bewildered citizens watch as entertainers fly private jets to host "environmental" rock concerts. Political figures commute to Congress in private limousines to testify about the risks of global warming. Prominent figures rail against nuclear power, but then oppose wind and alternative energy proposals when they may obscure the seafront view overlooking their mansions. Political figures who live in 30,000-square-foot mansions lecture the American people about the need to consume less energy and protect the environment.

We are all consumers. The choices we make as consumers are frequently at odds with our environmental beliefs. In describing his environmental schizophrenia, Sagoff writes,

> I speed on the highways; yet I want the police to enforce laws against speeding. I used to buy mixers in returnable bottles—but who can bother to return them? I buy only disposables now, but to soothe my conscience, I urge my state senator to outlaw one-way containers. I love my car; I hate the bus. Yet I vote for candidates who promise to tax gasoline to pay for public transportation. I send my dues to the Sierra Club to protect areas in Alaska I shall never visit ... I support almost any political cause that I think will defeat my consumer interests. This is because I have contempt for—although I act upon—those interests. I have an "Ecology Now" sticker on a car that leaks oil everywhere it's parked.[9]

Now consider another example. In 1969, the U.S. Forest Service approved a controversial plan that allowed Walt Disney Enterprises to develop a ski resort in a California wilderness area called Mineral King Valley. The proposal would have opened the area up to an estimated 14,000 visitors daily. The plan would also have involved construction of a 20-mile-long highway through Sequoia National Park. The Sierra Club filed suit, claiming that the Forest Service failed to take into account aesthetic and ecological factors, which outweighed economic interests.

Sagoff used this case in a class to illustrate the discrepancy between the citizen versus the consumer mindset. First, he asked how many students had visited or thought they would visit the Mineral King Wilderness area. A few hands were raised. Next, he asked how many would visit the site if it were developed into a Disney-run ski resort. This time, a lot more hands were raised. Finally, he asked them if they thought the Forest Service was right in granting Disney a lease to develop the resort. This time, the response was nearly unanimously opposed to the Forest Service.

Most students said the Forest Service's decision was despicable and that the agency had violated the public's trust decision. This example illustrates the conflicts that can arise between our choices as citizens and the personal choices that we may make as consumers.

A number of years ago, I took an ecological trip through the Philippines. I hired a guide with a four-wheel drive vehicle to travel into the mountains. Including the driver, there were five of us in the party. Along the road, we stopped at a fast-food restaurant and ordered lunch to go. We ate while we drove. I carefully stored the empty containers in a sack so that they could be disposed of when we returned. As we drove, our conversation turned to the state of the Filipino environment, which is degrading at an alarming rate. The driver was an introspective fellow and complained that the problem was that the Filipino people are not taught to protect and appreciate the environment. He was honestly concerned about the destruction that was taking place before our very eyes. At the summit of the mountain, I walked down the road and took some pictures. As I neared the truck, I could see the driver swinging something as if it were a slingshot. As I got closer, it dawned on me that he had just thrown the entire lunch bag out into a scenic ravine! Many would agree with Sagoff—perhaps, the biggest environmental problem we face is that of environmental schizophrenia.

2.3   Limits to growth

Policymakers peer into the future in an effort to detect adverse consequences of current trends and to craft policies that would prevent or reduce such consequences. This section describes various arguments for and against limits to growth. This topic is critical to understanding the sustainability concept (see Chapter 3) and global population policy (see Chapter 14). The limits to growth principle applies to ecological and economic systems.

In ecology, the principle means that as the population of a given species increases within a given geographic habitat, the supply of food, water, and other essentials eventually fails to meet the survival needs of the species. When this phenomenon begins to occur, members of the species that can migrate do so, while the others simply die out and become extinct, at least in that habitat. In today's world, there is concern that some key resources of the global commons will fail to sustain the needs of the growing human population.

In economics, the principle means that the supply of goods and services to a community becomes scarce in relation to the demand. When this happens, prices increase, and as the scarcity grows, increasing numbers of people will not be able to afford all the necessities of a minimum standard of living such as food, clothing, and medical care. The consequence is poverty and lower life expectancy (see Chapter 9).

At the start of the first millennium, when the Roman Empire prevailed, the world's population was somewhat less than 300 million. It did not reach 1 billion until the dawn of the twentieth century. It took only 80 years for the world to reach its second billion. Since 1850, the world population has increased from 2 billion to its present 6 billion (see Chapter 14). Over most of human history, the population was relatively stable. Births and deaths were nearly equal. However, throughout the Industrial Revolution the rate of population accelerated, enabled by agricultural technologies that increased affordable food production and by urban infrastructure that accommodated higher population densities.

The current U.S. population growth rate is about 1% (including immigration), with a doubling time of 70 years.[10] But many Third World countries experience significantly higher rates and shorter doubling times. The highest rate of national population increase, 3.7% (with a 29-year doubling period), currently occurs in the United Arab Republic and Burundi.[11] The United Nations projects that the current global population will increase from 6 to 9 billion by 2050 (see Chapter 14). This represents an annual growth rate of 1.02%, about the same as the United States, and a doubling in 50 years. While 1% seems like a modest growth rate, the effect of compounding on population growth is of great concern to many environmental policymakers.

2.3.1 Malthusian growth model

Economists and scientists have argued over limitations to growth ever since Thomas Robert Malthus (1766–1834) first popularized his hypothesis in the eighteenth century. Malthus was a pious and benevolent individual. A caring father, introspective scholar, and friendly clergyman, he lived an exemplary life. He was like Cassandra, a "green" mythological figure who prophesied doom. He has been referred to as the world's first professor of political economics.

Malthus examined census data from numerous countries and based his conclusions on a mound of statistics. He popularized his thesis on limits to growth published in 1798, *An Essay on the Principle*. He based the thesis on a simple principle: while the food supply grows at a linear rate (i.e., 1, 2, 3, 4, ...), population grows exponentially (i.e., 2, 4, 8, 16, ...). He concluded that exponential increase in human population would eventually outstrip the ability of society to feed itself.[12]

Malthus suffered a heavy penalty and was unmercifully attacked for his ideas. He countered that his theory was frequently misrepresented; he took pains to point out that his hypothesis did not necessarily predict future catastrophe if people were willing to take action to prevent it, pointing out[13]

> This constantly subsisting cause of periodical misery has existed ever since we have had any histories of mankind, does exist at present, and will for ever continue to exist, unless some decided change takes place in the physical constitution of our nature.

At the time, Malthus held that his Principle of Population could provide a sound basis for predicting our future. He believed it was critical for steps to be taken to control population growth.

Although controversial at the time, the Malthusian growth model has profoundly influenced the fields of socioeconomics and environmentalism. Prior to Malthus, many economists considered a high fertility rate an economic plus since it increased the number of workers available to contribute to the growth of the economy. Following Malthus, a generation of economists began to view fertility from a different perspective, arguing that while a large number of people might increase a nation's gross economic output, sheer numbers also tend to reduce the per capita output available to meet demand.

Malthus's concept continues to be the subject of lively debates to this day. For example, based partly on Malthusian concepts, Paul Ehrlich predicted in the late 1960s in *The Population Bomb*[14] that hundreds of millions of people would die of starvation and disease from overpopulation. He anticipated that this crisis would occur in the 1970s and that life expectancy in the United States would dwindle to only 42 years by the 1980s. Consistent with Malthus's premise, the Club of Rome published equally dire predictions in its 1972 bestseller, *The Limits to Growth*.[15]

2.3.2   Criticisms of Malthus

Today, the Malthusian growth model of population growth versus food supply is widely rejected as too simplistic; it can be demonstrated that for the past two centuries, global food supply has generally kept pace with population growth. For example, at least in developed nations, as population has increased, the price of food and many other resources has generally declined relative to wages. This is because of the stunning developments in science and technology that he had no basis of anticipating. Hence, his analysis was premised on two partially or completely flawed assumptions, discussed next.

2.3.2.1   *Constant rate of exponential increase*

Population statistics demonstrate that population growth is not necessarily exponential over the long term. Modern demographic analyses suggest that growth rates tend to flatten and then invert as a function of economic prosperity. Malthus developed his theory while England was experiencing a high population growth rate, which later began to flatten.

2.3.2.2   *Low rate of growth of food supply*

Modern studies reveal that the intensity of agricultural production rises in response to population increase and market demands. Modern science and technology enabled unprecedented growth in food production and delivery.

One of the key reasons Malthus's predictions did not materialize is that technology, combined with market economic incentives, significantly increased the Earth's carrying capacity. This is supported by *Jevon's Paradox*, which was quoted in a recent work citing Jevon's own 1866 work for the case of coal[16]:

> It is wholly a confusion of ideas to suppose that the economical use of fuel is equivalent to a diminished consumption. The very contrary is the truth ... As a rule new modes of the economy will lead to an increase of consumption ... Now if the quantity of coal used in a blast furnace be diminished in comparison with the yield, the profits of the trade will increase, new capital will be attracted, the price of pig iron will fall, but the demand for it increase, and eventually the greater number of furnaces will more than make up for the diminished consumption of each.

Translated to the food context described here, Jevon's example would read as follows:

> Now if the quantity of resources to produce food on the land is diminished in comparison with the yield, the profits of the food trade will increase, new capital will be attracted, the price of food will fall, but the demand for it will increase, and eventually the greater amount of land for food production will more than make up for the diminished consumption of each farm.

Jevon's phrase "more than make up" means that increased food will be needed to keep up with an expanding demand. In other words, the demand for food will ultimately outstrip the supply, which is consistent with Malthus's assertion. However, both Jevon and Malthus underestimated the power of technology and human ingenuity to increase the food supply. Many experts argue that it is becoming increasingly clear that the current trend of increased food supply may peak earlier than expected (see Chapter 12). In fact, Jevon and Malthus were right in principle, but made wrong assumptions about the limits of technology to continuously increase the efficiency of land and production. Such limits have become even more constrained by worldwide damages to land, water, and air resources by pollution and overuse.

Modern human population growth, however, has been based on finite resources such as petroleum, potable water, and agricultural land, and reliance on these scarce natural resources may yet prove to be unsustainable. Despite continued advances, crop production in some countries is no longer keeping pace with population growth. Increasing drought, protracted heat waves, intensified soil erosion, and the loss of remaining good arable land are all contributing to the problem. Few of the farmers in the world's poorest countries can afford the fertilizers needed to rejuvenate their soils, and considerable debate surrounds the subject of whether genetically modified (GM) crops will be able to contribute in the longer term to continued agricultural growth, to say nothing of the potentially calamitous impacts of GM foods.

Julian Simon was a leading economic optimist and one of the harshest critics of the predictions of environmental doom forecasted by Ehrlich and others. His 1984 book with Herman Kahn criticized the conventional theoretical limitations on population and economic growth.

Simon correctly noted that few of Ehrlich's 1968 predictions[17] about rising prices and famines had actually occurred. He expressed the belief that humans "are not just more mouths to feed, but possess productive and inventive minds that help find creative solutions to man's problems, thus leaving us better off over the long run." In other words, the more the population increases, the greater is the chance that another Einstein will be born who will develop new ways to improve and replenish the Earth's dwindling resources. In support of his thesis, Simon cited statistics showing that some countries with rapid population growth, such as Singapore and South Korea, foster more economic prosperity than other nations.

Environmentalists and social scientists are divided over the issue of environmental degradation and the limits that nature may place on development and population growth. Detractors have painted Simon as an arrogant optimist and argued that social scientists, in particular, have failed to place sufficient emphasis on the intrinsic limitations of technology and nature. At the same time, many scientists continue to warn that ultimate limits do exist on the number of people the planet can support.

However, there is a possibility that this ongoing debate may soon become muted as birthrates have been plummeting in many developed and some developing countries. This decline has led to projections that global human population growth might level off at somewhere around 10 billion people by the middle of this century. Even if this proves to be the case, one should not lose sight of the fact that with a current population of 6.5 billion people, much of the world is already overpopulated and it is frequently in the poorest countries with the weakest economic development and most corrupt governments where population growth continues unabated (see Chapter 14).

2.3.3   Neo-Malthusians

The premise that growth in food production would not keep pace with population has not been empirically confirmed by history. Particularly since the 1960s, world food production has systematically grown above demographic rates. Many argue that the rapid and continued increase in global population over the last century has complemented Malthus's predicted population patterns. *Neo-Malthusianism* is a doctrine that advocates population control. Although the term originally referred to limiting population growth, more recently it has been used to label those who are concerned that overpopulation may increase resource depletion or environmental degradation to a degree that is unsustainable and may even lead to ecological collapse. Fresh water scarcity is perhaps the most compelling concern contributing to the fear of a global food shortage in the near future. Neo-Malthusians typically believe that Malthus was right, but was simply a little ahead of his time. Many neo-Malthusians believe that Malthus is being proven correct, at least on a regional scale for the time being, in sub-Saharan Africa. This part of Africa is characterized by drought, shortage of arable land, and lack of technology, all of which limit food production. In support of their view, neo-Malthusians point out the following:

- The Earth's projected population of 9 to 10 billion people by the middle of this century could wreak huge ecological degradation and resource depletion.
- The Green Revolution, which has enabled India and other countries to feed themselves, has slowed.
- The amount of global farmland per person is steadily declining.
- Fisheries are becoming an increasingly scarce resource.
- Global warming degrades food supplies in unpredictable ways.

Neo-Malthusians assert that the population problem in many developing counties is an inevitable result of high reproductive rates. The theory of demographic transition rejects this view (see Chapter 14), and argues instead that population explosion is a transitory phenomenon occurring in the second stage of demographic transition due to a rapidly decreasing mortality rate without a corresponding fall in the birthrate. The theory of demographic transition maintains that every country passes through this demographic transition and that it can be proven empirically.

2.4   Gaia principle

According to this principle, all of the Earth's life is part of a single self-regulating living system or organism. This can be viewed as a religious and ethical idea in that Gaia was the ancient Greek goddess who protected life on Earth (also see Chapter 8). In the 1970s, the British scientist James Lovelock first proposed his *Gaia hypothesis,* named after the Greek goddess "Gaia" who drew the living world forth from chaos. Lovelock hypothesized that the Earth's life system functions as if it were a single self-regulating living system or organism.

Lovelock's hypothesis ranges between a spectrum of two widely opposing concepts: the almost undeniable (weak Gaia) to the much more sweeping (strong Gaia) hypothesis. Under the weak hypothesis lies the undeniable statement that life has dramatically altered planetary conditions. In contrast, the strong hypothesis goes much further, arguing that the Earth's biosphere effectively acts as if it is a self-organizing system, which works in such a way as to keep its systems in an approximate state of equilibrium that is conducive to life. However, geological history shows that the exact characteristics of this equilibrium have intermittently undergone rapid changes, which are believed to have caused the extinctions of many species.

Some policy critics have invoked the Gaia hypothesis as evidence that the change in environmental quality and limits to growth are either overstated or no threat at all. Others cite it to argue that we need to take immediate action to limit our global ecological footprint (see Section 3.6) before natural forces begin to react, perhaps in a harsh fashion, to bring the Earth back into balance.

On the extreme side of the Lovelock spectrum, some proponents state that the entire Earth is a single unified organism; under this strong hypothesis, the Earth's biosphere is consciously manipulating global processes to create conditions conducive to life. Most mainstream scientists contend that there is no evidence at all to support such a far-reaching hypothesis.

Scientists note that numerous global processes appear to be maintained by homeostatic mechanisms consistent with the Gaia principle. For instance, rising atmospheric carbon dioxide levels enhance plant growth because the increased carbon dioxide concentration increases the ability of organisms to extract this greenhouse gas from the atmosphere, restabilizing the atmosphere; however, this process might also be overwhelmed, leading to a chaotic response. Other examples include

- The atmospheric composition has remained relatively constant (79% nitrogen, 20.7% oxygen, and 0.03% carbon dioxide) over hundreds of millions of years. (However, some scientists argue that these concentrations have actually varied considerably during that time.) Lovelock maintains that this composition should be unstable, and the atmosphere's stability can only have been accounted for by the actions and effects of biological organisms.
- Lovelock has also observed that since life originated, the sun's energy output has increased by 25%–30%; yet the Earth's surface temperature has remained relatively constant over time. He believes that life and geological processes have maintained a reasonably stable climate conducive to life.
- The salinity of the world's oceans has been relatively constant over a long period of geological history. This has posed a long-standing mystery as rivers (carrying salts) should long ago have raised ocean salinity to a much higher level. Saline stability is vital as most life-forms cannot tolerate concentrations much higher than 5%. According to the Gaia principle, geological and biological forces must be working in unison to stabilize critical conditions in such a way as to maintain life.

Lovelock's hypothesis sparked almost instant controversy. For instance, Ford Doolittle argued in a scientific paper in 1981 that there was nothing in the genome of organisms that could explain the feedback mechanisms required by Gaia, and therefore the hypothesis was unscientific.[18]

Despite such criticisms, many supporters maintain there is much to be said for Lovelock's hypothesis. Echoing Lovelock's observations, Lewis Thomas, the author of *Lives of a Cell*, writes[19]:

> I have been trying to think of the Earth as a kind of organism, but it is no go. I cannot think of it this way. It is too big, too complex, with too many working parts lacking visible connections. The other night, driving through a hilly, wooded part of southern New England, I wondered about this. If not like an organism, what is it like, what is it most like? Then, satisfactorily for that moment, it came to me: it is most like a single cell.

The noted astronomer Carl Sagan is said to have joined the debate by suggesting that from an astronomical perspective, space travel and planetary probes appear to provide a perspective in which the Earth, as a "living" organism, may be on the verge of seeding other planetary systems.[20]

Many, and perhaps most, Earth scientists view the factors that stabilize the biosphere as an undirected aspect of the system; the combined actions resulting from competition among species, for example, tend to counterbalance environmental perturbations. Opponents of Gaia argue that there are many examples that show the effects of life have dramatically changed or even destabilized the biosphere (e.g., conversion of the Earth's atmosphere from a reducing environment to an oxidizing one), but proponents counter that in the long run, such changes promoted an environment that was even more suitable to life. Such intense scientific debates resulted in an international Gaia conference in 1988. A second international conference was held in 2000.

Throughout his career, Lovelock has generally been an adamant environmentalist. Yet in his recent book, *The Revenge of Gaia*, the potential effects of global warming have led him to strongly support nuclear power as the only practical technology that can meet the world's increasing energy demands while reducing climatic damage. Lovelock now believes that the global organism is sick, and drastic action must be taken.

Lovelock's pessimism about how climate change will affect the global community stems from his assessment of how Earth and life systems will respond in reestablishing the ecological balance. Earth will adjust to man-induced stresses, but it will do so with revenge. As a control system, Lovelock believes that counterbalancing forces that have generally worked in our favor are now beginning to turn against us. As human activity causes the global temperature to rise, these effects will turn harmful, perhaps, with disastrous consequences.

A number of noted scientists suspect the existence of a threshold set by temperature and carbon dioxide levels, past which the Earth's atmosphere will be irreparably harmed. Activities such as increasing atmospheric carbon dioxide levels, destroying wetlands and forests, and overfarming do not simply produce linear increases in temperature, but can also produce nonlinear effects that amplify the increase in temperature.

Lovelock believes we are now approaching one of these tipping points; our future is like that of passengers on a small raft quietly drifting toward the Niagara Falls. Like a raft going over the falls, the global climate may abruptly "flip" into an entirely new equilibrium state. As Lovelock views it, Gaia has no reason to favor the human species over any other life form. If global warming jeopardizes humanity or results in massive economic disruption, it will also presumably result in a reduction in the principal cause of global warming (i.e., human population). Just how Gaia would then react and "reset the thermostat" to maintain a new global ecosystem is problematic. Whether there is any validity to the strong hypothesis remains to be demonstrated. Regardless, the hypothesis arguably was one of the first serious attempts to show that the Earth's ecosystem is not merely a compilation of unrelated biological processes and chemical reactions that work independently of one another; instead, many processes appear to work in unison to maintain stable environmental conditions.

Part 2: Challenges to environmental policy

2.5 Interdependence of environmental and energy policies

Environmental and energy considerations are intimately interrelated. This section discusses some current issues in energy–environmental policy formulation.

Some policymakers have promoted solar power and renewable energy as if they are benign technologies with little or no detrimental consequences. One of the principal challenges facing many renewable-energy projects is that they tend to consume significantly larger tracts of land than conventional energy sources, such as nuclear or gas- and coal-fired power plants. This represents a major environmental impact that is frequently given appropriate attention in energy policy debates. One study concluded that solar power can consume up to 300 times as much land to produce 1 kW of energy as other conventional sources. The authors of this study predict that by the year 2030, U.S. solar energy production could occupy an additional 79,537 square miles of land.[21] The proposed U.S. SunZia electrical transmission line is a case in point. This line would connect sun- and wind-powered projects in New Mexico with cities in Arizona. Unfortunately, the line would disrupt large tracts of grasslands and skirt along two national wildlife refuges, potentially endangering sensitive species.

In the Southwest, surface water is scarce and groundwater levels are rapidly falling. Western cities are concerned that current water-use levels are not sustainable. New power plants that require water to cool condensers could impair the resource-based economies they are designed to support. However, commercially available air-cooled condensers have a growing market.[22] This technology can solve problems associated with the use of scarce water, while eliminating the thermal impacts of heated discharges on rivers and lakes. However, the environmental impacts of fossil-fuel emissions and high control costs remain. As a result, nonfuel power technologies (such as solar, wind, and geothermal) are challenged to continue lowering their costs if they are to increase their market share of power production.

2.5.1 National oil policy dilemma

By the middle of 2009, oil prices were hovering in the $60-per-barrel range, that is, a price drop of over 50% from their 2008 summer high of $147 per barrel. While consumers applauded the good news, many energy policymakers were less elated. According to many experts, prices fell for the wrong reasons— because of reduced demand resulting from a collapse in the global economy, not because of increased supplies or enhanced energy efficiency. The precipitous drop in prices, coupled with a shaky financial system, worried many policymakers because it also portends a decrease in down-line oil production investment projects.

Oil has always been, and will become increasingly more, expensive to extract (see Chapter 11). Future projects can only be financed if the price of crude stays above a certain level, which will make such projects profitable. When oil prices skyrocket, the public typically views large oil company profits with revulsion and distrust. Yet when prices plummet (i.e., in a volatile oil market) and some oil companies struggle to survive, the public views this favorably. What the public largely fails to understand is that in the end, a lack of sufficient revenue, resulting in lower investment, may come back to hurt the consumer by reducing future supply. Like a dog chasing its tail, this can lead to skyrocketing prices, and permanent high pricing of oil and other energy supplies.

Many U.S. politicians blame soaring gas prices on large American oil companies. But the days when American oil companies ruled the roost and set supply-and-demand have largely come to an end. Today, the Organization of Petroleum Exporting Companies (OPEC) has the power to manipulate supply and with it, the price of world crude. American oil companies have little say in how OPEC chooses to manipulate supply.

Some advocates argue that in order to prevent a calamitous situation, a painful, and perhaps even draconian, energy policy is urgently needed. But to date Congress has shown little stomach for sponsoring expensive and unpopular policies that would anger voters in the interim but could save the public from a potentially disastrous future (see Chapter 11). For example, placing a high tax on gasoline would reduce consumption, significantly extending proved reserves for many years into the future. Revenue generated from the tax could be used to fund alternative energy. Over the long term, this could reduce the amount of money spent on energy. It could also greatly reduce the level of imported oil, which would reduce the trade deficit and thereby strengthen the economy. Unfortunately, the pubic has shown little willingness to accept short-term pain as a means for preventing a festering and long-term calamity, which is almost sure to befall us. This is an example of where "risk communications" could help explain to the public the risk of not accepting some costly pain today (see Chapter 6).

Some advocates argue that all fossil fuels should be phased out in favor of nonfossil alternatives. Others argue for diversity of energy resources as among the fossil and nonfossil sources. Some advocates believe federal subsidy policies are essential to achieving their aims, while others insist energy technologies compete in a free market. There are even those who advocate against any environmental regulations that create costs for companies that extract, convert, and distribute energy.

2.5.2 *How failure to establish a comprehensive energy policy can lead to disjointed decision making*

As it stands, the United States and many other Western nations have failed to establish a comprehensive, science-based and programmatic energy policy. The lack of such a policy can lead to a hodgepodge of inconsistent and uncoordinated courses of action, resulting in poor decisions, wasted resources, and ineffective energy production. A coherent policy can only be accomplished by effective partnerships between the public and private sectors. The following example illustrates the types of problems that can result from uncoordinated polices and planning.

In 2009, gas drillers bid on leases to tap one of America's greatest energy resources, the Marcellus Sale gas deposit, which holds an estimated 480 trillion cubic feet of natural gas. In addition, a study by Department of Energy and Minerals Engineering at Penn Shale University, estimated that the project would generate nearly $14 billion in tax revenues and create 175,000 new jobs.[23]

Natural gas is arguably the cleanest, cheapest, and most abundant form of proven hydrocarbon fuels. The key to unlocking this great resource is hydraulic fracturing, in which engineers inject fluids to fracture the underlying rock so that the gas can be pumped out of the reservoir. This technology has been in use since the early 1950s. Yet an environmental group launched an unsubstantiated campaign alleging that this technology pollutes the underlying groundwater system. Studies by multiple organizations including the U.S. EPA have shown otherwise, concluding that this technology poses no significant danger to the underlying groundwater. The U.S. Secretary of Energy and Nobel Prize-winning physicists have also indicated that they see no reason not to use hydraulic fracturing. Congress is investigating the use of this technology, and some proposals are being circulated to regulate its use to the point that may make it economically impractical for commercial use.[24] Critics label this prospect an unscientific folly that could actually stimulate more detrimental alternatives, such as increased reliance on coal.

A sound national energy policy that includes a strategic implementation plan supported by a competent and science-based environmental impact study might substantively increase energy investment by making it less subject to controversy and unfounded criticism. It would enable a course for energy development that promotes clean and abundant energy resources while mitigating the effects of opposition claims that are not supported by the facts of case and scientific standards of evidence.

2.6   Pollution and its environmental impacts

According to Pojman, the EPA estimated in the year 2000 that perhaps 80% of all cancers are related to pollution. The World Health Organization (WHO) reports that approximately 1 billion urban people (nearly one-fifth of the Earth's inhabitants) are exposed to hazardous pollutants and that emphysema is widespread in the world's major cities.[25]

EPA studies indicate that indoor pollution is typically two to five times higher than pollution occurring outdoors.[26] The EPA estimates that the so-called sick building syndrome (SBD) affects 17% of the 4 million commercial buildings in the United States, including EPA headquarters. SBD is correlated with nausea, headaches, dizziness, sneezing, coughing, chronic fatigue, and so on, as is suspected in other more serious illnesses.

Three factors largely control the threat that environmental pollutants pose to modern society:

1. Concentration of the chemical (the amount measured by volume of mass)
2. Severity or toxicity of the chemical
3. Persistence (how long the chemical remains in an environmental media (water, air, body, etc.)

Ecologists describe environmental impact in terms of a formula. Although a gross oversimplification of the actual science, the formula nevertheless provides a simple first approximation to the pollution problem that can be easily understood by policymakers and the general public alike.

Environmental impact = (population) (resource use per person) (pollution per unit of resource)

Approximately 70,000 synthetic chemicals are in commercial use. The short-term effects of many of these chemicals are poorly understood. Even less is known about their long-term effects on affected humans and other species. But there is one rather chilling statistic: the EPA estimates that 80% of all cancers are caused by man-made pollution. Approximately half of all air pollution is caused by automobiles. The WHO estimates that more than 1 billion urban people (one-fifth of the Earth's total population) are exposed to the effects of air pollution. Air pollution is perhaps the principal factor causing emphysema (an incurable lung disease), which is rampant in major cities around the world. Smog, a mixture of fog and smoke, is responsible for thousands of respiration-related deaths each year. But the impact is not limited to pollution. It extends to resource usage, and in particular, nonrenewable resource usage.

Part 3: Additional policy principles and practices

2.7 Selected principles of international policy and law

International environmental policy and law, as it has evolved, embodies a number of principles whose applications are variously illustrated throughout this book.

2.7.1 Sustainability principle

A species, a population, or an economic system, in order to sustain itself indefinitely, cannot consume environmental resources at a rate that exceeds consumption (see Chapter 3 on sustainability and Part 4: Critical global environmental issues [Chapters 10–14] for additional case studies. Also, see Section 4.2).

2.7.2 Participation principle

This principle was first established by specific National Environmental Policy Act (NEPA) regulations that require public participation[27] (see Chapter 5).

2.7.3 Gaia principle

The Gaia principle (Section 2.4) has a strong ideological component that has inspired antipollution public advocates, including some who have resorted to violence. Nevertheless, policy must take this principle into account because such protests have had some political influence (see Sections 2.2 and 2.2.1 for an extreme example of Gaia-inspired protest).

2.7.4 Effective management principle

Governmental and private enterprise organizations are responsible for managing their respective programs and businesses in accordance with policies and laws that protect the environmental commons (see Chapters 6 and 7).

2.7.5 Polluter pays principle

Parties that damage the local or global commons with activities that pollute the air, water, land, and biotic environments are required to pay a fair share of the costs of preventing or ameliorating the damage (see Chapter 9 for the theory regarding the principle and Chapter 13 as a global case study). The following two principles apply to all of the other stated principles.

2.7.6   Precautionary principle

The Science and Environmental Health Network defines this principle as follows:

> When an activity raises threats of harm to human health or the environment, precautionary measures should be taken even if some cause and effect relationships are not fully established scientifically. In this context, the proponent of an activity, rather than the public, should bear the burden of proof. The process of applying the precautionary principle must be open, informed, and democratic, and must include potentially affected parties. It must also involve an examination of the full range of alternatives, including no action.[28]

NEPA states its purposes as follows[29]:

> The purposes of this Act are: To declare a national policy which will encourage productive and enjoyable harmony between man and the environment; to promote efforts which will prevent or eliminate damage to the environment and the biosphere and stimulate the health and welfare of man; to enrich the understanding of ecological systems and natural resources important to the Nation.

2.7.7   Procedural rights

These are rights that enable the implementation of the precautionary principle. They involve ways to access information and ensure participation in decision making. These are of particular importance in the application of environmental laws, as defined by the United Nations Economic Commission for Europe Convention on Access to Information, Public Participation in Decision-Making and Access to Justice in Environmental Matters, known as the Aarhus Convention (1998). The Aarhus Convention established

- The rights of everyone to receive environmental information that is held by public authorities (access to environmental information).

This can include information on the state of the environment, policies or measures taken, or on the state of human health and safety where this can be affected by the state of the environment. Applicants are entitled to obtain this information within 1 month of the request without having to disclose why they require it. In addition, public authorities are obliged, under the Convention, to actively disseminate the environmental information in their possession.

- The right to participate in environmental decision making. Arrangements are to be made by public authorities to enable the public affected and environmental nongovernmental organizations to comment on, for example, proposals for projects affecting the environment, or plans and programs relating to the environment; these comments are to be taken into due account during decision making, and information is to be provided on the final decisions and the reasons for making those decisions (public participation in environmental decision making).
- The right to review procedures to challenge public decisions that have been made without respecting the two aforementioned rights, or environmental law in general (access to justice).[30]

2.7.8 U.S. National Environmental Policy Act

The EPA places further requirements on U.S. federal agencies to establish procedural rights to persons concerned with the environmental (including economic and social) effects of proposed federal actions[31]:

Agencies shall:

a. Make diligent efforts to involve the public in preparing and implementing their NEPA procedures.
b. Provide public notice of NEPA-related hearings, public meetings, and the availability of environmental documents so as to inform those persons and agencies who may be interested or affected.
c. Hold or sponsor public hearings or public meetings whenever appropriate or in accordance with statutory requirements applicable to the agency.
d. Solicit appropriate information from the public.
e. Explain in its procedures where interested persons can get information or status reports

> on environmental impact statements and other elements of the NEPA process.
>
> f. Make environmental impact statements, the com-ments received, and any underlying doc-uments available to the public pursuant to the provisions of the Freedom of Information Act.

These NEPA principles have generally been accepted by the global community and are directly or indirectly reflected in the body of international environmental agreements (see Chapter 4).

2.7.9   Environmental impact assessments

These assessments are a critical tool of environmental policy. Starting with the U.S. National Environmental Policy Act of 1969, many nations have passed laws requiring that environmental impact assessments be conducted for major projects or programs that would significantly impact the environment (see Chapter 5).

2.8   Policy formulation process and its promulgation

Policy blueprints are usually implemented with imperfect knowledge concerning the ultimate effects or results. Most policies are not simply a list of static goals or laws. To the contrary, policies, particularly environmental ones, involve dynamic and changing goals.

2.8.1   Formulating an environmental policy

Typically, three broad steps are required to effectively formulate and implement a comprehensive policy. The first step provides the foundation and criteria for the next two:

1. A process to define the goals and priorities
2. A program(s) of initiatives and expenditures that can advance the policy goals
3. A method of monitoring (watchdog) progress, accessing the impact of the policy actions, and administrating aspects of the policy

2.8.2   Components of policy

Most well-conceived policies include the components indicated in Table 2.2. However, a formally published policy is only a statement of good intentions until it is actually implemented. When policies are intended to guide human behavior with respect to consumer preferences, the economy, the environment, and the role of the private sector in any aspect

Table 2.2 Basic Components of a Well-Constructed Policy

Background: Indicates the history and reasons, and intent that led to the formulation of the policy statement (i.e., a list of motivating factors).

Scope and applicability: Describes who the policy affects and what segments of society will be affected.

Purpose statement: Outlines why the policy is being issued, and what the desired effect or outcome of the policy is intended to be.

Policy statements: Specifies the particular laws, regulations, orders, or requirements to organizational behavior that the policy is intended to shape or create. Such statements are very diverse and depend on the type of policy being enacted, and the intent and entities involved.

Effective date: Dictates when the policy comes into force. In Western societies, retroactive policies are rare.

Enforcement and responsibilities: Specifies which entities are responsible for carrying out individual aspects of the policy.

Definitions: Provides clear and unambiguous definitions for terms and concepts found in the policy document.

of national or international life, many obstacles may be encountered that obstruct or derail the desired outcomes.

2.8.3 Example: Developing an environmentally responsible energy policy

The policy components listed in Section 2.8.1 are perhaps better explained with the help of the following example. The real world of rational national energy policy poses many challenges. These include integrating the policy with established cultural norms, customs, and lifestyles, and with existing laws, treaties, and agency missions. Development of a pragmatic policy can be problematic. An ill-formulated policy may cause detrimental economic supply–demand mismatches, which can severely harm a nation's economy. Even the best formulated policy will meet with some resistance. For a sovereign nation, development of an environmentally responsible energy policy might require many or all of the measures given in Table 2.3.

The issues cited in Table 2.1 pose challenges from a host of established economic, social, and political interests. Resistance can be expected for any proposal that affects a large class of diverse citizens (see Chapter 8).

2.9 Methodology aspects of policy formulation

This section presents a variety of common environmental policy theories, discusses their merits, and illustrates their applications. Additional theories employed in environmental economics are presented in Chapter 9.

Table 2.3 Scope and Measures That a Comprehensive National
Energy Policy May Need to Address

An explanation of the energy problem and the purpose for the energy policy
statement.

Defining the scope and extent of the energy policy and who will be subject to
its provisions.

When will the energy policy become effective and over what period (if any)
will it remain effective?

Who will be responsible for implementing and enforcing the energy policy?

A set of definitions that define the parameters, requirements, and constraints
used in the policy statement.

A set of specific objectives that will achieve the desired policy. A partial list of
potential policy objectives might include the following:

　Setting goals and standards for conservation and efficiency

　Defining market objectives for energy storage, trading, and fuel transport

　Establishing incentives for encouraging environmentally responsible drilling

　Defining objectives for limiting air emissions, water effluents, and other
　waste-energy-related waste products

　Establishing programs and incentives (such as taxes, exemptions, and
　subsidies) that encourage desired responses from consumers and producers

　Formulating an international initiative that promotes the stated policy goals
　(global, multilateral, and bilateral treaties that coordinate the multiple
　interests of the parties)

　Initiating alternative energy research, development, and commercialization

2.9.1　*Managing uncertainty*

Policy blueprints are usually implemented with imperfect knowledge concerning their ultimate results. Useful policies are not simply static goals or laws; they remain valid and flexible to accommodate geographic, political, and social changes over time. Generally, three broad categories of action are required in formulating global environmental policy:

1. Political and administrative processes that define the goals and priorities, and periodically assess and refine or replace them as appropriate
2. Programs and projects with funded budgets to advance the policy goals pursuant to the established priorities
3. A properly funded monitoring process to periodically assess the effectiveness of programs and projects against the goals and priorities

Althaus, Bridgman, and Davis[32] have advanced a policy model that is both *heuristic* and *iterative* in nature. In this context, the term "heuristic" refers to methods that are useful in solving problems. Commonly used in solving day-to-day problems, heuristics are simply educated guesses, common-sense approaches, rules of thumb, or intuitive judgments. Typically, heuristic methods involve trial-and-error techniques or simple experimentation. Such methods are widely used because they have been so successful in rapidly reaching a reasonable solution to a problem. They are particularly effective when low-cost trials reveal that an approach will not work.

Another common approach is simulation of a well-defined system whose elements can be captured in a computer model. In order to determine the economic ability of a power grid to absorb intermittent energy sources such as wind and photovoltaic energy, a utility organization can simulate the continuous operation of its grid and calculate the amount and value of coal, oil, and gas energy displaced, along with the rate of return on the investment in wind and photovoltaic. In doing so, the utility can assume a variety of policy scenarios related to uncertainties in prices, maintenance costs, and equipment performance. This approach is commonly applied in benefit–cost studies that assess return on investment within specified ranges of uncertainty, such as weather patterns and energy prices (see Chapter 6; also see Chapter 9, Section 9.3).

2.9.1.1 Example: Factors affecting an environmentally responsible energy policy

Experienced environmentally concerned policymakers will typically ask about and debate the following:

- What is the energy demand–supply forecast in the short, intermediate, and long term?
- How can we efficiently integrate the national energy policy with environmental policy along with state and local functions and with international commitments?
- What are the most appropriate mechanisms (e.g., incentives, subsidies, taxes, standards) for implementing the policy?
- What are the external economic, social, and environmental effects of alternative policy choices?
- How can we minimize dependence on unreliable external energy sources and prices, and move toward national self-sufficiency?
- How can we minimize the proportion of GDP represented by energy consumption?
- How can we anticipate and cope with the social impacts of energy prices that make energy unaffordable to less affluent consumers?
- How can we monitor and measure the many effects of the policy such as energy efficiency achieved, and environmental, economic, and social costs avoided?
- Will the bureaucracy cooperate in the policy?

2.9.2 Policy process cycle

The *policy process cycle* provides a useful model for analyzing policy development. One widely cited version of this cycle consists of the following five stages:

1. Agenda setting (identification of a problem to address)
2. Decision making
3. Policy analysis and evaluation (perform analyses to determine if the policy should continue or be terminated)
4. Policy formation
5. Policy implementation

The *Australian Policy Handbook* suggests an eight-step policy cycle[33]:

1. Issue identification
2. Policy analysis
3. Policy instrument development
4. Consultation (with ties through the entire process)
5. Coordination
6. Decision
7. Implementation
8. Evaluation

Defining such cycles is essential to managing the policy development process. In reality, there is constant feedback among these steps. Nevertheless, the cycle is helpful for defining expertise needed, budgets, and schedules, and for monitoring the success of the management and policy implementation. In the following discussion, you will learn about some of the many factors that can alter the intended outcome.

2.9.3 Five-phased policy fermentation process*

Modern society has amassed a long legacy of "solving" one problem by unwittingly creating another in its wake.[34] President Lincoln once announced that "those who fail to learn from history are condemned to relive it." Consistent with Lincoln's insight, environmental problems have a history of running full circle: from a solution, to a problem created by that solution, back to a solution that rectifies the problem generated from the earlier solution. One must wonder why society seems to so often for-

---

* This section is based on an article written by R. MacLean, "Environmental Leadership: The Road to (Environmental) Hell Is Paved with Good Intentions," *Environmental Quality Management* (Winter 2009): 93–99.

get Lincoln's cautionary warning and continues to struggle with relearning the age-old environmental lessons. As described in this section, many environmental problems follow a predictable five-phased process leading to a policy that finally mitigates the mistakes that were made years earlier. But first, we will examine the circular solution–problem–solution dilemma.

2.9.3.1 Around and around we go, and where we stop

Modern engineering and technological innovations are riddled with errors that all seemed good at the time. Consider the use of asbestos, chlorofluorocarbons (CFCs), polychlorinated biphenyls, and the addition of lead compounds to paint and gasoline. All share a common trait in that each was touted as significantly superior and sometimes even safer than older alternatives. Their replacements were often perceived as being less efficient or more hazardous. Even early efforts to centralize and control pollution led to problems. Sanitary landfills were promoted as a means of managing waste before it was recognized that leachate, seeping from these landfills, was contributing to large-scale groundwater pollution. Dichlorodiphenyltrichloroethane (DDT) was hailed as a wonder pesticide in the 1940s, and as a lifesaver for eradicating mosquitoes that carried the deadly malaria typhus (and this may well be the case). By the mid-1960s, however, its wondrous properties had come under intense scrutiny as accusations mounted that DDT was a toxin responsible for decimating bird populations and might potentially contribute to cancer.

Sometimes, the perceptions about an issue actually come full circle. Kudzu, an Asian vine, for instance, was imported to control erosion in earth-working projects. Although it controlled erosion, kudzu was also a highly invasive plant that began to spread out of control, decimating native plant species. Some projects are now underway to eradicate or at least mitigate the effects of its spread, as it has become an undesirable nonnative invasive plant.

Then there is the transportation arena. Beginning in the 1930s, many officials argued that streetcars and overhead wires were old, unsightly technology and began an effort to replace them, ushering in the era of gasoline-powered automobiles. Today, many of these same cities are spending millions to reestablish rail and mass transportation to offset congestion and air pollution created by vehicles.

The benefits of these technologies were all clearly visible at the time of their introduction, but their potential problems were unknown or appeared to be quite remote. This is no less the truth today as we struggle to understand the benefits and risks of an entirely new generation of technologies, such as GM foods and nanotechnology. Potential concerns about cellular phones and brain cancer is yet another controversial and poorly understood problem.

2.9.3.2   Five-phased problem-identification policy process

In terms of environmental policy, major issues tend to evolve over a predictable pattern of phases, beginning with an incipient phase (see Table 2.4).[35] In an ideal world, issues would be resolved in the early stages, leading to the best outcome for all parties involved. Thus, it is in society's interest to identify emerging issues early and deal with them directly rather than outright denying or debunking what may later turn out to be legitimate concerns.

One of the principal reasons such problems can fester for a long time is that powerful lobbying groups and economic interests have vested interest in maintaining certain industries. These industries often defend themselves by debunking the mounting evidence of harm. They also highlight the problems that would result from regulating a technology, in terms of lost jobs, higher costs, or losses to shareholders. Consider the tobacco industry. For decades, this industry has mounted sizable lobbying efforts and funded "research" with the objective of confusing the public and decision makers about the true risks of smoking. In some other cases such as the ban on CFCs, strict regulations have been enacted relatively swiftly, particularly when manufacturers realized these compounds could be replaced by others that might generate even larger revenues. Corporations can be quite proficient in building defensive shields to protect their vested interests, usually marketed in the name of good intentions.

Table 2.4 Five Phases Leading to Policy Formulation

Phase 1: The incipient phase frequently begins with "fringe" set of individuals, often portrayed as "crazed nuts," which initially identify the problem and attempt to put a "spotlight" on the problem by publicizing it. Critics often deny or debunk these early claims outright.

Phase 2: The next phase tends to be followed by scientific research and investigations to study and validate the problem.

Phase 3: The problem becomes more widely publicized and accepted; larger and established public interest groups often climb on aboard and start demanding that various policies and regulations be established to address the problem.

Phase 4: In this phase, some event, which is usually accompanied by widespread media attention, elevates the problem to that of a "crisis" that generates widespread public attention. Public frenzy can reach the point of panic and hysteria, at which point all reason is largely lost and the outcome may be driven more by fanaticism and non-science-based politics than by science and rational dialog.

Phase 5: Finally, as public concerns mount, politicians, policymakers, and regulatory agencies begin to address the problem by forging a public policy to resolve it. This is often the case, even where the actual risks are far less significant than the media attention it receives.

2.9.3.3 *Today's policymaking world*

Decades ago, a lack of scientific evidence and widespread ignorance on the part of policymakers and politicians may have explained the introduction of new and questionable technologies whose potential impacts were not well understood. But is the situation really any different today? Today's battles are often fought in the courtroom or in mass media by dueling experts who argue over nuances, resort to scare tactics, or attempt to elevate the debate to a level that confuses nonexperts.

Despite the enormous strides in modern science, the environmental headlines of today (nanotechnologies, GM foods, new replacement materials, and of course climate change) are frequently hedged in a minefield of endless and complex debates. It is not uncommon to find that rational discourse is replaced by money, politics, and ideology-driven demagogues.

Research funding may be redirected to issues receiving the highest media ratings or those that appear to be the most politically correct, regardless of their actual potential for undesirable consequences. This has led to enormous and expensive efforts that generate so much data (not necessarily results) that scientists complain of "analysis paralysis." Meanwhile, the public is left confused. The ultimate losers are the many serious issues that fail to make it onto the media's radarscope. True progress will probably be made possible only when policymakers are backed by systematic, objective, and scientifically based decision-making processes, such as environmental impact assessments.

2.9.4 *Discrepancies between formal policy and its implementation*

As Althaus, Bridgman, and Davis[36] view the process, policies are intentionally *normative* and not meant to be exacting or predictive. The term "normative" implies that laws, regulations, and other regulatory orders do not necessarily describe how the world is; instead, they prescribe how the world should be. There is a negative by-product for policy formulation. Regulations pursuant to policy sometimes result in unexpected effects, such as encompassing a much wider scope than is needed to address the problem. Thus, it is not uncommon to find a gaping crevasse between the originally stated policy and the specific implementing regulations that are eventually enacted. Sometimes, such differences arise as a result of political compromise between competing interests over how the policy should be interpreted and implemented. In other circumstances, such differences may be traced to a lack of policy implementation by those in enforcement roles who do not share the goals of the policymakers. Unexpected consequences may also arise as a result of selective or idiosyncratic enforcement of a policy mandate.

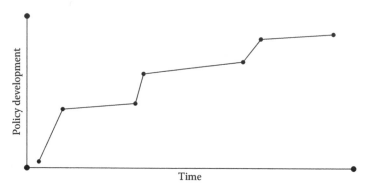

Figure 2.1 Punctuated equilibrium and environmental policy development.

2.9.5   Changes in policies and the theory of punctuated equilibrium

Punctuated equilibrium or *punctuated policy equilibrium* is a theory that seeks to explain the life development of policies. It is analogous to an aspect of the biological theory of evolution. Frank Baumgartner and Bryan Jones first described punctuated equilibrium as applied to political science in 1993.[37] This theory attempts to explain the observation that many policies remain relatively unchanged for a long period of time, before undergoing a sudden or rapid change in response to political factors. This process is represented by the step function depicted in Figure 2.1. Many policies experience abrupt changes in direction or goals; these sudden changes may be correlated with changes in political or party control and in public opinion. They may result from large shifts in how society perceives government, the economy, or the environment.

Baumgartner believes that punctuated equilibrium can be used to explain current trends in the environmental and energy policy arena.[38] The jury is still out, as more recent research with respect to U.S. gun control and state tobacco policies have found a weaker relationship between policies and supposed punctuated changes. For example, a study by Givel concluded that despite a substantial effort to change state tobacco policy, tobacco policymaking from 1990 to 2003 was largely characterized by limited and symbolic punctuation that favored the protobacco lobby.[39] The theory of punctuated policy equilibrium remains controversial and subject to revision.

2.9.6   How diverging interests affect environmental policy

Michael Thompson, who directs the Musgrave Institute in London, writes that four distinctly different schools of thought are interacting with one another to shape today's environmental policy[40]:

1. *Individualists*: Those who believe nature can basically absorb the impacts that are thrown at her. This group believes that regulation is a waste of time and money.
2. *Egalitarians*: Those who believe nature is exceedingly fragile. They favor a concerted effort at the local, grassroots level to protect the biosphere.
3. *Hierarchists*: Those who believe nature is stable within certain limits. They believe in regulating the environment from the top down.
4. *Fatalists*: Those who believe nature is capricious and has no clear principles. This group believes there is little hope and may even seek to escape what they view as the coming wrath of nature.

As Thompson views this framework, these four diametrically opposing forces and their interaction with society at large is actually healthy. An example of a successful outcome involved several villages in the Himalayas that were faced with an urgent need to protect their forests from avalanches. Their final solution involved switching from private ownership to an egalitarian model, which emphasized the need to take collective action to protect the forests. Later, as the forests again came under threat from the onslaught of mining and tourism, the villagers adopted a hierarchical approach, imposing laws against deforestation.

2.9.7 *Schoenbrod's delegation dilemma*

The English political philosopher, John Locke, argued that "the legislative cannot transfer the power of making law to any other hands." This is one reason why the U.S. founding fathers vested *all* federal legislative power in Congress. But in his book, *Power Without Responsibility: How Congress Abuses the People through Delegation*, David Schoenbrod argues that this is no longer the case.[41] Schoenbrod provides a fervent argument against one of the principal ways in which Congress yields it power, by delegating its lawmaking authority to administrative agencies, courts, and special interest groups.

As Schoenbrod views the problem, such delegation breaks the chain of accountability linking voters to their elected representatives. For this reason, legislators are not truly held accountable for rules they should have made but that were actually instigated by other largely non-elected entities. In its defense, Congress argues that it cannot be bothered with every detail or lay out highly prescriptive regulations for every statute it approves; it is argued, perhaps rightly, that disciplines such as environmental policy and law require technical expertise, which only administrative agencies such as the EPA—and not nontechnical legislators—possess. Thus, bureaucrats are delegated pervasive responsibility for dictating the rules that govern our society, with Congress only setting the programmatic policy and direction. To this end, bureaucracies are routinely entrusted with sweeping authority to regulate "in the public interest." Schoenbrod further argues that the U.S. Constitution imposes limits upon such delegation, but that the Supreme Court has largely ignored these restraints.

Schoenbrod attacks the congressional excuse that legislators delegate their power because they lack the time and technical capability to assess the details of their policies; in reality, Congress delegates its constitutional responsibilities as an "invisible" means of exercising its powers without being held accountable for the political consequences.

According to Schoenbrod, such congressional behavior perpetuates many problems. By delegating their powers, legislators can strike bargains with interest groups more easily and with less transparency. With respect to environmental regulation, Congress frequently prescribes absolute environmental goals such as "fishable and swimmable water" or air quality that protects public health "with an adequate margin of safety." The EPA is then authorized to implement such requirements by a certain date (often deadlines that are impossible to meet) with little regard to cost or technological feasibility. In Schoenbrod's view, it has become a highly effective way of institutionalizing irresponsibility.

In implementing their delegated responsibilities, agencies such as the EPA often inflict severe hardships on businesses, workers, and consumers. Not uncommonly, this eventually leads to a public backlash. Predictably, congressmen then rush to denounce the regulatory agency claiming the agency is at fault for the egregious acts. This may then be followed by corrupt practices such as trading campaign contributions for legislation that "rescues" constituents from the "arrogant, runaway" bureaucrats. In some cases, this has forged its own tortuous path, as when President Reagan's first EPA administrator, Ann Gorsuch, called Congress's bluff and threatened to rigidly enforce a nearly unworkable Clean Air Act; as retribution, irate lawmakers engineered her removal from office.

Conversely, if an agency like the EPA assumes a less rigid enforcement policy, lawmakers can simply point fingers at the "irresponsible" bureaucratic agency that has placated polluters and ignored the will of Congress. In the end, legislators hope to receive credit for popular environmental policies that have worked effectively while avoiding political blame for the adverse effects of such policies.

Many critics of our politically influenced lawmaking processes believe that they perpetuate a swelling, meddlesome bureaucracy; closed-room horse trading; and shady arrangements with special interests. Arguably, only sustained public outrage can lead to a substantial change; but delegation is far too arcane a subject for radio talk shows or call-in campaigns. As Schoenbrod states, "It's time for a little perestroika on the Potomac."

2.9.8 Iron Triangle model: Government and the corporations

James Swaney writes about the Iron Triangle,[42] which consists of (1) government bureaucrats in policy or decision-making positions, (2) legislative committees or commissions that have been delegated authority, and (3) corporations with common interests that develop mutually reinforcing attitudes that shape environmental policy. Swaney argues that a corporate elite controls hazardous technologies and that information concerning the associated environmental, safety, and health impacts are frequently shielded from public scrutiny. Consequently, these elites have the power to impose unacceptable risks to an uninformed, misinformed, or unsuspecting public who may actually support policies against their own interests. According to this view, special interests dominate the information that influences the public mindset.

In Chapter 8, we present two extreme case studies in which the international oil industry co-opts weak and corrupt Third World governments to accommodate petroleum extraction at great costs to the environment and indigenous populations. Global policy principles such as Agenda 21 (see Table 3.1 for Agenda 21 Summary and see Table 4.1 for Agenda 21 Statement of Principles) and policies such as those requiring environmental impact assessments (see Chapter 5) provide essential decision inputs designed to prevent such abuses.[43]

Discussions, problems, and exercises

1. What are the principal threats to the global commons?
2. Discuss possible scenarios for each stage of the Australian policy cycle, and comment on challenges likely to be confronted at each step.
3. Discuss how a heuristic policy process could work, along with the risk factors that would impede its success.
4. What did the economist Julian Simon argue?
5. Explain how you think the Iron Triangle might affect efforts to combat global climate change; explain how it may both promote and hamper regulations intended to control climate change.

Class project

The mining industry is notorious for its impacts in terms of pollution and degradation of environmental resources. Develop an environmental policy for managing a nation's mining industry. What legislation, controls, and measures would you include to mitigate pollution? What policy measures would you enact to foster a sustainable mining industry? (Note: See the Case Study given in Chapter 8 before answering.)

Notes

1. G. Hardin, "The Tragedy of the Commons," *Science* 162 (1968): 1243–8.
2. G. Hardin, "The Tragedy of the Commons," *Science* 162 (1968); G. Hardin, *The Immigration Dilemma: Avoiding the Tragedy of the Commons* (Washington, DC: Federation for American Immigration Reform, 1995), 13–30.
3. Hardin, "The Tragedy of the Commons."
4. L. Pojman, *Global Environmental Ethics* (Mountain View, CA: Mayfield Publishing Company, 2000).
5. L. Tanner, "Kids' Lower IQ Scores Linked to Prenatal Pollution." Associated Press, 2009, http://news.yahoo.com/s/ap/20090720/ap_on_he_me/us_med_pollution_iq (accessed July 26, 2009).
6. EPA, "Air Quality Trends," http://www.epa.gov/airtrends/aqtrends.html. (accessed October 8, 2009).
7. See note 4 above.
8. Unabomber Manifesto, *Washington Post*, September 19, 1997.
9. M. Sagoff, "At the Shrine of Our Lady Fatima, or Why Political Questions Are Not All Economic," *Arizona Law Review* 23 (1981).
10. Central Intelligence Agency Factbook, "Population Growth Rate," https://www.cia.gov/library/publications/the-world-factbook/fields/2002.html (accessed July 7, 2010).
11. Ibid.
12. T. Malthus, *Essay on the Principle of Population* (London: J. Johnson, 1798).
13. Ibid.
14. P. Ehrlich, *The Population Bomb* (New York: Ballantine, 1968).
15. Donella H. Meadows and others, *Limits to Growth* (Rome: Club of Rome, 1972).
16. John M. Polenimi, K. Mayumi and M. Giampietro, *The Jevon's Paradox and the Myth of Resource Efficiency Improvdements—Forward* (London: Earthscan, 2008), http://books.google.com/books?hl=en&lr=&id=nfHDSSqi4NQC&oi=fnd&pg=PR7&dq=jevons+paradox+population&ots=PVcLegZLt6&sig=uuu5ghyUX2UOJ96CgW-fKLvJugg#v=onepage&q=jevons%20paradox%20population&f=false (accessed July 7, 2010).
17. See note 14 above.
18. F. Doolittle, "Is Nature Motherly," *CoEvolution Quarterly* (1981).
19. L. Thomas, *Lives of a Cell* (London: Allen Lane, 1980).
20. Wikipedia.com, "Gaia Hypothesis," http://en.wikipedia.org/wiki/Gaia_hypothesis (accessed March 26, 2007).
21. A team of scientists, several of whom work for the Nature Conservancy, has written a paper that will appear in the journal *PLoS*.
22. *Electric Power Magazine*, "Air-Cooled Condensers Eliminate Plant Water Use," http://www.powermag.com/water/Air-cooled-condensers-eliminate-plant-water-use_1361.html (accessed July 7, 2010).
23. League of Women Voters of Pennsylvania, *Marcellus Shale Natural Gas Extraction Study 2009–2010* (2009).
24. *Washington Examiner*, January 20, 2010, P2.
25. L. P. Pojman, *Global Environmental Ethics* (Mountain View, CA: Mayfield Publishing Company, 2000), 249.
26. EPA, *Indoor Air Quality Basics for Schools* (Washington, DC: EPA, 1996).

27. National Environmental Policy Regulations 1506.6. Public "involvement" establishes specific requirements for meaningful public involvement.

28. Science and Environmental Health Network, "Precautionary Principle," http://www.sehn.org/precaution.html (accessed July 7, 2010).

29. National Environmental Policy Act, Part 1500: Purpose, Policy, and Mandate.

30. United Nations Economic Commission for Europe, "The Aarhus Convention, June 1998," http://ec.europa.eu/environment/aarhus/ (accessed July 7, 2010).

31. *National Environmental Policy Act Regulations* 1506.6, "Public Involvement," http://www.nepa.gov/nepa/regs/ceq/1506.htm#1506.6 (accessed July 7, 2010).

32. C. Althaus, P. Bridgman and G. Davis, *The Australian Policy Handbook*, 4th ed. (Sydney: Allen and Unwin, 2007).

33. Ibid.

34. D. D. Anderson, "Key Concepts in Anticipatory Issues Management," *Corporate Environmental Strategy* 5 (1997): 6–17.

35. Ibid.

36. See note 32 above.

37. F. R. Baumgartner and B. D. Jones, *Agendas and Instability in American Politics* (Chicago: University of Chicago Press, 1993).

38. F. R. Baumgartner, "Punctuated Equilibrium Theory and Environmental Policy," http://www.policyagendas.org/pdf/paper_Baumgartner1.pdf (accessed February 15, 2009).

39. M. Givel, "Punctuated Equilibrium in Limbo: The Tobacco Lobby and U.S. State Policy Making from 1990 to 2003," *Policy Studies Journal* 43, no.3 (2006): 405–18.

40. M. Thompson, "Understanding Environmental Values: A Cultural Theory Approach," http://www.cceia.org/resources/articles_papers_reports/710.html (accessed February 15, 2009).

41. D. Schoenbrod, *Power without Responsibility: How Congress Abuses the People through Delegation* (New Haven: Yale University Press, 1993), 255.

42. J. A. Swaney, "Social Economics and Risk Analysis," *Review of Social Economy* 53 (1995): 575–94.

43. See Chapter 1 for how the environmental impact assessment process first came about in the United States and then spread to European and other nations.

chapter three

Sustainability and environmental policy

> Suburbia is where the developer bulldozes out the trees, then names the streets after them.
>
> **Author unknown**

Case study

The world's first energy crisis was a progressive crisis, one that had no parallels in the annals of history. There was no distinct event that actually marked the crisis, although it was clearly beginning to fester around the year AD 1500. The crisis only intensified as demand increasingly outstripped supply. Even so, it was not initially a universal problem, but one that started principally in England and crept outward to other areas of Europe. Before it was over, it would come to jeopardize the very advancement of European civilization.

The demand for wood, which heated houses, cooked meals, was the lifeblood of ironworking, and provided building material for the great sailing ships of the time, soared to the point that it could no longer be supplied from within the British Isles. The price of this commodity began to rise sharply. Unsustainable wood harvests had decimated the forests of England and, to a lesser extent, much of Europe as well. Most of Europe found itself in the midst of a wood crisis by the end of the seventeenth century. Eventually, new wooden buildings were banned in London. The long era in which wood had fueled Western civilization was coming to a screeching halt. The world's first energy crisis was at hand.

Coal saves the Industrial Revolution

The world's first energy crisis was now threatening the Industrial Revolution and the reemergence of Western society. Western civilization was moving perilously close to a return to the Dark Ages.

Coal deposits, which occasionally outcropped along the land surface, had for centuries been relegated to a curiosity. This soft black rock would save everything. Inventors of the day learned how to harness the fantastic power of coal in the nick of time. The era of coal was born. People learned to heat homes, cook food, manufacture iron, and run machines that powered the Industrial Revolution with coal. James Watt would go on to invent the steam engine, which was fired by this shiny rock. It would spark the Industrial Revolution. Coal soon made its way to the United States, and the rest is history.

A crisis had been averted across Europe. But the transition from wood to coal was not an easy one. Coal was difficult and hazardous to mine. It was dirty. The skies in London turned dark with smog.

In the nineteenth century, coal began to be replaced by oil, an even more powerful energy source, and still later by nuclear energy. But you will recall, Europe's brush with disaster had been the result of placing unsustainable demands on its timber resource. Today, it is alarming that everything from oil (see Chapter 11) to minerals and our agriculture are being utilized at unsustainable rates (see Chapter 12). Will we dodge another disaster by pure luck, much as Europe did its timber industry crisis, or is a day coming when we will indeed have to pay the piper?

> **Problem:** After reading this chapter, return to this problem. Consider this case study in light of what you have read. This problem may be pursued either as an individual assignment or as a class project. In light of what you have just read, develop a sustainable resource policy for water, food, or energy, for some country or region.

Learning objectives

After you have read the chapter, you should understand the following:

- Definitions of sustainability
- The various definitions and meanings of sustainability
- The concepts of sustainable economies, energy, water, and food
- Key principles of sustainability
- The concept of sustainable populations
- How to measure sustainability
- Basic concepts related to achieving a sustainable energy policy

Traditional economists argue that as long as the price is right, the laws of supply and demand will provide needed resources while avoiding the harmful environmental externalities (see Chapter 9). Conversely, many environmental scientists are skeptical of the market's ability to supply goods and services over the long term and argue that the last generation has shown that sustained growth will result in continued environmental degradation or even catastrophe. Tillard claims that over the past 50 years, the global population has consumed more goods and services than the combined total of all previous generations.[1] Friedrich Schmidt-Bleek and others have taken this claim a step further, writing that developed nations account for just 20% of the global population yet consume 80% of the world's resources.[2]

It is worthwhile to examine how societies have responded to past environmental stresses. Jared Diamond's popular book, *Collapse: How Societies Choose to Fail or Succeed*, describes how the collapse of ancient societies such as the Anasazis, Maya, and Easter Islanders was due in large part to their failure to understand and mitigate the effects of their environment footprints. These societies were isolated, with only primitive knowledge and little ability to seek distant resources. Today, we have a much better understanding of our actions and an appreciation for sustainable development practices. Ecosystems do not necessarily change in a linear, predictable manner. As with many chemical reactions little change may occur until a critical activation energy is reached, after which things can proceed rapidly in nonlinear ways. For instance, discontinuities have occurred when species suddenly crashed instead of gradually declining. A possible scenario that could cause unexpected havoc (see Chapter 13) is that global warming might actually shut down the oceanic conveyor belt that distributes thermal energy, so that parts of the world become frigid.

Assuring the future and security of our grandchildren may sound quaint in an era of conspicuous consumption and instant oatmeal. Yet, history is littered with the remnants of societies that failed to plan ahead and use their resources wisely. Easter Island and the Maya civilization are two celebrated examples.

Although it may sound like a trendy buzzword, "sustainability" is not a novel concept. For instance, among the Iroquois Indian nations in North America, tribal elders were required to weigh the impact of important decisions for a full seven generations onward.

Is the term "sustainability" merely an overused and meaningless cliché, or does its application truly have the power to change the world? This term has been applied to everything from local communities, to agriculture and economic systems, to communities and automobiles. Yet, despite its almost inherent simplicity, this concept can be very deceptively difficult to define, communicate, and implement in practice.

3.1   Definitions of sustainability

The influential book, *Our Common Future*,[3] popularized the concept of sustainable development. The potential benefits of sustainable development are indeed alluring. However, it has also been said that nobody understands what the term sustainability really means. Of course, this is an exaggeration. Although as yet there is no universally accepted definition or concept of sustainability, various definitions have been proposed. Most definitions involve adopting a collection of economic, social, and environmental goals that are consistent with each other and mutually attainable. The modern use of the term sustainability is first attributed to its use in *Our Common Future*, in which the Brundtland Commission defined sustainable development as being

> Development which meets the needs of the present without compromising the ability of future generations to achieve their needs and aspirations.[4]

However, the scope of sustainable development can be viewed more comprehensively than by simply considering it in terms of natural resources. As a more comprehensive concept, sustainability can be defined as:

> Development that delivers basic environmental, social and economic services to all without threatening the viability of the natural, built and social systems upon which these services depend.[5]

The International Chamber of Commerce writes

> Sustainable development means adopting business strategies and activities that meet the needs of the enterprise and its stakeholders today while protecting, sustaining and enhancing the human and natural resources that will be needed in the future.[6]

Yet another definition based more on consumption of goods and services is that of "sustainable consumption." Under this definition,

> Sustainable consumption is the consumption of goods and services that have minimal impact upon the environment, are socially equitable and economically viable whilst meeting the basic needs of humans, worldwide. Sustainable consumption targets everyone, across all sectors and all nations, from the individual to governments and multinational conglomerates.[7]

In general, the concept of sustainable development or sustainability means that the consumptive use of renewable resources does not exceed the regenerative capacity of the environment.[8] Social progress, environmental protection and preservation, conservation of resources, and economic maintenance are all elements of sustainable development. Factors such as quality of life concerns, biological and cultural diversity considerations, and conservation, not to mention philosophical questions for humanity, are also within the aforementioned constraints. The welfare of future generations also fit into the sustainable development equation.

Sustainable yield can be thought of as the optimum level of production (e.g., timber, fisheries, or water) of a renewable resource that can be maintained indefinitely. In economic terms, it represents the maximum long-term level of income that can be derived from the use of a resource without causing eventual degradation or depletion of that resource.

It should be noted, however, that many ecologists largely reject the concept of maximum sustained yield that was promoted in the last century by commercial forestry, agricultural interests, and fishing interests; they reject this concept because it assumes a long-term stability in the underlying ecosystems that usually cannot be demonstrated to exist. That is to say, natural systems are usually more complex, more variable, and less stable in their response to disturbance than such a production strategy can account for.

3.1.1 Achieving sustainability

Neil Carter identifies principles essential to sustainable development[9]:

1. *Equity*: Our inability to promote the common interest in sustainable development is often a product of the relative neglect of economic and social justice within and amongst nations.[10]
2. *Democracy*: Sustainable development requires a political system that ensures effective citizen participation in decision-making.[11]
3. *The precautionary principle*: In order to protect the environment, the precautionary approach shall be widely applied by states according to their capabilities. Where there are threats of serious or irreversible damage, lack of full scientific certainty shall not be used as a reason for postponing cost-effective measures to prevent environmental degradation.[12]
4. *Policy integration*: The objective of sustainable development and the integrated nature of global environment/development challenges pose problems for institutions that were established on the basis of narrow preoccupations and compartmentalized concerns.[13]
5. *Planning*: Sustainable development must be planned. Only free market environmentalists believe that the unfettered market can of its own volition produce sustainable development.

Moreover, Principle 4 of the United Nations Agenda 21 declares (see Chapter 4),

> In order to achieve sustainable development, environmental protection shall constitute an integral part of the development process and cannot be considered in isolation from it.

Sustainable development is also one of the pillars of the United Nation's eight Millennium Goals pursuant to its population policy (see Chapter 14):

> Goal 7: Integrate the principles of sustainable development into country policies and programs and reverse the loss of environmental resources.

Specifically, Goal 7 declares the following four "targets" pursuant to that goal:

1. Integrate the principles of sustainable development into country policies and programs, and reverse the loss of environmental resources.
2. Reduce biodiversity loss, achieving, by the year 2010, a significant reduction in the rate of loss.
3. Halve, by the year 2015, the proportion of the population without sustainable access to safe drinking water and basic sanitation.
4. By the year 2020, achieve a significant improvement in the lives of at least 100 million slum dwellers.

3.1.2   Difference between sustainability and sustainable development

The terms "sustainability" and "sustainable development" are often used interchangeably. Before proceeding further, however, it is instructive to discuss the difference between these two terms. *Sustainability* refers to a long-term and perhaps unachievable goal. In contrast, *sustainable development* refers to the highly variable *process* used to move us closer to that goal of sustainability. More to the point, sustainability is the intended goal or outcome, whereas sustainable development refers to the developmental process or steps taken to achieve this goal.[14]

3.2 Impact, population, affluence, and technology equation

One of common precepts of sustainable development advocates is the idea that environmental effects can be mitigated with advances in technology despite increases in the number of consumers and their per capita income. This school holds that the human impact on ecosystems can conceptually be expressed and compared by an impact, population, affluence, and technology (IPAT) index:

$$\text{Impact} = \text{population} \times \text{affluence} \times \text{technology}$$

Obviously, population and technology play an important role in sustainable development. However, others argue that the role of affluence in this formula is much more significant than consumption patterns would indicate and is harder to describe and quantify. Moreover, increased wealth may overwhelm the ability of technology to mitigate the environmental impact of increased population.

3.2.1 Technological paradox

Optimists argue that increased technology will more than offset the resource depletion that results from affluence, through the invention of substitutes, discovery of new replacement materials, or improvement of efficiencies in reuse and recycling. Critics counter that the lure of sustainability has been obscured by the promises that technological advances will yield increased affluence forever.

The last couple of generations have indeed shown that resource prices (the ultimate measure of supply and demand) have generally decreased with time, while the standard of living has improved. Paradoxically, however, lower cost and improved technologies have actually reduced recycling and conservation measures and have caused accelerated resource depletion.

Although modern technology has moderated prices and made resources more abundant, at least in the short term, the number of proven reserves has not often kept pace with a global increase in demand. This observation applies to many nonrenewable ore deposits, as well as renewable resources like fresh water, topsoil, rainforests, and fish populations.

Researchers Michael and Joyce Huesemann analyzed the question, "Will technological progress avert or accelerate global collapse?" They came to the conclusion that technology alone is insufficient to achieve sustainable development. While technological progress has improved efficiencies, it has also increased both the number of consumers and their affluence. Without a major environmental policy change, they concluded that technological progress will ultimately cause an impending collapse.

The best available projections indicate that the Earth's population will stabilize at around 9 billion by the middle of the twenty-first century (see Chapter 14). The implication of this global rise in population is that individual consumption levels resulting from increased affluence are likely to grow disproportionately by a much larger amount compared to today's rates.

3.2.2   Marketing of "sustainable products" vs. the goal of profitability

Companies, industries, and government agencies have all reported numerous "sustainability success stories." But in reality, many of these success stories have done little more than reduce the rate of unsustainability. There is a significant difference between sustainability and reducing the rate of unsustainability—a difference that is rarely communicated. MacLean notes that in his experience, most sustainability mangers hired by large companies have limited backgrounds in environmental science and regulation and are, in fact, extensively trained in communication, marketing, and product development.[15] He concludes that they were principally hired to increase profits and build brand name recognition, with the goal of achieving truly sustainable products relegated to a distant priority.

Thus, marketing green products does not necessarily translate to a measurable environmental benefit or a commitment to sustainable development. Packaging an environmentally destructive product in a biodegradable wrapper may appeal to environmentally conscious consumers, but it is doing next to nothing for the environmental health of the planet.

By shifting the discussion to sustainable development and away from environmental compliance, many corporations have been able to change public opinion through the marketing of questionable "environmentally friendly" products. The problem is that both the public and companies may begin to believe their headlines and ignore the tough choices that must be made to achieve environmental stewardship.

3.3   United Nations' sustainability policy: Agenda 21

The concept of sustainability is the baseline goal for Agenda 21, the 40-chapter document that details the goals and programs resulting from the United Nations Conference on Environment and Development (informally known as the Earth Summit) held in Rio de Janeiro, Brazil, in

June 1992. The Rio Conference was the follow-up to the United Nations Conference on the Human Environment, the first global conference ever convened on the environment, held in Stockholm, Sweden, in 1972. Today, the United Nations remains committed to the global goal of sustainable development, a mutual challenge to societies, economies, and environments around the world.

Agenda 21 provides 27 principles for the implementation of its strategy (Table 3.1). Nearly half of these sustainable development principles focus on actions undertaken by national governments. The remainder focuses on actions undertaken by individuals and organizations.[16]

Table 3.1 Summary of the 27 Principles Contained in Agenda 21[a]

1. Human beings are entitled to a healthy and productive life in harmony with nature.
2. States have the right to exploit their own resources but without damage to others.
3. The right to development must meet the needs of present and future generations.
4. Environmental protection is an integral part of the development process.
5. People must eradicate poverty to decrease disparities in standards of living.
6. Needs of the least developed and the most environmentally vulnerable must be a priority.
7. States must cooperate to conserve, protect, and restore Earth's ecosystem.
8. States are to eliminate unsustainable patterns of production and consumption.
9. States are to improve scientific understanding to strengthen capacity building.
10. Environmental issues are best handled ensuring participation by all concerned.
11. States must enact effective environmental legislation.
12. States are to promote supportive, open economics for growth and development.
13. States must have laws to protect victims of pollution and environmental damage.
14. States are to cooperate in order to discourage and prevent severe environmental degradation.
15. The precautionary approach must be applied to threats involving serious damage.
16. Authorities are to promote "polluter pays" with due regard to the public interest.
17. Impact assessment must be undertaken for likely significant adverse impacts.

(Continued)

Table 3.1 *(Continued)*

18. States must notify others of disasters or emergencies likely to harm others.
19. States must notify others of transboundary environmental effects.
20. Women's full participation is essential to achieve sustainable development.
21. World youth partnership is essential to achieve sustainable development.
22. Indigenous people and communities have a vital role in sustainable development.
23. The environment and resources of people under oppression are to be protected.
24. Warfare is inherently destructive to sustainable development.
25. Peace, development, and environmental protection are interdependent and indivisible.
26. States must resolve environmental disputes peacefully.
27. States and people must partner for sustainable development.

[a] See Chapter 4, Table 4.1, for the complete 27 principles.

The complete first principle declares, "Human beings are at the centre of concerns for sustainable development. They are entitled to a healthy and productive life in harmony with nature."

In accordance with these principles, the United Nations promulgated a 40-chapter implementation plan (see Chapter 4, Table 4.3). Seven of these areas explicitly implement the sustainability principle. Moreover, 6 of the 27 principles are grounded on sustainability (as shown in Chapter 4, Table 4.1). Virtually all other elements in the policy are required for the goal of sustainable development to succeed.

In 2000, with Agenda 21 in mind, the United Nations identified eight Millennium Development Goals. In 2002, the United Nations World Summit on Sustainable Development held in Johannesburg, South Africa, benchmarked the goals and agenda to the world, generating an improved implementation plan.

Sustainable development is also embedded within ISO 14001, an increasingly popular international standard for environmental management systems (see Chapter 7). The ISO 14001 standard embraces Agenda 21 from the Earth Summit along with strategies outlined by the International Chamber of Commerce business charter for sustainable development.

3.4 Common principles of sustainability

Hargroves and Smith (2005) have identified a number of common principles that can be found in most pragmatic sustainable development programs.[17] These principles are depicted in Table 3.2.

Table 3.2 Common Principles Found in Most Pragmatic Sustainable
Development Programs

Integration of economic, social, and environmental goals in policy formulations

Commitment to best practices of sustainable development

Promoting continuous improvement

Dealing transparently and systemically with risk and uncertainty

Ensuring appropriate valuation, appreciation, and restoration of natural
 resources and environs

Conserving biodiversity and ecological integrity

Avoiding the loss of human capital as well as natural capital

Providing opportunities for community or stakeholder participation

Being cognizant of intergenerational equity

3.4.1   Sustainable development, the National Environmental Policy Act, and the Environmental Protection Agency

Although the National Environmental Policy Act (NEPA) predates the modern concept of sustainable development, the rudimentary concept is nevertheless embedded in the Act. Consider the following two excerpts from NEPA:

> *Productive and enjoyable harmony* between man and his environment; to promote efforts which will *prevent or eliminate damage* to the environment and biosphere and *stimulate the health and welfare* of man ...[18] (Emphasis added.)

> It is the continuing policy of the Federal Government to use all practicable means and measures, including financial and technical assistance, in a manner calculated to foster and promote the general welfare, to create and *maintain conditions* under which man and nature can exist in *productive harmony*, and fulfill the social, economic, and other requirements of present and *future generations* of Americans.[19] (Emphasis added.)

The U.S. Environmental Protection Agency (EPA) has taken its own path toward describing sustainable development. The EPA's Center for Sustainability promotes the linking of environmental, economic, and social goals to "enhance" quality of life and encourage livable communities that can someday realize a "new American Dream." It recommends protecting vital resource lands, conserving energy and nonrenewable resources, and reversing unsustainable transportation trends.

The concept of sustainability has also received significant international attention, particularly within Europe and among other industrialized nations. For example, in 1991, the Resource Management Act of New Zealand was enacted. This act blazed a new precedent by articulating what some experts have called the world's first legislative statement promoting the principle of sustainability.

3.4.2   Executive order on sustainability

A new executive order (EO), EO 13514, Federal Leadership in Environmental, Energy, and Economic Performance, directs federal agencies to establish an integrated strategy for sustainability and make reduction of greenhouse gas (GHG) emissions a federal agency priority.[20] This EO directs agencies to enhance other aspects of sustainability by reducing water consumption, minimizing waste, supporting sustainable communities, and using federal purchasing power to promote environmentally responsible products and technologies.

In implementing this EO, each agency must submit an annual Strategic Sustainability Performance Plan, subject to approval by the Office of Management and Budget (OMB) director, to address among other topics

- Sustainability policy or goals, including GHG reduction targets
- Integration with agency strategic planning or budgeting
- Schedules or milestones for activities covered by the EO
- Evaluation of past performance based on net lifecycle benefits
- Planning for adaptation to potential climate change

3.4.3   Adoption of sustainability policies

To achieve its goal, sustainability requires that a proactive approach encompass economic development and preservation of environmental quality. The overlapping of governance with ecological, social, and economic needs generates the interdisciplinary nature of sustainability science. How we identify such overlaps depends partly on how we frame the environmental issues. If policymakers perceive resource depletion as the prevailing issue, they may emphasize recycling policies; if they believe there is an unlimited amount of extractable resources or perceive new technologies and substitutions on the horizon, they may have little incentive to protect such resources. Similarly, if the public and politicians believe that automobile emissions are the cause of global warming, they may promote more fuel-efficient cars or seek other alternatives.

Citizens must also feel empowered and must believe that their efforts have some meaning. Individual perceptions mold the collective will, and hence shape policies, which can institutionalize what may have begun as simple lifestyle choices. With respect to automobiles, monetary incentives can encourage people to do the right thing (i.e., purchase more fuel-efficient vehicles). Conversely, prohibitions are more heavy-handed policy instruments. In forging a sustainable development policy, the beliefs of key policymakers, the system of governance, and individuals and organizations all play a role.

3.4.3.1 Basic requirements for global sustainability

Three major elements are required to develop and achieve an international sustainability strategy that can protect the global commons:

1. Commitment
2. International cooperation (see Chapter 4)
3. Ability to assess global impacts and develop comprehensive plans for mitigating their effects (see Chapter 5)

3.4.3.2 Sustainability hierarchy

A *sustainability hierarchy* was first proposed by Marshall and Toffel to provide a framework around a variety of issues that have been associated with sustainability.[21] This hierarchical system categorizes actions as unsustainable based on their direct or indirect potential to

1. Endanger the survival of humans.
2. Impair human health.
3. Cause species extinction or violate human rights.
4. Reduce quality of life or have consequences that are inconsistent with other values, beliefs, or aesthetic preferences.

Marshall and Toffel argue that for sustainability to become a more meaningful concept, the various worthy issues in the fourth category (values, beliefs, and aesthetic preferences) should not be considered sustainability concerns.

3.4.3.3 Economic costs of sustainability

The rising costs of fossil energy and material commodities have much to do with motivating environmentally sustainable policies. The principal reason American automakers are currently in economic trouble is that they have pushed high-profit, gas-guzzling sport utility vehicles (SUVs) and trucks to the detriment of more fuel-efficient vehicles. When the price of crude oil skyrocketed in 2008, the market for big cars crashed and has not yet recovered, even though gas prices have plummeted. Over the long run, rising energy prices will stimulate research into more fuel-efficient and sustainable technologies.

Sometimes, ensuring sustainability does not require radically new technologies but merely a new way of looking at an existing business model. The biggest challenge to marketing electric cars involves the development of a new generation of batteries that can power cars over a longer or perhaps even unlimited range. Clearly, range limitation is a significant reason many people shy away from electric vehicles. But what if the range problem can be overcome without developing a new generation of electric batteries?

Shai Agassi, an Israeli entrepreneur, wants to do just that—electrify the world's car fleet—not by developing a new battery that gets 250 miles on a charge, but by providing a system that allows drivers to travel as far as they want without stopping for a time-consuming recharging. He proposes to establish a system of highway battery exchange stations, analogous to the gas canister exchanges people now use for barbecue grills. Drivers would simply pull in to a station and the drained battery would be swapped for a fully charged one within a few minutes.[22]

3.4.3.4   Sustainable dilemmas

A policy that may at first appear to be sustainable can be anything but on closer inspection. Consider policies that promote the production of fuel ethanol from corn. Corn is a renewable resource; in fact, it is a solar-based product. What sounds cleaner and more sustainable than a national ethanol program? It seems a great idea until one performs a more thorough lifecycle analysis, which reveals the high energy intensity involved in cultivating and harvesting corn and its subsequent conversion into ethanol fuel. An ethanol program is only marginally more efficient and sustainable than producing gasoline. But the story does not end there. Diversion of corn to manufacture ethanol can only mean less corn will be left to feed livestock and people, which will drive the cost of food up. In some cases, rainforests in locations such as Brazil are being destroyed to produce farmland to grow crops that produce fuel. The loss of valuable rainforests in turn increases carbon dioxide concentrations, which negates one of the most powerful arguments for manufacturing ethanol in the first place. We cannot confidently declare any practice as sustainable until a full lifecycle analysis of its environmental costs has been performed.

3.4.4   Critics of sustainability

Critics argue that sustainability has become such an overused buzzword as to have lost much of its meaning. More to the point, no sustainable blueprint of Shangri-la will bring Shangri-la to existence. There are volumes of discussions about impending doom if sustainable paths are not followed.

Some of these dire warnings are based on rigorous scientific analysis, and some are pure hyperbole.

Julianne L. Newton and Eric T. Freyfogle argue that the term has no clear goal confusing sustaining natural systems with other human goals and values.[23] The key point derived from reading much of the literature critical of sustainable development is that it is a movement that is more hype than substance.

Stanley Temple believes that the term "sustainable" has no real meaning. He cites proposals circulating for "sustainable development," "sustainable growth," "sustainable economies," "sustainable societies," "sustainable healthcare," "sustainable agriculture," and "sustainable energy"—it seems everything is sustainable. He questions the very use of the words "sustainable development" and so on. For Temple, the word sustainable has been used in too many situations, and "ecological sustainability" is just one of those terms that confuses people. In his view, the term has come to mean too much and nothing at all, at the same time.[24] Paul Reitan points out that, "Surely no one would want simply to sustain the maximum number of humans organized into societies, knowing that that would mean existence at the barest, meanest survival level."[25]

3.5   Theory of sustainable population growth

We live in a global community of 6.8 billion citizens and growing. It is all but certain that at least another 2 billion people will be added to the population of this planet by 2050 (also see Chapter 14). The developing world envies the lifestyle and consumption habits of Westerners and believe they also have a right to such prosperity. Until the world's population stabilizes, there will be no end to the spiraling increase in the demand for fossil fuels and other natural resources. The mounting ecological imprint will be on a scale too large to sustain human lifestyles at today's rates of resource consumption. In the short term, we can only reduce environmental pressures by significantly reducing individual environmental footprints through improvements in technology that radically reduce consumption rates by increasing efficiency, reducing waste, and lowering our expectations for our standard of living. However, such reductions in individual consumption may be overwhelmed by an increase in population. Ultimately, the key to long-term sustainability is to lower consumption and stabilize population simultaneously. This raises two overarching questions:

1. Can any feasible reduction in population growth actually result in a more sustainable path?
2. If so, are there measures that policymakers can take to encourage such changes?

Historical data reveals that population grows the fastest when per capita consumption is low. Later, consumption tends to explode when a population is large, but by then the population is growing more slowly. For example, throughout the nineteenth century, the U.S. population grew at a rate that typifies Africa today; later, it slowed as per capita wealth increased.

This same population growth pattern followed by consumption growth is now occurring in China. But much of the Third World has yet to pass through this transition phase. In many regions, environmental problems may be so dire that the world simply cannot wait for their populations to stabilize (see also Chapter 14).

3.5.1   Approaches to sustainable population policies

One strategy for dealing with escalating environmental problems involves shifting to sustainability practices as quickly and inexpensively as possible. The most direct solution appears to be population control. But what types of population policies should be adopted? Should we advertise the benefits of small families? Should tax codes be restructured to favor small families? Should free birth control or even sterilization procedures be provided? Each of these policy options might have an effect, but none has proved to affect demographic trends over the long term. Perhaps more importantly, none of these aforementioned policies has gained widespread public support.

Consider India's experience (see also Chapter 14). In 1976, healthcare workers were rewarded for meeting sterilization quotas; the zeal of some of these workers contributed to the downfall of Indira Gandhi's government in 1977, which ended this program. Enough said. Arguably, after the issue of climate change, no other environmental issue generates as much passion or controversy as population control.

Fifteen years ago, most of the world's nations agreed to a new approach for population control—one that bases positive demographic outcomes on decisions individuals make of their own accord and in their own self-interest. Some evidence suggests that what most women actually want is not so much as to have *more children* but instead to have *more* for a smaller number of children.

A policy along this line was signed in 1994 at a United Nations conference in Cairo. The policy was simple: the government does not concern itself with personal childbirth decisions; instead, it helps every woman bear a child in good health when she wants one. Left to their own devices, women would collectively "control" population while acting in their own personal self-interests. Where women are provided a choice of contraceptives, they tend to have two or fewer children. This tends to be true for both rich and poor nations.[26] For more information on population, see Chapter 14.

3.6 Measuring sustainable development

It is now generally recognized that sustainable development does not focus entirely on the environment. John Elkington coined the term "triple bottom line," consisting of three very broad criteria to measure an organizational or societal success: economy, society, and environment.[27] Environmental policy analysts have long relied upon economic models that seek to convert physical environmental impacts on air, water, land, and biota into surrogate marketplace dollars. In Chapter 7, we review some of the principal approaches that have been taken. But many critics of dollar measures for nonmarket environmental effects have proposed alternative indexes to guide environmental decision making. In this section, we describe two significant efforts to develop quantitative methodology tools that implement this idea.

3.6.1 Methods for assessing ecological footprint

This section describes two methods in use for assessing ecological footprints (EFs).[28]

3.6.1.1 Global Footprint Network

In 1990, Mathis Wackernagel and William Rees of the University of British Colombia developed the concept of an "ecological footprint." Since that time, the Global Footprint Network (GFN), an organization that promotes sustainable development, has matured the concept into a standard methodology that has been adopted by many public and private-sector organizations as a research and decision tool. The term is defined in the next section.

3.6.1.1.1 Ecological footprint (EF)
This is a measure of how much biologically productive land and water an individual, population, or activity requires in order to produce all the resources it consumes and to absorb the waste it generates using prevailing technology and resource management practices. EF is usually measured in global hectares.

Thus, EF is a measure of the impact of different kinds of human activities that use air, water, land, and biotic products that support a given lifestyle tied to a level of technology and economic activity. GFN has tabulated generic per capita footprint for regions, cities, and corporate enterprises. This enables those who analyze policy and propose development plans for regions, cities, and corporations to compare the EF values of any number of alternative development schemes. When incorporated into environmental assessment reports, EF values can contribute to the best decision choice from among these alternatives.

3.6.1.1.2   Biocapacity (shorthand for biological capacity)   This is the ability of an ecosystem to produce useful biological materials and to absorb wastes generated by humans. For an ecosystem to sustainably meet human needs (food, fiber, energy, minerals, water, and so on), its biocapacity measured in hectares must be equal to or exceed its rate of consumption by a given population. The result of the calculation is defined as follows:

Ecological (deficit) or reserve = EF of consumption − total biocapacity

Table 3.3 shows how GFN compared the EF impacts of three broad classes of nations: high-income, middle-income, and low-income countries. It shows the results of applying the equation.

The messages of Table 3.3 are as follows:

- The world's population, technologies, and economies are currently exceeding the world's biocapacity by 0.8 hectares (1 hectare = 2.471 acres).

Table 3.3 Example of Global Ecological Footprint Application[a]

	High-income countries	Middle-income countries	Low-income countries	World
Population (million)	1022.1	4281.1	1277.0	6592.9
Footprint components	Ecological footprint in global hectares/capita			
Ecological footprint of consumption	1.0		1.8	2.6
Cropland footprint	–	–	–	
Grazing footprint	–	–	–	
Forest footprint	–	–	–	
Fishing ground footprint	–	–	–	
Carbon footprint	–	–	–	
Built-up land	–	–	–	
Total biocapacity	3.4	1.7	1.0	1.8
Cropland	–	–	–	
Grazing land	–	–	–	
Forest	–	–	–	
Fishing ground	–	–	–	
Built land	–	–	–	
Ecological deficit or reserve	(2.7)	(0.1)	(0)	(0.8)

[a] A number within parentheses means a deficit; otherwise, it is a reserve.

- The developed nations, representing only 15.5% of the world's population, when averaged with the EF of the other 84.5%, are responsible for about 0.7 EF or 87% of the world's ecological deficit.
- This situation is not sustainable even if the world's population does not increase.
- This suggests that global policymakers need to carefully project and assess the impacts of this unsustainable situation in terms of ecological resource depletion and its ultimate consequences for the quality of human life.
- One can expect a steady deterioration of the average human condition, punctuated by a series of catastrophic events as though nature itself is taking revenge for the abuses heaped on it.

How can policymakers use this information? After all, a simple index like the EF by itself conveys no causal information that informs actions that can make a difference. The answer resides in the far more complex and detailed analysis that lies behind the EF calculated for each element of Table 3.3.

Each individual EF computation is backed by research into the causal factors that may be amenable to policy. For example, to usefully estimate the cropland footprint for a given population, we must evaluate how the entire food chain, starting with consumer demand, responds to real-world economic, technological, cultural, and political factors. The reason for doing this is not to generate an EF number but to enable useful analysis of how policy can alter the food-chain system to reduce EF.

For the most part, policymaking organizations do not maintain the research capabilities required to construct and maintain complex policy metric tools like the EF. GFN is one of many research organizations that provide such services.

3.6.1.2 Ecological rucksack

The ecological rucksack (ER) is another index approach that has been used as an environmental policy metric.[29] ER is defined as the total quantity (typically in kilograms) of materials removed from nature to create a product or service minus the actual weight of the product. That is, an ER resembles a hidden material flow assessment. ERs take a lifecycle approach and signify the environmental strain or resource efficiency of the product or service manufactured or produced. More specifically, rucksacks measure the amount of materials not directly used in the product, but displaced because of the product. That is, ERs represent the materials necessary to produce, use, recycle, and dispose of a product, but not the materials specifically used in the product.

ERs use a cradle-to-grave approach. They focus attention on the entire lifecycle of a product or service and the environmental and resource impacts of producing the product or service.

3.6.1.2.1 Simplified methodology Friedrich Schmidt-Bleek from the Wuppertal Institute for Climate, Environment and Energy (Germany) first proposed the ER concept. ERs are calculated by subtracting the weight (W) of the product from the material intensity (MI):

$$ER = MI - W$$

The material intensity is found using the following equation:

$$MI = SUM (Mi \times Ri)$$

where Mi is the weight of a material removed from nature in kilograms and Ri is the rucksack factor for each respective material used to make the product.

The material input is calculated from five main categories:

1. Abiotic raw materials
2. Biotic raw materials
3. Moved soil (agriculture and forestry)
4. Water
5. Air

The rucksack factor is the quantity (in kilograms) of materials moved from nature to create 1 kg of the resource. For example, the rucksack factor for aluminum is 85:1 (85 kg of materials moved for every 1 kg of aluminum obtained), for recycled aluminum it is 3.5:1, and for diamond the rucksack factor is 53,000,000:1.

Each material used in the production of the good or service is ultiplied by its rucksack factor and then each normalized value is summed to produce the material intensity of that product or service. As far as possible, all materials used for the production, use, and disposal of the product or service, whether directly or indirectly, are included in the calculation.

The ER value of some materials will change over time as they become rarer or as technology makes their extraction or processing more efficient. For example, copper has moved from an ER of 1:1, when copper nuggets were easy to find, to 500:1, as copper is now being extracted from sulfide ores.

3.6.1.3 Case studies examples

1. ER values: The Association of Cities and Regions for Recycling provides some rucksack values for various products. A 5-g gold ring, for example, was found to have an ER of 2000 kg. An aluminum drink can was found to have an ER of 1.2 kg and a 20-kg computer was found to have an ER of 1500 kg.
2. Reducing the ER of a watch: By conducting a material intensity per unit service (MIPS) analysis for the production phase of a new

watch, opportunities to reduce the ER of the watch were revealed. By changing some of the materials used in the watch, the rucksack value was reduced.

3.6.2 Policy options for reducing the environmental impact or footprint

The EF and ER both take into account the relationship between a population's use of natural resources and the ability of the environment to sustainably supply those resources. Whichever method is used, controlling population is perhaps the most direct policy means for reducing environmental impact. There are generally two options available: (1) reduction of population fertility and (2) reduction of immigration (legal and illegal). In Chapter 14, we discuss the difficulties of depending on these strategies alone. However, the policy options discussed next can help meet society's needs while reducing or eliminating the need for population control.

3.6.2.1 Increasing the carrying capacity with science and technology

This has been summarized as "increasing the size of the pie." Efficiency could play a central theme under this policy avenue. The cornucopians would argue that science and technology innovations can be tapped to increase food yields, to recycle water, to find alternatives to scarce materials, and for a host of other prescriptions. Many (but certainly not all and perhaps not even a majority) of them, such as Simon, would argue that population should remain unconstrained.

3.6.2.2 Reduction in lifestyle or living simpler lives

Limit conspicuous consumption. Reduce our dependency on material and energy-intensive gadgets and things. Ride bikes instead of cars. Incorporate recycling and conservation into our daily lives.

3.6.2.3 Reforming and redistributing resources

Assist poor countries in developing green technology. Provide food, educational programs, and technological assistance to developing countries. Compensate developing countries for preserving their rainforests.

The chapters that follow emphasize policy concepts, principles, and applications that address these fundamental strategies.

3.7 Ecological modernization

Ecological modernization emerged in the early 1980s, and has gained increasing attention among scholars and policymakers in recent decades. It is an optimistic, reform-oriented school of environmental social science.

While its name does not intuitively imply its meaning, Neil Carter clearly explains the concept as a more practical approach to sustainable development that he characterizes as "weak sustainability."[30]

> Ecological modernization concedes that environmental problems are a structural outcome of capitalist society, but it rejects the radical green demand for a fundamental restructuring of the market economy and the liberal democratic state. The political message of ecological modernization is that capitalism can be made more environmentally friendly by the reform (rather than overthrow) of existing economic, social, and political institutions.

One of its basic tenets is the environmental adaptation of economic and industrial development. That is, an enlightened populace, a developed economy, and the concept of sustainable development can be combined to produce a sustainable society. Under this view, sustainable development provides a source of future growth and development in the same way as labor and capital have traditionally done. New approaches and technological innovations can be developed to actually reduce resource consumption and pollution.[31]

Some advocates believe that the scope of ecological modernization is mainly focused on economic reorientation and techno-industrial adaptation, rather than on cultural factors (value reorientation, changes in behavior, and lifestyle adaptations). Similarly, there are varying opinions as to whether it requires political and government intervention, entrepreneurship and market implementation, or social implementation. However, there is a common understanding that ecological modernization requires innovative and structural changes throughout society.[32] In recent years, research has focused on the "sustainable household," that is, environmental reshaping of lifestyles and consumption patterns.[33]

Some forms of ecological modernization have been embraced by business interests because they meet the triple bottom line of economics, society, and environment that underpins sustainable development, yet do not challenge free-market principles. In contrast, many environmental movements regard free enterprise and its notion of business self-regulation as part of the problem.

Critics assert that ecological modernization will not protect the environment and that it does little to alter the impulses within a free-enterprise system. They also question whether technological advances alone can achieve resource conservation and environmental protection, particularly if left to business self-regulation practices. They also argue that the theory has limited global promise because it applies primarily to its countries of origin—Germany and other European nations—and has much less acceptance in developing countries and other parts of the world. Arguably, the harshest criticism is that the theory is predicated on the notion of sustainable growth, and in reality this is not possible because growth entails the consumption of natural and human capital, which degrades the ecosystem.[34]

Some related key paradigms are described in the following sections.

3.7.1 Societal–environmental dialectic

In 1975, Allan Schnaiberg proposed a theory that significantly influenced the field of environmental sociology. He proposed a *societal–environmental dialectic*. This theory is based on the following dialectical or opposing forces:

- The desire for economic expansion will prevail over ecological concerns. Policies will tend to maximize economic growth at the expense of environmental degradation.
- Governments will formulate policies that are focused on controlling only the most severe of problems to prevent environmental disasters. This results in the illusion of environmental action more than on actual environmental benefits.
- Such policies will generate scenarios in which environmental degradation is so severe that political forces will eventually be forced to respond with sustainable policies.

These conflict-based and opposing forces may lead to several potential outcomes. For instance, the most powerful economic and political forces may preserve the status quo and bolster their dominance.

3.7.2 Human exemptionalism paradigm and the new ecological paradigm

The human exemptionalism paradigm argues that humans are such a uniquely superior species that they are not subject to environmental forces to the extent that other species are. Human dominance is determined by our culture, which is more adaptable than our biological traits and limitations. Our culture allows us to adapt and innovate, yielding a nearly unbounded resource capable of solving natural limitations and problems. Because humans are largely exempt from natural conditions, they have control over both the environment and their destiny. This was the popular paradigm from the Industrial Revolution through the 1960s.

The new ecological paradigm was developed in the 1970s, with assumptions contrary to the human exemptionalism paradigm. This new paradigm recognizes the innovative capacity of humans but also states that humans are still ecologically interdependent. While it accounts for the power of social and cultural forces, it also states that humans are still constrained by environmental limitations.

3.8   Applications of the sustainable resource development principle

This section provides two contrasting examples of how the sustainability principle has been applied to specific global and national issues. The first provides a global perspective on a sustainable global economy. The second is a business view of sustainable development.

3.8.1   Application 1: Report by the Worldwatch Institute

A report by the Worldwatch Institute, *State of the World: Innovations for a Sustainable Economy,* begins by citing the following global phenomena that raise questions about the continued sustainability of the human economy given the scale of its environmental impacts on vital natural resources essential to that economy[35]:

- The highest atmospheric levels of carbon dioxide in the last 600,000 years and projections of an ice-free Arctic Ocean as a consequence
- The possible collapse of current fish populations
- The rapid increase of oxygen-depleted dead zones in the oceans from 149 to 200 in the past 2 years, posing threats to fish species
- Urban air pollution estimated to cause 2 million premature deaths each year
- Declines in the populations of bees, bats, and other vital pollinators across North America, threatening crops and ecosystems
- A potential peaking of the world's oil production (see Chapter 11)

Actions promoted in these and other sections of the Worldwatch report come together in the report's Chapter 13, "Investing in Sustainability," summarized here.[36] Specifically, investment policies are divided into four categories as follows:

1. *Socially responsible investments*: These are investments that focus on environmental and social sustainability and include investment in certain values not directly related to sustainability.
2. *Project finance*: This involves funding for major infrastructure or extractive projects such as dams and mines that add value to social and environmental sustainability.
3. *Private equity and venture capital*: The focus here would be on green energy and products.
4. *Microfinance*: This involves very small loans widely distributed to small-scale artisans and craftsmen to serve local markets. The objective is to increase income among the poorest populations and alleviate poverty.

All of these strategies are actually pursued by various United Nation's environmental and social programs and by a range of international donor organizations, often funded by major corporations that select specific issues or areas of geographic focus for their work.

3.8.2 Application 2: Business and Dow Jones sustainability concepts

The World Business Council for Sustainable Development is a coalition of 180 companies from some 35 countries. Each member shares a commitment to sustainable development via three pillars—economic growth, ecological balance, and social progress.

The Dow Jones Corporation, known for its business and financial indexes and other publications, measures sustainability in three dimensions: economic, environmental, and social responsibility (Table 3.4). It publishes a

Tables 3.4 Dow Jones Measurements of Sustainability

Environment
    Policy/management
    Performance (eco-efficiency)
    Reporting (content and coverage)
Economic
    Codes of conduct, compliance, corruption, and bribery
    Corporate governance
    Customer relationship management
    Investor relations
    Risk and crisis management
    Brand/supply chain/marketing practices criteria
    Innovation/research and development/renewable energy criteria
Social
    Citizenship/philanthropy
    Stakeholder engagement
    Labor practice
    Human capital development
    Social reporting
    Talent attraction and retention
    Product quality/recall management
    Global sourcing
    Occupational health/safety
    Healthy living
    Bioethics

family of indexes to track the performance of companies in terms of corporate sustainability as defined by the Dow Jones.

The Dow Jones concept of corporate sustainability is a business approach

> to create long-term shareholder value by embracing opportunities and managing risks deriving from economic, environmental and social developments. Corporate sustainability leaders harness the market's potential for sustainability products and services while at the same time successfully reducing and avoiding sustainability costs and risks.[37]

3.9   Examples of a sustainable energy policy

The maximum worldwide power consumed at any given moment is about 12.5 trillion watts (terawatts or TW). The U.S. Energy Information Administration projects that by 2030 the world will require 16.9 TW. Is it possible to develop a global, clean, and sustainable energy system? This section investigates two plans for achieving a sustainable energy infrastructure within two decades using only proven technologies or those that are close to working on a large scale today.

This section describes two dramatically different approaches for achieving a national sustainable energy policy. The first plan provides a comprehensive approach for the way energy is both generated and used. It has been characterized by many critics as being more hype than reality. Still, it is useful for illustrating the rudiments of sustainable development, such as a national energy sustainability plan. The second plan provides a different and radically innovative approach for generating and allocating energy, principally for use in the transportation industry.

3.9.1   Plan 1: Jacobson and Delucchi energy plan

In developing their energy plan, Jacobson and Delucchi[38] assumed that most fossil-fuel heating (e.g., ovens) could be replaced by electric systems and most fossil-fuel transportation could be replaced by battery and fuel-cell vehicles. Wind, water, and sunlight are collectively referred to as WWS. The renewable technologies considered in this plan have a low generation of greenhouse and air-polluting emissions over their entire lifecycle (an assertion that has been disputed by some critics), including mining, milling, construction, operation, and decommissioning. Electrolysis (powered using WWS electricity) would produce hydrogen for use in fuel cells. This hydrogen would also be burned in airplanes and used by industry.

They chose a mix of different technologies, emphasizing wind and solar, with nearly 10% of power produced using mature water-related methods.[39] Wind provided by 3.8 million large wind turbines (each rated at 5 MW) would supply about 50% of the worldwide demand. An additional 40% of the power would come from photovoltaics and solar power plants, with 30% of the photovoltaic output coming from rooftop panels on homes and buildings. Nearly 90,000 photovoltaic and concentrated solar power plants, averaging 300 MW, would need to be constructed. The plan also includes 900 hydroelectric stations worldwide, 70% of which have already been constructed.

Compared to fossil fuels, electrification tends to be a more efficient way to use energy. For instance, approximately 17–20% of gasoline energy is actually used to move a vehicle (the rest is wasted as heat). In comparison, 75–85% of the electricity delivered to an electric vehicle goes into generating motion.

3.9.1.1 Materials scarcity

One of the most problematic materials required for such a mammoth undertaking involves rare-earth metals such as neodymium used in turbine gearboxes. Photovoltaic cells rely on materials such as amorphous or crystalline silicon and cadmium telluride. Limited supplies of materials such as indium could reduce the prospects for some types of thin-film solar cells. Recycling old cells might help ameliorate material difficulties.

Components for building millions of electric vehicles include rare-earth metals for electric motors, lithium for lithium-ion batteries, and platinum for fuel cells, all of which are scarce resources. Again, recycling could help ameliorate some of the material challenges.

3.9.1.2 Cost

The authors of this plan claim that their mix of WWS sources can reliably supply a nation's entire energy demand. According to their plan, the cost of wind, geothermal, and hydroelectric are all less than 7 ¢/kW·h (per kilowatt hour). Wave and solar are higher, but by 2020 and beyond, wind, wave, and hydro are expected to be equal to or less than 4 ¢/kW·h. In contrast, the authors claim that the average cost of U.S. conventional energy transmission is about 7 ¢/kW·h, and is projected to rise to 8 ¢/kW·h by 2020. Wind turbine energy already costs about the same or less than energy from a new coal or natural gas plant.

Today, solar power is relatively expensive, but it should be competitive by as early as 2020. One study concluded that within 10 years, photovoltaic system costs could drop to about 10 ¢/kW·h, including transmission costs and the cost of compressed-air storage of power for use at night.

Worldwide WWS construction costs over a 20-year period might run on the order of $100 trillion, not including transmission costs. Each nation would need to invest in a robust long-distance transmission system that could carry WWS power from remote regions, where it is often most abundant (such as the Great Plains for wind and the desert Southwest for solar in the United States), to major population centers.

3.9.1.3   Reliability

The authors claim that WWS technologies typically have less downtime than traditional power sources. The average offline time for a coal plant, for example, is 12.5% of the year for maintenance. In comparison, offshore wind turbines have a downtime of less than 2% and 5%. Photovoltaic systems average about 2%.

The principal challenge faced by WWS is that the wind does not always blow and the sun does not always shine. However, the WWS authors argue that intermittency problems can be mitigated by balancing the power sources. For example, base electricity supply could be generated from steady geothermal or tidal power while relying on wind at night (when it tends to be more plentiful), solar by day, and hydroelectricity to turn power on and off quickly in order to smooth out supply or meet peak demand. Smart electric meters can be used in homes to automatically recharge electric vehicles when demand is low.

3.9.1.4   Time table

With sensible policies in place, nations could initially set a goal of generating 25% of their new energy supply using WWS sources in 10–15 years. This goal would be gradually raised to the point where nearly 100% of new supply would be generated by WWS in 20–30 years. Under a very aggressive policy, all existing fossil-fuel capacity might be retired and replaced during the same period; under a more modest and likely policy, full replacement might be completed in 40–50 years.

3.9.1.5   Plan 1: Phenomenal or fantastic?

Critics have responded that the study is riddled with errors, inconsistencies, and frequent use of very optimistic assumptions. For example, with only 6 billion people in the world, there may not be enough rooftops in the entire world to house the 1.7 billion rooftop solar systems called for in this study.

One critic calculated that approximately 15 square miles of land are needed to generate 1000 MW with solar thermal and voltaic plants. The 90,000 solar plants (rated at 300 MW apiece) would require some 450,000 square miles, nearly the size of Texas and California combined. Moreover, the solar mirrors and panels must be washed once a week, which would require massive quantities of water and call for huge maintenance costs. Furthermore, a wind turbine generating 5 MW would probably take up an area equal to about the length of two football fields and tower perhaps

80 stories; the 4 million windmills (rated at 5 MW apiece) required by the plan would extract a huge EF, hardly a benign source of energy.

What is interesting is that the WWS authors barely mention the use of nuclear power, the one proven technology that can produce relatively clean energy. One critic observed that the authors of this proposal barely make any distinction between nuclear and fossil fuels, which were lumped together as "old" technologies. Nuclear energy's advantage is its tremendous energy yield per pound of resource consumed. Simply put, a pound of uranium contains 2000 times as much energy as a pound of coal. This translates into a 110-car train loaded with coal arriving every 30 hours to fuel a plant, versus six tractor trailers loaded with uranium fuel rods arriving once every 18 months.

Another criticism is that the authors of this study claim that nuclear power emits 25 times more carbon over its lifecycle compared to wind. This assertion is debatable. A nuclear reactor contains about 500,000 cubic yards of concrete and 120 million pounds of steel. In contrast, a single 45-story wind turbine stands on a base of 500 cubic yards of concrete and contains as much metal as 120 automobiles. As 2000 of these are equal to one nuclear reactor, that adds up to twice as much concrete and steel and this translates into significant quantities of greenhouse emissions produced over the construction phase (mining, concrete curing, forging, transportation, and construction).

Every form of electrical generation has a "capacity factor," which is the percentage of time, on average, the plant is up and running. All plants go periodically on- or offline (due to maintenance, refueling, or simple unavailability). Of *all* energy sources, nuclear energy has the highest capacity factor (greater than 90%), while solar has the lowest (20%). Although the WWS authors were factually correct in stating that windmills are only offline for maintenance 2% of the time, they neglected to mention that wind only blows about 30% of the time, resulting in a very low capacity factor.

3.9.2 Plan 2: Pickens sustainable energy generation plan

We will now consider an alternative sustainable energy policy, the "Pickens plan," which is less comprehensive in scope compared to the one portrayed in Section 3.9.1. However, the authors of this work believe that the Pickens plan is more pragmatic and subject to less uncertainty than the WWS plan.

In recent years, wind power has experienced exponential growth in the United States. Oil tycoon T. B. Pickens has proposed a U.S. national sustainable energy policy, widely referred to as the Pickens plan.[40] He specifically wants to strengthen U.S. national energy security by increasing the exploitation of domestic energy sources while decreasing the country's dependence on vulnerable imported foreign oil. Import of foreign petroleum and natural gas has caused a significant negative trade deficit that is decreasing national wealth. According to one estimate, the Pickens plan could reduce the amount the country spends on foreign oil imports by $300 billion annually (i.e., 43% decrease).

3.9.2.1   Description of the Pickens plan

Pickens wants to reduce America's dependence on imported oil by investing $1 trillion in enormous renewable-energy wind turbine farms that would produce electricity. This would allow natural gas currently burned in electrical power plants to be shifted to fuel trucks and heavy vehicles. Specifically, his plan would

1. Involve private-industry funded construction of thousands of wind turbines in the Great Plains region of the United States (i.e., the U.S. "wind corridor"), because of the region's favorable wind patterns. He estimates these turbines could provide 22% or more of the country's electricity supply.
2. The U.S. government would fund and construct electric power transmission lines to connect the turbine farms to the nation's power grid. This would provide electrical energy to the Midwest and southern and western regions of the United States. The new transmission lines would cost between $64 billion and $128 billion.[41] Testifying before the U.S. Senate Homeland Security and Government Affairs Committee, Pickens said that the transmission lines should have the same status priority that President Eisenhower assigned, while declaring priority emergency, to building the interstate highway system in the 1950s and 1960s.
3. With wind energy providing a large portion of the nation's electricity, natural gas, currently used to fuel electrical power plants, would be diverted and used as fuel for thousands of trucks and heavy vehicles. The plan places emphasis on developing a national fleet of trucks and buses that would burn this relatively clean resource. This would significantly reduce the demand for imported petroleum, currently used in manufacturing gasoline or diesel fuel.

Many geologists and economists have voiced concerns that U.S. natural gas production may have peaked or is nearing its maximum production rate, a point known as "peak gas" (see Chapter 11). It is possible that the plan would help natural gas efficiency by conserving U.S. natural gas supplies.

According to Pickens, the plan could provide 22% of the nation's electrical power supply from wind power and would provide for the conversion of vehicles from gasoline to natural gas in less than 10 years. The plan calls for increasing the installed wind power capacity by at least a factor of 10 by the year 2018 from its 2008 level.

Pickens plan would significantly reduce CO_2 emissions by shifting a large percentage of electricity production from natural gas combustion to carbon-neutral wind power generation. While natural-gas-powered vehicles would still produce CO_2, they would produce about 25% less when compared to the same amount of energy derived from gasoline or diesel.[42]

Pickens does not believe that electric batteries are likely to power the nation's large trucks, which is why his plan focuses on diverting natural gas to power large vehicles. However, wind-generated electricity could be used to power electric-battery-powered automobiles.

3.9.2.2   Intermittency problem

A key challenge to the Pickens plan is that the wind energy used to replace natural gas is available only intermittently, as wind does not blow much of the time. But this may not pose a significant obstacle. Modern technology could ensure that several wind farms from varying regions of the country are connected together on electrical grid. While some are idle, others could make up the difference.[43]

Others have argued that to produce power when the wind is not blowing, backup natural-gas-powered plants may be needed. These backup plants could be brought online to help supply peak capacity during periods of low wind or peak demand.

Another suggestion is that instead of producing additional electricity, energy could be conserved. For instance, the Smart Power Grid, an intelligent electrical distribution network currently under development, could reduce power consumption during peak hours and thereby lessen dependence on gas-fired plants. Still another option, one which would be more expensive, involves storing energy by methods such as pumping water uphill when the wind is blowing and releasing it through turbine generators when the electricity is needed.

3.9.2.3   Cost

According to a study performed by the Cato Institute, the wind energy component would result in higher retail electric rates because wind power is twice as expensive as natural-gas-generated power. However, the American Wind Energy Association counters that the cost of wind power has dropped by 90% over the last 20 years and further declines can be expected. Also, the cost of economic "externalities" (see Chapter 9) of gas- or diesel-powered vehicles is greater than that of natural-gas-powered ones.

3.9.2.4   Endorsements and criticisms

President of the United States Barack Obama has publicly stated that he supports many elements of the Pickens plan. Representatives of the Sierra Club and Center for American Progress have also endorsed this plan. The Institute for Energy Research, an organization funded by the oil industry that advocates offshore drilling, claims that the Pickens plan relies on government subsidies and that producing large amounts of wind power is not a viable option.[44]

3.10   Conclusion

Many policymakers and corporate marketing executives would have us believe that if sacrifices are needed, they will be akin to a transition from gas-guzzling SUVs to compact hybrids with properly inflated tires. Many companies will advertise their sustainable practices and products. People will use recycled bottles, carry unbleached cloth grocery sacks, and buy fuel-efficient cars. But will these measures be sufficient? They are very unlikely to make a significant dent on the world's ecosystems.

Citizens in developed countries do not show a significant inclination toward a decline in their lifestyles. Those in developing nations appear to be unsatisfied with anything less than what developed nations have become accustomed to, and those in undeveloped countries are struggling to simply maintain the little that they have. However, we have the opportunity to do much more than choose between plastic and paper sacks. We can decide what to build, how and where to live, and how and where to travel. Our children will know if our pursuit of sustainability was a saving grace or merely an expedient advertisement to satisfy critics.

Discussions, problems, and exercises

1. List all the elements you can think of that must be assured as part of a global policy and plan for sustaining the human species at a reasonable minimum standard of quality living.
2. What are the factors that threaten this sustainability?
3. Cite some critics of the sustainability program and support or refute their allegations as you see fit.
4. Is there hope for the future of humanity? Make the case one way or another in the light of possibilities for technology and the global political community to make a difference?

Notes

1. Sustainable Consumption, http://www.gdrc.org/sustdev/concepts/22-s-consume.html (accessed January 23, 2010).
2. Friedrich Schmidt-Bleek and others, (Germany's Wuppertal Institute for Climate, Environment and Energy).
3. United Nations World Commission on Environment and Development (Brundtland Commission), "Our Common Future," http://www.un-documents.net/wced-ocf.htm (accessed July 8, 2010).
4. Ibid., 43.

5. International Council for Local Environmental Initiatives (Germany: ICLEI, 1994).
6. Definition adopted by the International Chamber of Commerce from the book published by the International Institute for Sustainable Development, *Business Strategies for Sustainable Development: Leadership and Accountability for the 90s* (Winnipeg, Manitoba, Canada: International Institute for Sustainable Development, 1992).
7. See note 1 above.
8. B. Sadler, "Towards the Improved Effectiveness of Environmental Assessment," Executive Summary of Interim Report Prepared for IAIA'95, Durban, South Africa, 1995.
9. N. Carter, "Sustainable Development and Ecological Modernization," in *The Politics of the Environment,* 1st ed. 2001, 2nd ed. 2007, chap 8. (Cambridge UK: Cambridge University Press)
10. World Commission on Environment and Development, *Our Common Future* (New York: Oxford University Press, 1987), 49.
11. World Commission on Environment and Development, *Our Common Future* (New York: Oxford University Press, 1987), 15.
12. Agenda 221, Principle 15.
13. World Commission on Environment and Development, *Our Common Future* (New York: Oxford University Press, 1987), 9.
14. J. W. Handmer and S. R. Dovers, "A Typology of Resilience: Rethinking Instructions for Sustainable Development," *Industrial and Environmental Crisis Quarterly* 9 (1996): 482.
15. R. MacLean, "I Say Green, You Say Sustainable," *Competitive Environment,* September 2009.
16. United Nations, Report of the United Nations Conference on Environment and Development, Agenda 21, Rio Conference 27 principles, Annex I, Rio Declaration on Environment and Development, (Rio de Janeiro, June 3–14, 1992).
17. K. Hargroves and M. Smith, eds., *The Natural Advantage of Nations: Business Opportunities, Innovation and Governance in the 21st Century* (London: Earthscan/James and James, 2005).
18. *The National Environmental Policy Act of 1969,* as amended (Pub. L. 91-190, 42 U.S.C. 43214347, January 1, 1970, sec. 2).
19. *The National Environmental Policy Act of 1969,* as amended (Pub. L. 91-190, 42 U.S.C. 43214347, January 1, 1970, title 1, sec. 101(a).)
20. EO 13514, "Federal Leadership in Environmental, Energy, and Economic Performance," 74 FR 52117; October 8, 2009, http://www.archives.gov/federal-register/executive-orders/2009-obama.html (accessed July 8, 2010).
21. J. D. Marshall and M. W. Toffel, "Framing the Elusive Concept of Sustainability: A Sustainability Hierarchy," *Environmental Science and Technology* 39, no.3 (2005): 673–82.
22. S. Agassi, "In Denmark Ambitious Plan for Electric Cars," *New York Times* (December 1, 2009) http://www.nytimes.com/2009/12/02/business/energy-environment/02electric.html?scp=1&sq=S.%20Agassi,%20New%20York%20Times,%20December%202,%202009&st=cse (accessed August 5, 2010).
23. J. L. Newton and E. Freyfogle, "Conservation Forum: Sustainability—A Dissent," *Conservation Biology* 19, no.1 (2005): 36–8, p. 29.
24. S. Temple, interview remarks, "Old Issue, New Urgency?" *Wisconsin Environmental Dimension* 1, no. 1 (Spring 1992): 1.

25. P. Reitan, "Sustainability Science—and What's Needed beyond Science," *Sustainability: Science, Practice, and Policy* (2005), online at http://ejournal.nbii .org/archives/vol1iss1/communityessay.reitan.html (accessed July 8, 2010).
26. R. Engelman, "Population and Sustainability: Can We Avoid Limiting the Number of People?" *Scientific American*, June 2009.
27. J. Elkington, "Towards the Sustainable Corporation: Win-Win-Win Business Strategies for Sustainable Development," *California Management Review* 36, no.2 (1994): 90–100.
28. This section is based on information provided by The Global Footprint Network, http://www.footprintnetwork.org/en/index.php/GFN/ (accessed July 8, 2010).
29. This description was provided by the Global Research Development Center in Wuppertal, Germany, "Ecological Rucksacks," http://www.gdrc.org/ sustdev/concepts/27-rucksacks.html (accessed July 7, 2010).
30. See note 9 above.
31. S. C. Young, *The Emergence of Ecological Modernisation: Integrating the Environment and the Economy?* (London: Routledge, 2000).
32. J. Huber, *New Technologies and Environmental Innovation* (Cheltenham, UK: Edward Elgar, 2004).
33. OECD, ed., *Towards Sustainable Household Consumption? Trends and Policies in OECD Countries* (Paris: OECD Publication, 2002).
34. R. York and E. A. Rosa, "Key Challenges to Ecological Modernization Theory," *Organization and Environment* 16, no.3 (2003): 273–88.
35. The Worldwatch Institute, *State of the World: Innovations for a Sustainable Economy* (Washington D.C.: The Worldwatch Institute, 2008).
36. The Worldwatch Institute, "Investing in Sustainability," in *State of the World: Innovations for a Sustainable Economy*, (Washington D.C.: The Worldwatch Institute, 2008), chap. 13.
37. Dow Jones, *Dow Jones Sustainability North America Index Guide Book* (New York: Dow Jones Inc., 2006), section 3.1, version 2.0.
38. M. Z. Jacobson and M. A. Delucchi, "A Path to Sustainable Energy by 2030," *Scientific American*, November 2009, 58–65.
39. Ibid.
40. Pickens Plan, http://www.pickensplan.com/theplan/ (accessed January 24, 2010); Pickens Plan, http://en.wikipedia.org/wiki/Pickens_plan (accessed January 24, 2010).
41. D. R. Baker, "Experts Wary of Pickens' Clean-Energy Plan," *San Francisco Chronicle*, September 1, 2008, http://www.sfgate.com/cgi-bin/article.cgi?f=/ c/a/2008/09/01/MNO512K43O.DTL&type=printable (accessed October 4, 2008).
42. U.S. Department of Energy, "Natural Gas Vehicle Emissions," http://www.eere .energy.gov/afdc/vehicles/natural_gas_emissions.html (accessed July 8, 2010).
43. The Texas Energy Report, "Wind Energy," http://www.window.state.tx.us/ specialrpt/energy/renewable/wind.php (accessed July 8, 2010).
44. R. Bradley, *Pickens Plan Leaves U.S. Energy Security Blowing in the Wind* (Washington, DC: Institute for Energy Research, 2008), available at http:// www.instituteforenergyresearch.org/2008/07/11/pickens-plan-leaves-us-energy-security-blowing-in-the-wind (accessed July 7, 2010).

chapter four

Environmental policy treaties and their implementation

> Ancient Rome declined because it had a Senate; now what's going to happen to us with both a Senate and a House?
>
> **Will Rogers**

Case study

Following is the table of contents of an actual treaty in the United Nations (UN) System that resulted from processes such as described in Chapter 2. It exemplifies the result of such a process implemented by the UN.

Montreal Protocol on Substances That Deplete the Ozone Layer[1]

Preamble
Article 1: Definitions
Article 2: Control measures
Article 2A: CFCs
Article 2B: Halons
Article 2C: Other fully halogenated CFCs
Article 2D: Carbon tetrachloride
Article 2E: 1,1,1-trichloroethane (methyl chloroform)
Article 2F: Hydrochlorofluorocarbons
Article 2G: Hydrobromofluorocarbons
Article 2H: Methyl bromide
Article 2I: Bromochloromethane
Article 3: Calculation of control levels
Article 4: Control of trade with non-parties
Article 4A: Control of trade with parties
Article 4B: Licensing
Article 5: Special situation of developing countries
Article 6: Assessment and review of control measures

Article 7: Reporting of data
Article 8: Noncompliance
Article 9: Research, development, public awareness, and exchange of information
Articles 10–20 Administrative matters
Annex A: Controlled substances
Annex B: Controlled substances
Annex C: Controlled substances
Annex D: A list of products
Annex E: Controlled substance

Problem: Access a copy of this treaty[1], print it, and review it to obtain a sense of the various issues involved and how the treaty addressed them.

1. Identify elements of treaty development described in Chapter 2 and relate them to the issues in the Montreal Protocol.
2. Participate in a class (or small group) discussion of what was probably involved in the process that developed this treaty.

Learning objectives

After you have read the chapter, you should understand the following:

- The major global environmental policy instruments
- The principles of Agenda 21
- The strategy for implementing Agenda 21
- The programs of the UN Environmental Program that implement the strategy
- How global environmental policy is connected to the global population policy

4.1 Framework for international environmental treaties and programs

In Chapter 2, we described the principles, concepts, processes, and dilemmas that underlie global environmental policy. In this chapter, we focus on the actual content of the policies as implemented through a wide variety of agreements and international programs, mainly through various UN organizations.

4.1.1 Taxonomy of international instruments

The UN maintains the world's most important and comprehensive list of treaties.[2] International instruments that create mutual obligations on the parties go by a variety of names such as treaties, protocols, agreements, charters, conventions, declarations, exchange of notes, memoranda of understanding, modus vivendi, signatories, and parties. These terms have variable meanings, and there is no single definition of what each usually includes or excludes. While the UN summarizes the most general characteristics of these terms, it treats all such instruments in the same way[3]:

> Although these instruments differ from each other by title, they all have common features and international law has applied basically the same rules to all of these instruments. ... The 1969 Vienna Convention on the Law of Treaties (1969 Vienna Convention), which entered into force on 27 January 1980, contains rules for treaties concluded between States. The 1986 Vienna Convention on the Law of Treaties ... (which has still not entered into force) added rules for treaties with international organizations as parties.

Thus, one cannot assume the nature of the instrument from the taxonomy alone but must be guided by its actual content. Virtually all important UN business is conducted pursuant to treaty-like instruments. Its very charter is in effect a treaty. While the charter did not specifically include environmental policy, it established the legal framework for the evolution of several important environmental policy instruments.

4.2 Agenda 21: Principles of global environmental policies

Global environmental policy has evolved in the post–World War II era as the UN has forged a general international consensus framework of environmental policies as an integral element of its population policies. The U.S. National Environmental Policy Act (see Chapter 1) has served as a model for the environmental policies of many other countries, and indirectly for the first comprehensive global environmental policy document of the UN, commonly referred to as Agenda 21 (depicted in Table 4.1). Agenda 21 lays out 27 key principles that essentially implement the many policy dimensions that flow from Principle 1, namely that "Human beings are at the centre of concerns for sustainable development. They are entitled to a healthy

Table 4.1 Agenda 21: Rio de Janeiro Declaration on Environment and
Development

The United Nations Conference on Environment and Development, having met
at Rio de Janeiro from June 3 to 14, 1992, reaffirming the Declaration of the
United Nations Conference on the Human Environment, adopted at
Stockholm on June 16, 1972, and seeking to build upon it, with the goal of
establishing a new and equitable global partnership through the creation of
new levels of cooperation among States, key sectors of societies and people,
working towards international agreements which respect the interests of all
and protect the integrity of the global environmental and developmental
system, recognizing the integral and interdependent nature of the Earth, our
home, proclaims that:

Principle 1: Human beings are at the centre of concerns for sustainable
development. They are entitled to a healthy and productive life in harmony
with nature.

Principle 2: States have, in accordance with the Charter of the United Nations
and the principles of international law, the sovereign right to exploit their
own resources pursuant to their own environmental and developmental
policies, and the responsibility to ensure that activities within their
jurisdiction or control do not cause damage to the environment of other
States or of areas beyond the limits of national jurisdiction.

Principle 3: The right to development must be fulfilled so as to equitably
meet developmental and environmental needs of present and future
generations.

Principle 4: In order to achieve sustainable development, environmental
protection shall constitute an integral part of the development process and
cannot be considered in isolation from it.

Principle 5: All States and all people shall cooperate in the essential task of
eradicating poverty as an indispensable requirement for sustainable
development, in order to decrease the disparities in standards of living and
better meet the needs of the majority of the people of the world.

Principle 6: The special situation and needs of developing countries,
particularly the least developed and those most environmentally
vulnerable, shall be given special priority. International actions in the field
of environment and development should also address the interests and
needs of all countries.

Principle 7: States shall cooperate in a spirit of global partnership to
conserve, protect, and restore the health and integrity of the Earth's
ecosystem. In view of the different contributions to global environmental
degradation, States have common but differentiated responsibilities. The
developed countries acknowledge the responsibility that they bear in the
international pursuit of sustainable development in view of the pressures
their societies place on the global environment and of the technologies and
financial resources they command.

Table 4.1 *(Continued)*

Principle 8: To achieve sustainable development and a higher quality of life for all people, States should reduce and eliminate unsustainable patterns of production and consumption and promote appropriate demographic policies.

Principle 9: States should cooperate to strengthen endogenous capacity-building for sustainable development by improving scientific understanding through exchanges of scientific and technological knowledge, and by enhancing the development, adaptation, diffusion and transfer of technologies, including new and innovative technologies.

Principle 10: Environmental issues are best handled with the participation of all concerned citizens, at the relevant level. At the national level, each individual shall have appropriate access to information concerning the environment that is held by public authorities, including information on hazardous materials and activities in their communities, and the opportunity to participate in decision-making processes. States shall facilitate and encourage public awareness and participation by making information widely available. Effective access to judicial and administrative proceedings, including redress and remedy, shall be provided.

Principle 11: States shall enact effective environmental legislation. Environmental standards, management objectives and priorities should reflect the environmental/developmental context to which they apply. Standards applied by some countries may be inappropriate and of unwarranted economic/social cost to particular developing countries.

Principle 12: States should cooperate to promote a supportive and open international economic system that would lead to economic growth and sustainable development in all countries, to better address the problems of environmental degradation. Trade policy measures for environmental purposes should not constitute a means of arbitrary or unjustifiable discrimination or a disguised restriction on international trade. Unilateral actions to deal with environmental challenges outside the jurisdiction of the importing country should be avoided. Environmental measures addressing trans-boundary or global environmental problems should, as far as possible, be based on an international consensus.

Principle 13: States shall develop national law regarding liability and compensation for the victims of pollution and other environmental damage. States shall also cooperate in an expeditious and more determined manner to develop further international law regarding liability and compensation for adverse effects of environmental damage caused by activities within their jurisdiction or control to areas beyond their jurisdiction.

Principle 14: States should effectively cooperate to discourage or prevent the relocation and transfer to other States of any activities and substances that cause severe environmental degradation or are found to be harmful to human health.

(Continued)

Table 4.1 Agenda 21: Rio de Janeiro Declaration on Environment and Development (*Continued*)

Principle 15: In order to protect the environment, the precautionary approach shall be widely applied by States according to their capabilities. Where there are threats of serious or irreversible damage, lack of full scientific certainty shall not be used as a reason for postponing cost-effective measures to prevent environmental degradation.

Principle 16: National authorities should endeavor to promote the internalization of environmental costs and the use of economic instruments, taking into account the approach that the polluter should, in principle, bear the cost of pollution, with due regard to the public interest and without distorting international trade and investment.

Principle 17: Environmental impact assessment, as a national instrument, shall be undertaken for proposed activities that are likely to have a significant adverse impact on the environment and are subject to a decision of a competent national authority.

Principle 18: States shall immediately notify other States of any natural disasters or other emergencies that are likely to produce sudden harmful effects on the environment of those States. Every effort shall be made by the international community to help States so afflicted.

Principle 19: States shall provide prior and timely notification and relevant information to potentially affected States on activities that may have a significant adverse transboundary environmental effect and shall consult with those States at an early stage and in good faith.

Principle 20: Women have a vital role in environmental management and development. Their full participation is therefore essential to achieve sustainable development.

Principle 21: The creativity, ideals and courage of the youth of the world should be mobilized to forge a global partnership in order to achieve sustainable development and ensure a better future for all.

Principle 22: Indigenous people and their communities and other local communities have a vital role in environmental management and development because of their knowledge and traditional practices. States should recognize and duly support their identity, culture and interests and enable their effective participation in the achievement of sustainable development.

Principle 23: The environment and natural resources of people under oppression, domination and occupation shall be protected.

Principle 24: Warfare is inherently destructive of sustainable development. States shall therefore respect international law providing protection for the environment in times of armed conflict and cooperate in its further development, as necessary.

Principle 25: Peace, development and environmental protection are interdependent and indivisible.

Table 4.1 (Continued)

Principle 26: States shall resolve all their environmental disputes peacefully and by appropriate means in accordance with the Charter of the United Nations.

Principle 27: States and people shall cooperate in good faith and in a spirit of partnership in the fulfillment of the principles embodied in this Declaration and in the further development of international law in the field of sustainable development.

Source: United Nations, "Report of the United Nations Conference on Environment and Development," Agenda 21, Rio Conference 27 principles, Annex I, Rio Declaration on Environment and Development, (Rio de Janeiro, June 3–14, 1992), http://habitat. igc.org/agenda21/rio-dec.htm.

and productive life in harmony with nature." This declaration is closely tied to the UN's Universal Declaration of Human Rights, which mainly deals with political and social rights. However, its Article 25 states[4]:

> Everyone has the right to a standard of living adequate for the health and wellbeing of himself and of his family, including food, clothing, housing and medical care and necessary social services.

Clearly, the standard of living depends upon a well-managed and sustainable natural and social environment to assure adequacy of food, clothing, housing, medical care, and many other amenities of life. Principle 2 recognizes the rights of each nation to define and enforce its respective environmental policies pursuant to these principles. The Rio Declaration is the most remarkable international achievement in recognizing the intimate connection between human community well-being, the quality of the environment on which that well-being depends, and the policies and laws that enable the achievement of values such as social and economic equity and basic human rights.

There is an immense gap between the humanistic concerns of the declaration and the reality of the human condition evidenced by the poverty that afflicts the majority of the world's population. Part of the challenge lies in the fact that environmental policy by itself cannot solve the deep-rooted problems of overpopulation, political and economic corruption, rebellions and wars motivated by power hungry leaders and institutionalized corruption. In Section 4.3, we review the general structure of global environmental policy, the international legal framework that has evolved in an effort to fulfill its principles, and the institutional methodology of international policy organizations. We also summarize the current status of the international law available to the global community in implementing various aspects of Agenda 21.

4.3   International law as environmental policy instruments

International environmental law is concerned with the protection of the global environmental commons on behalf of the global community. The UN has evolved an effective leadership role in this regard as evidenced by agreements to respect principles such as those shown in Table 4.1.[5] With 27 current members and likely to grow, the European Union (EU) has evolved its own internal system of environmental legislation, and is a major player in the development of UN environmental policy (see Chapter 13).

As with most international law, one of the guiding principles is that of sovereignty. Each nation has complete power to do as it pleases in its own territory, while being subject to the international laws it has agreed to, as emphasized in Principle 2 of Table 4.1. Sometimes there are sanctions for violations, but in general, adherence to international environmental laws depends on the good faith of the nations to implement them within their own policies, laws, and strategies.

Table 4.2 lists the major global environmental policy instruments developed to date. The most comprehensive document on this list is Agenda 21. However, there are also many other agreements not specifically about environmental management that raise a whole host of environmental issues.

4.3.1   Customary international law

Customary international law embodies the norms and rules that nations follow as a matter of custom. They are so pervasive that nearly all states of the world agree. For example, Principle 2 of the Rio de Janeiro Declaration is sometimes called the "environmental good neighbor policy." Its substance was already customary for most nations.

In accordance with the UN Charter and the principles of international law, states have the sovereign right to exploit their own resources pursuant to their own environmental and developmental policies. However, they also have the duty to ensure that activities within their jurisdiction or control do not damage the environment of other states.

The duty to promptly warn other states about emergencies of an environmental nature and environmental damages to which another state(s) may be exposed is also generally accepted. The point at which such principles become "customary law" is not clearly defined. Many arguments can be made against such "laws" by states not wishing to be bound.

Table 4.2 Programmatic Global Environmental Policy Instruments

General

The Stockholm Declaration on the Human Environment, 1972[12]

Agenda 21: The Rio Declaration on Environment and Development, 1992[13]

Johannesburg Summit commitment to full implementation of Agenda 21, 2002

Biological resources

The Convention on Biological Diversity, Rio de Janeiro, 1992

The Conservation of Antarctic Seals in 1972

The Conservation of Antarctic Marine Living Resources in 1980

Global air environment

The Kyoto Protocol to the United Nations Framework Convention on Climate Change, Kyoto, 1997

The Montreal Protocol on Substances That Deplete the Ozone Layer, 1999[14]

Global water environment

The United Nations Convention on the Law of the Sea (UNCLOS) 1994[15] (see Chapter 2, Section 2.4 for details)

International Convention for the Prevention of Pollution from Ships, 1973, as modified by the Protocol of 1978 and implemented by the UN International Maritime Organization (IMO)

United Nations Fish Stocks Agreement[16] in force 2001

Convention on the Law of the Nonnavigational Uses of International Watercourses, 1997

Global land environment

The United Nations Convention to Combat Desertification in Those Countries Experiencing Serious Drought and/or Desertification, Particularly in Africa. Paris, 1994[17]

1992 United Nations Conference on Environment and Development (UNCED) Forest Principles[18]

Chapter 11 of Agenda 21 Strategy: Combating Deforestation[19]

Outer space environment

There are five treaties respecting the outer space environment.[20]

Global population

Global Environmental Policy is an essential element of UN Population Policy (see Section 14.4).

Source: http://www.un.org/cyberschoolbus/treaties/index.asp.

4.3.2 Judicial framework

The body of international environmental law includes judgments and opinions by international courts or tribunals, including the entities outlined in Sections 4.3.2.1 through 4.3.2.3.

4.3.2.1   International Court of Justice

The International Court of Justice (ICJ) is the principal judicial organ of
the UN. It was established in June 1945 by the UN Charter and began
work in April 1946. The seat of the court is at the Peace Palace in the
Hague, the Netherlands. The court has a dual jurisdiction, which means
that it decides, in accordance with international law, disputes of a legal
nature that are submitted to it by states (jurisdiction in contentious
cases) and it gives advisory opinions on legal questions at the request
of the UN organs or specialized agencies authorized to make such a
request (advisory jurisdiction).[6] The court is composed of 15 judges, who
are elected to terms of office for 9 years by the UN General Assembly
and the UN Security Council, assisted by a registry, its administrative
organ. Its official languages are English and French. The ICJ acts as a
world court. However, the courts have heard only a few environmental
cases.[7]

4.3.2.2   European Court of Justice

Since the establishment of the Court of Justice of the European Communities
in 1952, its mission has been to ensure that "the law is observed in the
interpretation and application" of the treaties. As part of that mission, the
Court of Justice

- Reviews the legality of the acts of the institutions of the EU.
- Ensures that the member states comply with their obligations under
  community law.
- Interprets community law at the request of the national courts and
  tribunals.

The court, thus, constitutes the judicial authority of the EU and in cooperation with the courts and tribunals of the member states, ensures the application and uniform interpretation of community law. The Court of Justice
consists of three courts: the Court of Justice, the Court of First Instance
(created in 1988), and the Civil Service Tribunal (created in 2004). Since
their establishment, approximately 15,000 judgments have been delivered
by the three courts.[8] Of these, about 27 have involved applications of environmental law between 2001 and 2009.[9]

4.3.2.3   International Tribunal for the Law of the Sea

The International Tribunal for the Law of the Sea is an independent judicial
body established by the convention to adjudicate disputes arising out of the
interpretation and application of the convention pursuant to the provisions
of its statute. The tribunal has formed the Chamber of Summary Procedure,
the Chamber for Fisheries Disputes, the Chamber for Marine Environment
Disputes, and the Chamber for Maritime Delimitation Disputes. At the

request of the Chilean and the European Community, the tribunal has also formed a special chamber to deal with the case *Conservation and Sustainable Exploitation of Swordfish Stocks in the South-Eastern Pacific Ocean (Chile/ European Community)*.

Disputes relating to activities in the International Seabed Area are submitted to the Seabed Disputes Chamber of the Tribunal, consisting of 11 judges. Any party to a dispute over which the Seabed Disputes Chamber has jurisdiction may request the Seabed Disputes Chamber to form an ad hoc chamber composed of three members of the Seabed Disputes Chamber. Unless the parties otherwise agree, the jurisdiction of the tribunal is mandatory in cases relating to the prompt release of vessels and crews under Article 292 of the convention and to provisional measures, pending the constitution of an arbitral tribunal under Article 290, Paragraph 5, of the convention.[10]

4.4   Agenda 21: Strategic implementation program

The UN Strategic Program for Agenda 21 specifically implements environmental policy as elements of the Population Policy, as identified in Table 4.3[22] (see also Section 14.4).

4.4.1   Programmatic elements of Agenda 21

Programmatic elements of implantation mainly derive from the body of treaties implemented through the UN and its network of allied organizations. Table 4.2 lists the most environmentally relevant instruments.[11]

The Stockholm Declaration listed in Table 4.2 preceded Agenda 21. It has 7 articles and proclaims 26 principles. It is essentially a code of ethics and a philosophical statement of the values guiding global environmental policy. Although it does not establish any rules, its ethical stance strongly resembles that of the U.S. National Environmental Policy Act.

The foundational international environmental policy instrument is Agenda 21 that followed the Stockholm Declaration. The entire edifice of international environmental treaties and programs is linked directly or indirectly to Agenda 21. Table 4.3 shows Agenda 21 Strategic Program Areas that constitute implementatation strategy. The Strategic Program lists the contents of what is essentially Agenda 21's implementation strategy. It consists of 40 chapters, which are readily accessible from the UN Web site referenced in the table. Agenda 21 defines the comprehensive scope of the Strategic Program. It covers all environmental areas. It emphasizes a central concern for humans in framing environmental programs and calls for the fullest possible engagement of affected communities in decisions and actions that limit the degree of adverse environmental impacts. This environmental strategy is integrally linked to the UN's economic and

Table 4.3 Agenda 21: Strategic Program Areas

Chapter	Environmental issue
1. Preamble	
2. Introduction	
3. Combating Poverty	Population
4. Changing Consumption Patterns	Population
5. Demographic Dynamics and Sustainability	Population sustainability
6. Protection and Promotion of Human Health	Population
7. Promoting Sustainable Human Settlement Development	Sustainability
8. Integrating Environment and Development in Decision Making	Management
9. Protection of the Atmosphere	Air
10. Integrated Approach to the Planning and Management of Land Resources	Land
11. Combating Deforestation	Land
12. Managing Fragile Ecosystems: Combating Desertification and Drought	Land
13. Managing Fragile Ecosystems: Sustainable Mountain Development	Land sustainability
14. Promoting Sustainable Agriculture and Rural Development	Land sustainability
15. Conservation of Biological Diversity	Biological
16. Environmentally Sound Management of Biotechnology	Biological
17. Protection of the Oceans, All Kinds of Seas, Including Enclosed and Semi-enclosed Seas and Coastal Areas and the Protection, Rational Use, and Development of Their Living Resources	Water
18. Protection of the Quality and Supply of Freshwater Resources: Applications of Integrated Approaches to the Development, Management and Use of Water Resources	Water
19. Environmentally Sound Management of Toxic Chemicals Including Prevention of Illegal International Traffic in Toxic and Dangerous Products	Toxic substances
20. Environmentally Sound Management of Hazardous Wastes Including Prevention of Illegal International Traffic in Hazardous Wastes	Hazardous wastes
21. Environmentally Sound Management of Solid Wastes and Sewage-related Issues	Water pollution

Table 4.3 (*Continued*)

Chapter	Environmental issue
22. Safe and Environmentally Sound Management of Radioactive Wastes	Radioactive wastes
23. Preamble to Section III	Management
24. Global Action for Women towards Sustainable and Equitable Development	Management
25. Children and Youth in Sustainable development	Management
26. Recognizing and Strengthening the Role of Indigenous People and Their Communities	Management
27. Strengthening the Role of Non-governmental Organizations: Partners for Sustainable Development.	Management
28. Local Authorities' Initiatives in Support of Agenda 21	Management
29. Strengthening the Role of Workers and Their Trade Unions	Management
30. Strengthening the Role of Business and Unions	Management
31. Scientific and Technological Community	Management
32. Strengthening the Role of Farmers	Management
33. Financial Resources and Mechanisms	Management
34. Transfer of Environmentally Sound Technology, Cooperation and Capacity	Management
35. Science for Sustainable Development	Sustainability
36. Promoting Education, Public Awareness and Training	Management
37. National Mechanisms and International Cooperation for Capacity Building	Management
38. International Institutional Arrangements	Management
39. International Legal Instruments and Mechanisms	Management
40. Information for Decision Making	Management

Source: http://www.unep.org/Documents.Multilingual/Default.asp?documentID=52.

social development policies. In fact, the Johannesburg Summit simultaneously commits the UN to the Millennium Development Goals that address quality of life issues for the global population.[21]

It should be clear that global environmental policy is not viewed as a set of "stovepipes" that have their own environmental domain. UN Population Policy stresses that human development programs must integrate the requirements of any and all environmental policy areas as they are essential to protect the global population from harm and create numerous opportunities for the benefit of the people.

4.5 Programs that implement global environmental strategy

Environmental policy is directly and indirectly implemented through a wide variety of organizations and programs. The only way in which UN policy is created or modified is through treaties and related instruments agreed to by a sufficient number of UN members. Table 4.4 lists those activities that directly implement environmental policy and those with environmental factors or those likely to have environmental impacts.[23] The UN General Assembly has oversight responsibility for almost

Table 4.4 United Nations Environmental Policy Implementation

Environmental programs

United Nations Environmental Program (UNEP)
United Nations Development Program (UNDP)
Commission on Sustainable Development
United Nations Forum on Forests
Committee of Experts on the Transport of Dangerous Goods and on the Globally Harmonized System of Classification and Labeling of Chemicals
International Maritime Organization (IMO)

Programs dealing with environmental factors or impacts

General
Commission on Science and Technology for Development

Social, economic and health
Commission on Population and Development
United Nations Population Fund (UNFP)
United Nations Human Settlements Program (UN-HABITAT)
Permanent Forum on Indigenous Issues
Commission for Social Development
Committee on Economic, Social and Cultural Rights
International Trade Center (ITC)
Joint United Nations Programme on HIV/AIDS (UNAIDS)

Regional development commissions
Economic Commission for Africa (ECA)
Economic Commission for Europe (ECE)
Economic Commission for Latin America and the Caribbean (ECLAC)
Economic and Social Commission for Asia and the Pacific (ESCAP)
Economic and Social Commission for Western Asia (ESCWA)

Table 4.4 (Continued)

Specialized agencies
United Nations Industrial Development Organization (UNIDO)
United Nations Educational, Scientific and Cultural Organization (UNESCO)
International Maritime Organization (IMO)
World Bank Group
Food and Agriculture Organization (FAO)
International Fund for Agricultural Development (IFAD)
World Health Organization
World Meteorological Organization (WMO)
World Tourism Organization (UNWTO)

Other United Nations international law entities
Oceans and Law of the Sea (in force 1994)
International Law Commission (1949–present)
International Atomic Energy Agency (IAEA)
United Nations Office for Outer Space Affairs

Source: Compiled from http://www.un.org/aboutun/chart_en.pdf.

all of the programs and organizations listed. Details about each of the organizations or programs are readily accessible from the UN Web site search feature.

We will briefly describe the elements of two of these programs as particularly relevant to the environmental policy concerns developed in this book: the environmental program, including environmental assessment, and climate change, which we address in Chapter 13.

4.5.1   United Nations Environmental Program

The United Nations Environmental Program (UNEP) mission is "To provide leadership and encourage partnership in caring for the environment by inspiring, informing, and enabling nations and peoples to improve their quality of life without compromising that of future generations." UNEP has six priority program areas that focus on what the UN considers the most challenging environmental threats in the twenty-first century[24]:

- Climate change
- Disasters and conflicts
- Ecosystem management
- Environmental governance

- Harmful substances
- Resource efficiency

4.5.2　Climate change: Kyoto Protocol and the European Union

The climate change issue is the subject of Chapter 13 of this book. The *Kyoto Protocol* became effective in 2005. It established a cap-and-trade system for the six major greenhouse gases. Quotas on emissions were agreed to by each participating country. The goal of the original treaty was to reduce overall emissions by approximately 5% from their 1990 levels by the end of 2012. Under the treaty, nations that emit less than their quota would be allowed to sell emissions credits to other nations that exceed their quota. Developed countries can also sponsor carbon projects (e.g., maintenance of natural forest) that provide a reduction in greenhouse gas emissions, as a means of generating tradable credits.

The Intergovernmental Panel on Climate Change (IPCC) projected that the financial compliance cost through trading would be "limited" between 0.1% and 1.1% of gross domestic product among trading countries.[25] According to the Stern report, the costs of doing nothing were estimated to be 5–20 times higher,[26] but these figures are controversial and widely disputed.

All EU states have ratified the Kyoto Protocol. The EU Emission Trading Scheme (EU ETS) is the largest multinational greenhouse gas emissions trading system in the world; it was created in conjunction with the Kyoto Protocol. This program caps the carbon dioxide emissions from large installations, such as power plants and carbon-intensive factories. It regulates nearly half of the EU's carbon dioxide emissions.[27] Phase I of this system permitted participants to trade among themselves and in validated credits from developing countries. Phase 1 was widely criticized for providing an oversupply of allowances and for the distribution method of assigning allowances. Carbon dioxide trading currently makes up the bulk of emissions trading. It provides one means through which countries can meet their obligations under the Kyoto Protocol. In recent years, carbon trading has been steadily increasing. Phase II has attempted to address some of the criticism of Phase I.

4.5.3　Additional scope of the United Nations Environmental Program

The UNEP has 24 "Thematic Areas," each of which has supporting programs as listed in Table 4.5.[28] The reader can readily access each of these activities from the cited Web site. Among these, the environmental assessment activity is of particular interest in this book, as it relates to Chapter 1.

Table 4.5 Additional United Nations Environmental Program Activities

Art for the environment	Governments
Biodiversity	Health and environment
Biosafety	Indigenous knowledge
Business and industry	Indigenous peoples
Youth	Land
Children	Ozone
Civil society	Poverty and environment
Education and training	Regional seas
Energy	Scientists
Environmental assessment	South cooperation
Freshwater	Sports and environment
Gender	Urban issues

Source: http://www.unep.org/.

4.5.3.1 Environmental assessment

The UN describes its assessment program as follows:

> UNEP's strategic approach is to undertake and sup-
> port timely, participatory, and scientifically credible
> environmental assessments that are legitimate and
> relevant to decision-making processes, based on
> best available scientific expertise, knowledge, data,
> and indicators. Environmental assessments are key
> vehicles for promoting the interaction between sci-
> ence processes and the various stages of the policy
> and decision-making cycle. They underpin decision
> making by UNEP's Governing Council, multilat-
> eral environmental agreements, regional ministe-
> rial environmental forums, the private sector, and
> national and local authorities.[29]

The UN provides an index to environmental assessments organized
by means of the following topics[30]:

- Global environmental outlook (GEO) process
- Ecosystems
- Biodiversity
- International Assessment of Agricultural Science and Technology

Discussions, problems, and exercises

1. How do the international environmental policies cited here relate to issues of human rights?
2. Select six principles or concepts from Chapter 2 and briefly describe how they appear to be implemented in the system of treaties.
3. What factors constrain or limit the current framework of international environmental agreements for adequately protecting the future of the global commons?
4. In what ways and to what extent would the aggressive growth of renewable energy contribute to enabling the fuller implementation of Agenda 21?

Notes

1. United Nations Environment Programme. 2005. *The Montreal Protocol on Substances that Deplete the Ozone Layer.* http://www.unep.org/OZONE/pdfs/Montreal-Protocol2000.pdf (accessed August 5, 2010).
2. The United Nations. *Treaty Collection.* http://untreaty.un.org/English/treaty.asp (accessed August 5, 2010).
3. The United *Nations. Treaty Reference Guide.* http://untreaty.un.org/English/guide.asp (accessed August 5, 2010).
4. United Nations Universal Declaration of Humans Rights, http://www.un.org/en/documents/udhr/(accessed August 5, 2010).
5. United Nations, "Report of the United Nations Conference on Environment and Development," Agenda 21, Rio Conference 27 principles, Annex I, Rio Declaration on Environment and Development, (Rio de Janeiro, June 3–14, 1992), http://habitat.igc.org/agenda21/rio-dec.htm (accessed August 5, 2010).
6. International Court of Justice, http://www.icj-cij.org/court/index.php?p1=1 (accessed August 5, 2010).
7. Also see *Guide to the Court,* http://www.icj-cij.org/information/en/ibleubook.pdf
8. European Court of Justice, General Presentation: http://curia.europa.eu/jcms/jcms/Jo2_6999/general-presentation (accessed August 5, 2010).
9. Search for "environment" at http://europa.eu/jcms/jcms/Jo2_6999/general-presentation
10. International Tribunal for the Law of the Sea, http://www.itlos.org/start2_en.html (accessed August 5, 2010).
11. The United Nations Core Treaties: The Cyberschoolbus Treaty Project. http://www.un.org/cyberschoolbus/treaties/index.asp (accessed August 5, 2010).
12. United Nations Environment Programme. *Declaration of the United Nations Conference on the Human Environment.* 1972. http://www.unep.org/Documents.Multilingual/Default.asp?DocumentID=97&ArticleID=1503 (accessed August 5, 2010).
13. United Nations Department of Economic and Social Affairs. Agenda 21: Section II, Conservation & Management of Resources for Development, Chapter 21. http://www.un.org/esa/dsd/agenda21/res_agenda21_21.shtml (accessed August 5, 2010).

14. See note 1 above.

15. United Nations Convention on the Law of the Sea, http://www.un.org/ Depts/los/convention_agreements/convention_overview_convention.htm (accessed August 5, 2010).

16. The United Nations Agreement for the Implementation of the Provisions of the United Nations Convention on the Law of the Sea, http://www.un.org/ Depts/los/convention_agreements/convention_overview_fish_stocks.htm (accessed August 5, 2010).

17. United Nations Convention to Combat Desertification, http://www.unccd. int/convention/menu.php (accessed August 5, 2010).

18. United Nations Forum on Forests. *History and Milestones of International Forest Policy.* http://www.un.org/esa/forests/about-history.html (accessed August 5, 2010).

19. United Nations Department of Economic and Social Affairs. Agenda 21: *Section II, Conservation & Management of Resources for Development,* Chapter 11. http://www.un.org/esa/dsd/agenda21/res_agenda21_11.shtml (accessed August 5, 2010).

20. United Nations Office for Outer Space Affairs. *United Nations Treaties and Principles on Space Law.* http://www.oosa.unvienna.org/oosa/SpaceLaw/ treaties.html (accessed August 5, 2010).

21. See Table 14.3, "Summary of Key Millennium Goals and Targets for 2015," in this book.

22. United Nations Environment Programme. *Agenda 21: Environment and Development Agenda.* http://www.unep.org/Documents.Multilingual/ Default.asp?documentID=52 (accessed August 5, 2010).

23. United Nations. *The United Nations System: Principal Organs.* 2007. http:// www.un.org/aboutun/chart_en.pdf (accessed August 5, 2010).

24. United Nations Environment Programme, http://www.unep.org/ (accessed August 5, 2010).

25. G. Hinsliff, *£3.68 trillion: The price of failing to act on climate change.* 2006. *The Observer.* http://www.guardian.co.uk/money/2006/oct/29/greenpolitics. politics (accessed December 10, 2007).

26. Ibid.

27. The Washington Times. *Britain, California to join forces on global warming.* 2006. http://www.washtimes.com/business/20060731-011601-7934r.htm (accessed August 5, 2010).

28. United Nations Environment Programme, http://www.unep.org/themes/ thematic_areas.asp (accessed August 5, 2010).

29. United Nations Environment Programme. *UNEP Activities in Environmental Assessment.* http://www.unep.org/themes/assessment/ (accessed August 5, 2010).

30. United Nations Environment Programme. *Assessments.* http://www.unep. org/DEWA/assessments/index.asp (accessed August 5, 2010).

Sustainability, environmental impact assessment, and decision making

chapter five

Environmental impact assessment

> I think that all good, right thinking people in this country are sick and tired of being told that all good, right thinking people in this country are fed up with being told that all good, right thinking people in this country are fed up with being sick and tired. I'm certainly not, and I'm sick and tired of being told that I am.
>
> **Monty Python**

Case study

For generations, Niagara Falls, New York, has been a tranquil retreat for newlyweds. Love Canal was a quiet neighborhood located near Niagara Falls. As Niagara Falls' population expanded, the local school board was desperate for land. The district sought to purchase a property from the Hooker Chemical Company. Lacking strict environmental regulations and inspections at that time, Hooker had routinely buried toxic waste at this site. In total, 21,000 tons of toxic waste was buried beneath this property.

Because of safety concerns, Hooker declined to sell this property. The company even took school board members to the canal and drilled boreholes, showing that there were toxic chemicals below the surface. However, the board wanted the property and refused to capitulate. Faced with the property being expropriated, Hooker agreed to sell the property on condition that the board purchase it for $1. The agreement, signed in 1953, contained a 17-line provision written by Hooker, which explained the dangers of this site. This provision essentially released Hooker from all legal obligations should future lawsuits be filed.

In 1978, events took a turn for the worse. Lois Gibbs, a local mother and president of the Love Canal Homeowners Association, began to question why there were many health problems in the area. Her son, Michael Gibbs, developed epilepsy and suffered from asthma and other health problems. Residents began reporting strange odors and substances that were surfacing in their yards. There were high rates of unexplained illnesses, miscarriages, and mental retardation in the surrounding neighborhoods. Many basements were covered with a seeping thick black substance.

Vegetation in yards was dying. City officials investigated the area but took no action to solve the problem.

A survey by the Love Canal Homeowners Association indicated that 56% of the children born from 1974 to 1978 had birth defects. By 1978, Love Canal had become the stage for a national media event. At first, scientific studies did not conclusively prove that the chemicals were responsible for the residents' illnesses. Over time, however, the case for severe health effects mounted. The school that was built over this property was closed and later was demolished. However, both the school board and the former Hooker Chemical Company refused to accept liability.

The U.S. government relocated more than 800 families and reimbursed them for their lost homes. The U.S. Congress eventually passed the strict Comprehensive Environmental Response, Compensation, and Liability Act (CERCLA). Also known as the Superfund Act, CERCLA holds polluters accountable for damages caused by them and cleanup costs that are incurred.

> **Problem:** After reading this chapter, return to this problem. Consider this example in light of what you have read. This problem may be pursued either as an individual assignment or as a class project. How much of an effect did events such as these have on the passage of U.S. National Environmental Policy Act (NEPA) and other international environmental impact assessment (EIA) processes? Imagine if a process similar to that of NEPA had been used to assess the potential impacts and alternative sites for the proposed school ground. Had such a process been used, perhaps the Love Canal catastrophe would have been avoided. How could an EIA planning process be used to avoid such disasters in the future?

Learning objectives

- How EIAs are used in formulating environmental policies
- How the NEPA has affected international EIA processes
- General concepts underlying virtually all EIA processes
- Benefits of an EIA process
- Limitations and problems encountered with implementing an EIA process

5.1   How the National Environmental Policy Act's environmental impact assessment process has influenced international policy

A rigorous and scientifically based planning process can be instrumental in developing a successful environmental policy. At a minimum, the final policy should be based on a comparative analysis of the impacts of various policy alternatives. A practical plan may also require that other key factors (e.g., capital and operating costs, cost–benefits, social disruptions) be assessed and weighed in the decision-making process.

An EIA process provides nations with a valuable tool for improving decision making and advancing superior projects. EIA can be applied to proposals as diverse as land-use development projects, power generation and transmission infrastructure facilities, waste management facilities, and transportation infrastructure projects.*

This chapter examines the basic elements that are common to nearly all international EIA processes. The International Association for Impact Assessment (IAIA; www.iaia.org) was founded in 1980, and has become the premier international professional association for EIA scholars and practitioners. It has over 2500 members from 125 countries, publishes a professional journal, holds annual meetings and training workshops (English is its primary working language), and helps spread and improve the practice of EIA around the world. The IAIA, in cooperation with the Institute of Environmental Assessment of the United Kingdom, defines EIA to mean[1]:

> The process of identifying, predicting, evaluating and mitigating the biophysical, social, and other relevant effects of development proposals prior to major decisions being taken and commitments made.

In scientific jargon, a prediction is a categorical statement. It says that X will happen. In environmental science, it is often difficult to make *predictions* with absolute certainty. Often, the best that an analyst can do is to indicate that X is "likely" to happen. Terms such as "projection" or "forecast" tend to be used when making a "prediction" that is less than complete certainty.

Following the historic adoption of NEPA, the international community began to realize that environmental degradation was not only an American problem but also a global one. Other nations began to appreciate how the EIA elements inherent in the NEPA process could facilitate sound economic development while providing the methodology to establish plans and policies that would enhance public participation and help their own decision makers avoid making costly environmental mistakes. Throughout the 1970s, and continuing into the decades that followed, many countries moved quickly to adopt their own versions of national environmental policies and assessment procedures.[2] So many, in fact, that today some have postulated that NEPA may be one of the most emulated statutes in the world.

---

* For additional information on selected EIA topics, see *Preparing Environmental Impact Assessments - A Guide to Best Professional Practices for Preparing: Greenhouse Emissions, Cumulative Impact, Risk, Socioeconomic, and International Environmental Impact Assessments*, CRC Press (2010).

According to the U.S. Council on Environmental Quality (CEQ), the NEPA process has been patterned or copied in one form or another by more than 80 countries worldwide.[3] However, the number of nations that have adopted some form of an EIA process may actually be much higher than this estimate. Canter, for example, reports that over 100 countries have instituted some form of EIA measure.[4] A book by John Cronin and Robert F. Kennedy, Jr., cites an even larger number—over 125 nations.[5] By one account, developing Asian countries alone have already performed more than 15,000 EIA studies.[6] For detailed instruction on NEPA and environmental planning processes, the reader is referred to the author's companion text, *NEPA and Environmental Planning*, which describes tools, techniques, and approaches for planning actions and assessing environmental impacts.[7] For a detailed review of environmental impact statements (EIS), the reader is directed to the author's book, *Environmental Impact Statements*.*

5.1.1   How the environmental impact assessment process promotes democracy

Because of NEPA, American citizens are now able to participate in and influence proposed federal actions that may affect their lives during the early planning process, before a final decision has been made. Arguably, no other single U.S. law has contributed so much toward opening up the federal planning and decision-making process to its citizens. Adoption of an international EIA processes similar to NEPA has opened up government decision-making processes to tens of millions of citizens around the world. A book by John Cronin and Robert F. Kennedy, Jr., points out[8]:

> NEPA, which has now been adopted in some form
> by over 125 countries, has become one of the great
> promoters of democracy around the world.

These EIA processes have opened up what was once a largely closed decision-making process (often instrumented by entrenched interests) to the scrutiny of millions of citizens around the world.

5.2   Status of environmental impact assessment legislation

It is remarkable how many foreign EIA processes have come to mirror NEPA's original model, which remains virtually unchanged after more than 40 years. Some countries that did not initially incorporate U.S. EIA

---

* Eccleston, C. H. 2001. *Environmental Impact Statements: A Comprehensive Guide to Project and Strategic Planning.* New York: John Wiley & Sons Inc.

principles, such as considering alternatives, encouraging public participation, or investigating cumulative impacts, eventually revised their EIA
processes to include such elements. A chronological outline of some representative nations that have followed the NEPA footprint is presented in
Table 5.1.[9]

While some developing countries, such as the Philippines, required
EIAs to be prepared for major development projects as early as the
1970s, a few of the leading industrialized countries, such as Japan and
the Federal Republic of Germany, adopted similar requirements relatively recently.[10] A former senior policy advisor to the CEQ, Ray Clark,
has written that this is indeed a tribute to the vision forged in NEPA by
U.S. Congress.[11]

While the process used in reaching environmentally significant decisions is not and has never been a smooth or uniform one, the traditional
lack of prominence devoted to environmental considerations is changing.
As depicted in Table 5.2, this process has undergone a series of evolutionary changes to reach its current state.[12]

5.2.1 Organizations that have adopted environmental impact assessment processes

The Organization of Economic Cooperation and Development (OECD)
recommended that its member states adopt EIA processes as early as 1974.
The OECD now uses an EIA process similar to that of NEPA in granting aid to developing nations.[13] Throughout the 1980s, many developing
countries continued to establish EIA processes as an essential element of
environmental policy and project planning. The requirement to implement
EIA studies for activities that are likely to significantly affect the environment is reflected in

- Principle 17 of the Rio Declaration on Environment and Development[14] (see Chapter 4, Table 4.1 for Principle 17)
- Article 5 of the Legal Principle for Environmental Protection and Sustainable Development[15]
- The 1987 United Nations Environmental Program (UNEP) Goals and Principles of Environmental Impact Assessment[16]

The European community now requires its members to comply with
an environmental process similar in nature to NEPA. Another example
involves the North American Free Trade Agreement (NAFTA), which
incorporates an EIA process modeled after NEPA for evaluating extrater-
ritorial impacts that cross the borders of the United States, Canada, and
Mexico.

Table 5.1 Representative Nations That Have Adopted an Environmental Impact
Assessment Process Patterned on the Model of NEPA

Nation	Year EIA process was adopted	Notes
United States of America	1969	National Environmental Policy Act
Canada	1973	Environmental Assessments Review Process (EARP).
Australia	1974	Environmental Protection (Impact of Proposals) Act, 1974
Malaysia	1974	EIA required under Section 34 A, Environmental Quality Act, 1974
France	1976	National Environmental Assessment Legislation
Philippines	1978	As per Presidential Decree No. 1586
Japan	1984	Environmental Assessment implemented via a cabinet resolution
United Kingdom	1985	Town and Country Planning (Assessment of Environmental Effects) Regulations 1988 (SI. No. 1199)
Indonesia	1986	AMDAL (EIA) process established by law through Government Regulation No. 29 of 1986
The Netherlands	1986	Environmental Protection (General Provisions) Act transformed into Environmental Management Act of 1993
New Zealand	1986	Environmental Act of 1986 and Resource Management Act of 1991
Sri Lanka	1988	National Environmental Act No. 47 of 1980 was amended to include an EIA provision
CEC	1988	EU Directive on Environmental Assessment for 12 Member States
Norway	1989	Under the Planning Act of 1989
Germany	1990	National Environmental Assessment Legislation
Thailand	1992	Sections 46 & 47 under National Environmental Quality Act 1992
Nepal	1993	In the form of National EIA Guidelines issued by National Planning Commission Secretariat

Nation	Year EIA process was adopted	Notes
India	1994	Before January 1994, obtaining Environmental Clearance from Central Ministry was only an administrative requirement intended for mega projects but from 1994 the EIA notification was issued

EIA = Environmental impact assessment; CEC = Commission for Environmental Cooperation; EU = European Union.

Table 5.2 Evolution of the EIA Process

1970s	Beginning with NEPA, early EIA processes focused primarily on the natural environment.
1980s	Socioeconomic assessments were eventually accepted as elements of the processes.
1990s	Integrated environmental management promotes principles of transparency, accountability, and informed decision making throughout the life cycle of a project.
1990s to present	Strategic environmental assessment emerged as a proactive tool for addressing the environment in plans and policies.

5.2.1.1 United Nations

The UNEP developed guidelines for performing EIAs and since has strongly encouraged member states to establish EIA processes.[17] According to the UNEP, EIA provisions now exist in the environmental legislative framework of 55 developing countries.

5.2.1.2 North American Free Trade Agreement

Under the 1994 North American Agreement on Environmental Cooperation, the United States, Canada, and Mexico agreed to develop recommendations covering proposed projects "likely to cause significant adverse transboundary effects." The Commission for Environmental Cooperation (CEC) is a tri-national organization created at the time when NAFTA was signed to address regional environmental concerns, prevent potential trade and environmental conflicts, and promote enforcement of environmental law. In 1997, the CEC agreed to begin developing a transboundary EIA agreement.

Presidential Executive Order 13141, issued during the Clinton administration, directed responsible agencies to assess and consider environmental impacts of trade agreements through a process of ongoing assessment and evaluation.[18] A provision of the Executive Order designates the U.S.

trade representative and the chair of the CEQ to develop procedures for conducting environmental reviews in consultation with appropriate foreign policy and environmental and economic agencies.

5.2.1.3   World Bank

In 1989, the World Bank ruled that an EIA process should normally be prepared for those projects for which it provided funding. In 1991, the World Bank published a three-volume EIA sourcebook that provides practical guidance for the preparation of EIA documents for various types of development projects.[19] The bank's most recent procedure clarifies the need for the nature of environmental assessment that needs to be prepared for various investment projects. However, this procedure does not apply to macroeconomic adjustment lending.[20]

5.3   Typical environmental impact assessment process

Paoletto details the principal steps that an EIA process should typically include (Table 5.3).[21]

5.3.1   Guiding principles

The IAIA has gone beyond this fundamental guidance by producing a set of 14 *basic principles* for all EIA processes (Table 5.4).[22] These principles apply to all stages of an EIA or strategic environmental assessment (SEA) process.

Table 5.3 Typical Steps Followed during a Project-Specific EIA Process

1. *Impact identification*: The EIA process typically involves a broad analysis of the impact of project activities with a view to identifying those which are worthy of a detailed study.
2. *Baseline study*: This involves collection of detailed information and data on the condition of the project area prior to the project's implementation.
3. *Impact evaluation*: This is performed whenever possible in quantitative terms and should include the working-out of potential mitigation measures.
4. *Assessment*: The environmental losses and gains, with economic costs and benefits for each analyzed alternative, are assessed.
5. *Documentation*: A document is prepared with details of the EIA process and conclusions regarding the significance of potential impacts.
6. *Decision making*: The document is transmitted to the decision maker, who will either accept one of the project alternatives, request further study, or reject the proposed action altogether.
7. *Post audits*: Reviews and audits are made to determine how close to reality the EIA predictions were.

EIA = Environmental impact assessment.

Table 5.4 Governing EIA Principles

Purpose	Support informed decision making that results in appropriate levels of environmental protection and well-being of the community.
Rigorous	Apply "best practicable" science, employing methodologies and techniques appropriate to address the problem under investigation.
Systematic	Ensure full consideration of all relevant information on the affected environment, proposed alternatives and their impacts, and measures necessary to monitor and investigate residual effects.
Interdisciplinary	Provide appropriate techniques and experts in the relevant biophysical and socioeconomic disciplines are employed, including use of relevant traditional knowledge.
Practical	Apply information and outputs that assist with problem solving and that are both acceptable to proponents and can be implemented by the practitioners.
Participative	Provide appropriate opportunities to inform and involve interested and affected public and ensure that their inputs and concerns are addressed explicitly in the documentation and decision-making process.
Relevant	Provide sufficient, reliable, and useable information for development planning and decision making.
Cost-effective	Implement a process that achieves the EIA objectives within the limits of available information, time, resources, and methodologies.
Efficient	Impose minimum cost burdens in terms of time and finance on proponents and participants, which are consistent with meeting accepted requirements and EIA objectives.
Focused	Concentrate on significant environmental effects and key issues; that is, issues that need to be considered during the decision-making process.
Adaptive	Can be adjusted to the realities, issues, and circumstances of the proposals under review without compromising the integrity of the process and is iterative, incorporating lessons learned throughout the proposal's life cycle.
Credible	Can be carried out with professionalism, rigor, fairness, objectiveness, impartiality, and balance, and be subject to independent checks and verification.
Integrated	Address the interrelationships of social, economic, and biophysical aspects of the environment.
Transparent	Present clear, easily understood EIA requirements; ensure public access to information; identify factors that are to be taken into account in decision making; and acknowledge limitations, problems, and difficulties.

Source: International Association for Impact Assessment.

The IAIA has also identified 10 *operating principles* applicable to all EIA processes (Table 5.5).[23] These operating principles describe how the basic principles outlined in Table 5.4 should be applied to the main steps and specific activities of the EIA process (i.e., screening, scoping, identification of impacts, and assessment of alternatives).

Table 5.5 IAIA Governing Operating Principles Underlying an EIA Process

Screening	Determine whether a proposal should be subject to an EIA and, if so, at what level of detail.
Scoping	Identify the issues and impacts that are likely to be important.
Examination of alternatives	Establish the preferred or most environmentally sound and benign option for achieving the proposal's objectives.
Impact analysis	Identify and predict the likely environmental, social, and other related effects of the proposal.
Mitigation and impact management	Establish measures necessary to avoid, minimize, or offset predicted adverse impacts and, where appropriate, to incorporate these into an environmental management plan or system.
Evaluation of significance	Determine relative importance and acceptability of residual impacts (i.e., impacts that cannot be mitigated).
Preparation of environmental impact statement or report	Document clearly and impartially the impacts of the proposal, the proposed measures for mitigation, the significance of the effects, and the concerns of the interested public and the communities affected by the proposal.
Review of the EIS	Determine whether the report: meets its terms of reference; provide a satisfactory assessment of the proposal(s); and contain the information required for decision making.
Decision making	Approve or reject the proposal and establish the terms and conditions for its implementation.
Follow-up	Ensure that the terms and conditions of approval are met; monitor the impacts of development and the effectiveness of mitigation measures; strengthen future EIA applications and mitigation measures; and, as necessary, undertake environmental audit and process evaluation to optimize environmental management. Note: when monitoring or evaluating, and when management plan indicators are designed, it is desirable, whenever feasible, that they also contribute to local, national, and global monitoring of the state of the environment and to sustainable development.

Source: International Association for Impact Assessment.
EIA = Environmental impact assessment.

Table 5.6 Basic Elements Addressed in a Typical Environmental
Impact Assessment Document

A brief nontechnical summary of the significant issues and findings

Any uncertainties and gaps in information

Description of the proposal

Description of the affected environment

Description of reasonable alternatives

Assessment of potential environmental impacts of the proposal (proposed
action and alternatives), including short-term and long-term effects, and the
direct, indirect, and cumulative impacts

Assessment of whether the environment of any other state/province or areas
beyond national jurisdiction is likely to be affected by the proposal

Description of practical measures (including their effectiveness) for mitigating
significant adverse environmental impacts of the proposed activity and
alternatives

5.3.1.1   *Minimal document requirements*

Paoletto suggests a set of minimal elements that a typical EIA document
should contain (Table 5.6).[24]

5.3.2   *Comparison of the National Environmental Policy Act with other environmental impact assessment processes*

Generally, a final decision to pursue an action cannot be made before the
EIA process has been completed. This requirement protects the environment and ensures the integrity of the EIA process. Unlike NEPA, in some
countries, a final decision can be made before the EIA process has been
completed.

A U.S. federal agency generally formulates a proposal and is responsible for preparing the EIS. In practice, consultants are often used to assist
the agency in preparing the EIS. Allowing the proposing agency to prepare the EIS has been criticized as potentially biasing the analysis. To promote a more objective analysis, some countries have their EIAs prepared
by independent agencies.

A NEPA EIS is only required for U.S. *federal* actions that may significantly affect environmental quality; this restriction also includes private
actions that are enabled by a federal agency (funded, authorized, and
approved by a federal agency). However, most U.S. states have their own
State Environmental Policy Act (SEPA), some of which have requirements
that are more rigorous than those of NEPA. These state SEPAs often encompass private actions that occur within those states. In contrast, European

Union (EU) assessments can be triggered by private or public actions that may significantly affect the environment; categories of actions requiring some level of EIA analysis have been established. Under both the U.S. and EU processes, private applicants generally supply data for the analysis. Both, the United States and the EU have well-defined public participation requirements.

In the United States, a programmatic EIS can be prepared for programs, policies, and plans. The EU does not recognize a programmatic EIA. Instead, it prepares strategic EISs, which are somewhat similar to a programmatic EIS (see Section 5.5).

The NEPA implementing regulations establish basic requirements that each U.S. agency must follow. Each federal agency is required to supplement these basic requirements with their own specific implementing regulations. Similarly, the EU establishes basic EIA requirements that each member state must follow. Each nation may supplement these basic requirements with its own specific implementing requirements. In the United States, the federal court system is used to litigate legal challenges involving NEPA compliance issues. In the EU, the European Court of Justice decides legal challenges involving the EIA process. Neither the United States nor the EU impact assessment processes require that one choose an alternative that protects environmental quality. Both processes are largely founded on the premise that the analysis will lead to more informed decision making and ultimately to decisions that will protect the environment.

Of late, NEPA lawsuits increasingly involve climate change issues. For proposals that could significantly affect the climate, U.S. courts appear to be moving in the direction of requiring some level of analysis. As applicable, EIAs in the EU must address the EU's emission trading system.

5.3.2.1 Comparison with the World Bank

NEPA recognizes only one instrument for investigating proposals that are deemed to result in a significant environmental impact—the EIS. The World Bank recognizes a number of different instruments, which depend on the nature of the problem; these instruments include EIAs, regional or sectoral environment assessments, environmental audits, risk or hazard assessments, and environmental management plans.

In the United States, a Notice of Intent (in addition to other notices that an agency's NEPA implementing regulations require) must be published in the U.S. Federal Register. The public is kept abreast of the status and is afforded an opportunity to participate over the course of the EIS, beginning with the public scoping process.

The World Bank uses various mechanisms for publicizing its EIA process, including publishing a status of all projects, by country, on its Web site. EISs are circulated, normally for 120 days, to the potentially affected parties in their local language. The bank directly engages people for their views and comments.

5.4   Programmatic and strategic environmental assessments

A relatively recent innovation involves the concept of programmatic environmental assessments and SEAs. The term "strategic environmental assessment" can be defined as[25]:

> A process of anticipating and addressing the potential environmental consequences of proposed initiatives at higher levels of decision-making. It aims at integrating environmental considerations into the earliest phase of policy, plan or program development, on a par with economic and social considerations.

SEAs can provide policymakers with a valuable tool in formulating policies and plans. In essence, an SEA extends the application of EIA to the level of policies, plans, and programs (PPPs). A key distinction between a project-level EIA and an SEA is that an SEA can be applied to PPPs at an earlier stage than individual projects. Thus, an SEA allows for environmental considerations and objectives to be viewed proactively as inherent elements of the planning process, rather than just as problems to be mitigated after other development decisions have been made.

Increasingly, SEAs are being used to shape the initial stages of decision making to assess the consequences of PPPs. Countries such as the United States, Australia, Canada, Denmark, and New Zealand have already applied mechanisms similar to SEAs in developing plans and policies.[26] Some aid agencies in Africa have also started to use them.[27] SEAs are also being recognized as a proactive tool for promoting sustainable development that may also serve to reduce the number of required project-specific EIAs. Planners may use it as a method to assess different ways for accomplishing sustainability policies (Chapter 3 provides an overview of the concept of sustainability).

Much of the world, including the EU, does not formally recognize a "programmatic" EIA, which is a concept defined in the NEPA regulations; the concept of the strategic EIA is recognized throughout the EU and is gaining acceptance in many other countries. Conversely, the United States and many other nations do not formally recognize the concept of the "strategic" EIA. In practice, the two concepts are similar. While there is no universally accepted criterion for differentiating between the two concepts, the two concepts are most commonly applied in the following manner.

A programmatic assessment tends to be prepared for proposals that involve the development of a definite policy, plan, or program. In contrast, a strategic assessment tends to denote a scope of analysis that is a level higher than that of a programmatic assessment. More to the point, a strategic assessment defines a high-level direction or strategy for a nation (e.g., national energy, agricultural, or water strategy). In theory, once this direction or strategy is defined, a "programmatic" assessment could then be prepared to consider a specific policy, plan, and in particular, a program for implementing the strategy defined in the SEA; the programmatic EIA can be tiered from the SEA. Once the programmatic EIA has defined a program-level course of action (such as a specific nuclear energy or solar program for implementing the energy strategy), more standard EIAs can then be tiered off the programmatic EIA to assess site-specific impacts of an individual project(s), such as construction of a nuclear power plant or solar array farm.

5.4.1   Goals of strategic environmental assessment

According to Sadler, SEAs should be prepared to[28]:

- Focus project-specific EIA by ensuring that issues of need and alternatives are addressed at the appropriate policy, plan, or program level.
- Improve the scope and assessment of cumulative impacts, particularly where large projects stimulate secondary development and where many small developments not requiring EIAs may occur.
- Facilitate the application of sustainable principles and guidelines, for example, by focusing on the maintenance of a chosen level of environmental quality, rather than by minimizing individual impacts.

5.4.2   Performance criteria

As depicted in Table 5.7, the IAIA has established performance criteria that SEA analyses should meet.[29]

5.4.3   Relationship between environmental impact assessment and strategic environmental assessment

The difference between EIA and SEA processes is evident in the scale of their frameworks. When compared to a project-specific EIA, the scope of SEA tends to be broader, both temporally and geographically, and allows consideration of alternatives and a higher "programmatic" view of the "bigger picture."

Table 5.7 Strategic Environmental Assessment Performance Criteria

Integrated	Ensure adequate environmental assessment of all strategic decisions relevant for the achievement of sustainable development. Address interrelationships of biophysical, social, and economic aspects. Tier to policies in relevant sectors and (transboundary) regions and, where appropriate, to project EIA and decision making.
Sustainability led	Facilitate identification of development options and alternative proposals that are more sustainable.
Focused	Provide sufficient, reliable, and usable information for development planning and decision making. Concentrate on key issues of sustainable development. Customize analysis to the characteristics of the decision-making process. Cost- and time-effective.
Accountable	Is carried out with professionalism, rigor, fairness, impartiality, and balance. Is subject to independent checks and verification. Document and justify how sustainability issues were taken into account in decision making.
Participative	Inform and involve interested and affected public and government bodies throughout the decision-making process. Address their inputs and concerns in documentation and decision making. Provide clear, easily understood information and requirements and ensures sufficient access to all relevant information.
Iterative	Ensure availability of the assessment results early enough to influence the decision-making process and inspire future planning. Provide sufficient information on the actual impacts of implementing a strategic decision to judge whether this decision should be amended and to provide a basis for future decisions.

Ideally, a project-specific EIA should be prepared once a policy has been established via SEA. The EIA provides information about the likely environmental impacts of an individual project and is useful in implementing mitigation measures. For example, if a government agency decides to develop a national wind power program, EIAs can be used to minimize the environmental damage from building specific power stations, but it cannot practically address the more fundamental questions regarding design of the national wind power program. In contrast, an

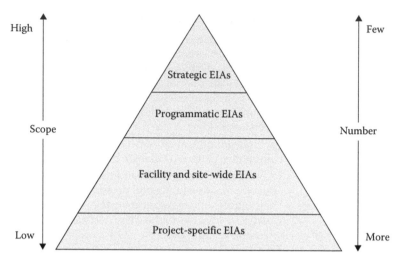

Figure 5.1 Relationship between different levels of environmental impact assessments.

SEA could effectively lay out the overall policy and investigate the programmatic impacts associated with such a policy, but is not an appropriate tool for evaluating site-specific impacts, alternatives, or mitigation measures. Figure 5.1 depicts the relationship between various levels of EIA assessments (note: the types of assessments shown in this figure are not recognized by many nations).

5.4.3.1   Comparison of strategic environmental assessment and environmental impact assessment

McDonald and Brown have written that[30]:

> EIA tends to focus on the mitigation of impacts of proposed activities rather than determining their justification and siting.

Perhaps SEA differs significantly from EIAs in that an SEA is a proactive tool for environmental management, whereas an EIA is used to assess specific development proposals. Some fundamental differences between SEAs and EIAs are summarized in Table 5.8.

SEAs and EIAs also tend to be applied at different stages of plans and policies and to different levels of decision making. Such a *tiered approach* is employed in New Zealand, the EU, and the United States.[31] Under a tiered approach, SEAs are used to formulate strategies and policies in a proactive way. These policies and strategies create a framework against which specific development proposals and projects can then be assessed

Table 5.8 Comparison of EIA with SEA

Attribute	PEA/SEA	EIA
Decision making	Formulates a high-level direction; supports site selection, but may not support site-specific construction and operational activities; tends to support multiple decisions	Supports detailed site selection decisions; supports site-specific decisions such as construction/operational activities
Timing	Analysis is performed before project-specific proposals are formulated; proactive; informs stakeholders about future development proposals	Analysis is performed after SEA/PEA is prepared; reactive; applied to the development of a site-specific proposal
Spatial bounds	Considers large areas, regions, or national development; focuses on impacts of national or regional significance	Focuses on a specific project site; focuses on site-specific impacts
Temporal bounds	Longer "shelf life;" tends to be a continuing process over a life cycle, which is aimed at providing information at the right time.	Shorter "shelf life;" tends to have a defined beginning and endpoint
Level of uncertainty	Greater degree of uncertainty	Less uncertainty
Impact analysis	Evaluates broad assessment of cumulative impacts and identifies issues for sustainable development	Tends to focus more on site-specific cumulative impacts, and direct and indirect impacts
Degree of quantitative analysis	Tends to be more descriptive and qualitative	Tends to be more quantitative
Focus of analysis	Focuses on maintaining a chosen level of environmental quality	Focuses on mitigating impacts
Level of detail	Has a broader perspective and a correspondingly lower level of detail, providing an overall vision	Has a narrow (project specific) perspective with a higher level of detail
Mitigation	Considers general mitigation measures	Considers more specific mitigation measures
Interim actions	Frequently involves interim actions (actions related to the EIA that need to occur while the analysis is being prepared)	Less likely to involve interim actions. Interim actions (actions that need to be taken before a final decision is made) are often prohibited or are severely restricted

using the EIA. Swedish planners have used SEA to ensure that plans and environmental goals encourage sustainable development.[32]

5.5   Challenges and limitations

In practice, application of the EIA process tends to be focused on project-level planning (although it has also been infrequently applied to programs and strategic planning). A project in the context of EIA is "an individual development or other scheme as distinct from a suite of schemes or a strategy for development of a particular type or in a particular region."[33] When the EIA process is applied to broader programs or regional planning, it is often done through the related analytical process of SEA (see Section 5.5).

While the EIA process does not necessarily prevent a project from having an impact on the environment, it frequently does minimize the severity of its adverse impacts.[34] Nonetheless, there are some fundamental problems associated with most processes. For example, some EIA processes address alternatives to the proposed project in a *limited* manner; that is, by the project assessment stage, a number of options having potentially different environmental consequences from the chosen ones are likely to already have been eliminated.

5.5.1   Optimism bias and the planning fallacy

Optimism bias is a demonstrated and systematic tendency for people to be overoptimistic about the outcome of an action. This includes overestimating the likelihood of positive events and underestimating the likelihood of negative ones. It is one of several kinds of positive illusions to which people are generally susceptible.

5.5.1.1   Planning fallacy
The *planning fallacy* is the tendency to underestimate the projected time required to complete a task. In one study, 37 college students were asked to estimate the completion time for their senior theses. The average estimate was 33.9 days. Only about 30% of the students completed their thesis in the predicted amount of time; the average actual completion time was 55.5 days.[35]

Lovallo and Kahneman expanded this concept from being a tendency to underestimate task-completion times to a more general tendency to underestimate the time, costs, and risks of future actions and, at the same time, overestimate the benefits of the same actions.[36] According to their definition, the planning fallacy results not only in time overruns but also in cost overruns and benefit that do not reach original expectations. Popular examples include construction of the Sydney Opera House and Boston's Big Dig project, both of which ran many years past their original schedule.

Flyvbjerg argues that what appears to be optimism bias may, on closer examination, be strategic misrepresentation.[37] He believes that planners frequently deliberately underestimate costs and overestimate benefits in order to get their projects approved, especially when projects are large and when organizational and political pressures are high. Kahneman and Lovallo maintained that optimism bias is the main problem.[38] Many studies have shown that this bias can cause people to take imprudent or even unacceptable risks.[39] This effect, however, is not universal; for instance, some people tend to overestimate the risks of low-frequency events, particularly negative ones.

5.5.2 Disadvantages of project-specific environmental impact assessments

Sadler describes some disadvantages of project-specific EIAs[40]:

- Restricted opportunities for effective public participation in planning or decision-making processes
- Restricted ability to address cumulative impacts, particularly for large development projects where secondary development could occur
- Limited analysis in a "stand-alone" process, which may be poorly related to the project cycle

Because a project-level EIA often precludes consideration of alternative strategies, locations, and designs, at least one EIA practitioner argues that, in effect, "An EIA at the project level is essentially damage control."[41] Application of EIAs at a more strategic level can promote a more effective assessment of alternatives and cumulative impacts at an earlier stage in the decision-making process. It can also facilitate consideration of a wider range of actions over a greater area.[42]

5.6 Mathematical modeling

No discussion of the EIA process would be complete without a consideration of computer modeling. Increasingly, policies are being developed based on the results of computer models. While computer models provide a powerful tool in one's policy arsenal, they likewise suffer from some significant weaknesses. Models can condense large amounts of difficult data into simple representations, but it is not a given that this model will provide an accurate answer, predict correct scenario consequences, or account for all possible variables, particularly human behavior. Even if all variables could be included in a complex dynamical model, chaos theory implies that the predictions will frequently be unreliable.

In their book, *Useless Arithmetic: Why Environmental Scientists Can't Predict the Future*, environmental scientists Orrin Pilkey and Linda Pilkey-Jarvis present cases that have demonstrated the misuse of mathematical models.[43] Some of these examples include the collapse of Atlantic cod fishing, the rise of sea levels due to global climate warming, shoreline erosion, toxicity of abandoned open-pit mines, and the spread of nonindigenous plants.

The Pilkey's assert that when models are used to study complex dynamical systems, the predictions are inherently (for many technical and mathematical reasons) unreliable. Nevertheless, policymakers, some with their own agendas, may offer them to the public as "proof" that a particular policy is the correct one, which is often a gross misrepresentation of the "prediction." As such, models offer no guarantee to policymakers that the right actions will be set into policy. Nor are the results of these realities necessarily benign for the environment. There are noted instances where modeling predictions resulted in actual severe harm to the environment.

Consider one noted example, the collapse of North Atlantic cod stocks. The results of computer simulations were misused by the fishing industry and the Canadian government to push for a fishing policy that essentially destroyed cod fisheries and the cod fishing industry in this region. The mathematical model was a simple dynamical system based on the population dynamics of a single species, the Atlantic cod. The model ignored interacting populations within the ecosystem. The predictions grossly overestimated the safe catch, resulting in the collapse of the Grand Banks cod fishery in 1992.

Computers have a mystique that can sometimes dazzle the public and policymakers alike. The public tends to believe that computer models are accurate. They also accept the standard lame excuses when things go astray as they did with the cod fishing industry. Years of optimistic claims by scientists about the accuracy of computer model predictions are not easy to turn around.

One alternative is adaptive management, a form of trial-and-error management.* For instance, adaptive management could be applied to the storage of high-level nuclear waste at Yucca Mountain, in Nevada. A policy could be formulated, given the best available science and with the understanding that the waste would be monitored. The storage facilities might be modified years and even centuries later, based on new circumstances and data that become available. Similarly, harvesting limits on a certain species such as cod could be set and modified over time depending on annual field observations.

---

* Eccleston, C. H. 2001. *NEPA and Environmental Planning: Tools, Techniques, and Approaches for Practitioners.* New York: John Wiley & Sons Inc.

Discussions, problems, and exercises

1. What is the closest equivalent under NEPA to a "strategic environmental assessment"?
2. Does the World Bank require a process similar to NEPA for funding international development projects?
3. Why has it been said that NEPA promotes international democracy?
4. You are the secretary of the environment for the island state of Pacific Shangri-La. Your state desires to develop sustainable resorts and eco-tourism industry. You need to develop a cost-effective and flexible EIA process tailored to plan a sustainable tourism industry. Prepare a report that outlines the key provisions of this EIA process. This report will discuss how it will be implemented, its scope, who is responsible for executing it, and what trigger mechanisms will ensure that the EIA process is initiated for new proposals. Also, define the scope and process for preparing a strategic EIA that will set the overall direction and goals for this proposal. Define the programmatic impacts and issues that this strategic EIA will evaluate. Define the programmatic alternatives that this EIA will investigate.

Notes

1. International Association for Impact Assessment and Institute of Environmental Assessment, *Principles of Environmental Impact Assessment Best Practice*, January 1999 (UK: IAIA, 1999).
2. A. K. Biswas and S. B. C. Agarwala, eds., *Environmental Impact Assessment for Developing Countries* (Oxford, United Kingdom: Butterworth-Heinemann, 1992).
3. House of Representatives, Committee on Resources, *Problems and Issues with the National Environmental Policy Act of 1969*, Oversight hearing before House Committee on Resources, 105th Cong., 2nd session, March 18, 1998, (Washington DC: U.S. Government Printing Office, 1998).
4. L. W. Canter, *Environmental Impact Assessment*, 2nd ed. (New York: McGraw-Hill Inc., 1996).
5. J. Cronin and R. F. Kennedy Jr., *The Riverkeepers* (New York: Scribner, 1997), 37, 175.
6. M. Prasad and A. K. Biswas, *Conducting Environmental Impact Assessment in Developing Countries* (New York: United Nations University Press, 1999).
7. C. H. Eccleston, *NEPA and Environmental Planning: Tools, Techniques, and Approaches for Practitioners* (Boca Raton, FL: CRC Press, 2001).
8. See note 5 above.
9. See note 6 above, p. 375.
10. See note 6 above.
11. R. Clark and L. Canter, eds., *Environmental Policy and NEPA: Past, Present, and Future* (Boca Raton, FL: St. Lucie Press, 1997), chap. 7.
12. CSIR, *Strategic Environmental Assessment (SEA): A Primer*, CSIR Report, ENV/ S-RR 96001, September 2, 1996.
13. OECD, *Good Practices for Environmental Impact Assessment of Developing Projects* (Paris: Development Assistance Committee, 1992).

14. United Nations. A/CONF.151/26 (vol. 1), August 12, 1992, *Report of the United Nations Conference on Environment and Development* (Rio de Janeiro, June 3–14, 1992).
15. Adopted by the Experts Group on Environmental Law of the World Commission on Environment and Development.
16. UNEP Working Group on Environmental Law, adopted by the UNEP Governing Council at its 14th session, Available at http://www.unep.org/Documents.Multilingual/Default.Print.asp?DocumentID=100&ArticleID= 1 658 (accessed July 8, 2010).
17. UNEP, *Environmental Impact Assessment: Basic Procedures for Developing Countries* (Bangkok: Regional Office for Asia and the Pacific, 1988).
18. Executive Order 1314, *Environmental Review of Trade Agreements,* 64 FR 63167, November 18, 1999.
19. World Bank, *Environmental Impact Assessment Sourcebook,* vol. 1–3, (Washington, DC: World Bank, 1991), http://web.worldbank.org/WBSITE/EXTERNAL/TOPICS/ENVIRONMENT/EXTENVASS/0,, contentMDK:20282864~pagePK: 148956~piPK:216618~theSitePK:407988,00.html (accessed September 1, 2010).
20. Operational Policy (OP)/Bank Procedures (BP) 4.01, January 1999.
21. G. Glen Paoletto, "Environmental Impact Assessments (EIA)" (lecture notes) http://www.gdrc.org/uem/eia/lecture-notes.html (accessed July 8, 2010).
22. International Association for Impact Assessment.
23. International Association for Impact Assessment, "Principles of Environmental Impact Assessment Best Practice," http://www.iaia.org/publicdocuments/special-publications/Principles%20of%20IA_web.pdf (accessed September 16, 2010).
24. See note 21 above.
25. B.Sadler and R. Verheem, *Strategic Environmental Assessment: Status, Challenges and Future Directions* (Amsterdam, the Netherlands: Ministry of Housing, Spatial Planning and the Environment, 1996).
26. Ibid.
27. R. Goodland, J. -R. Mercier and S. Muntemba, eds., "Environmental Assessment (EA) in Africa: A World Bank Commitment," in *Proceedings of the Durban (South Africa) Workshop,* June 25, 1995 (Washington, DC: The World Bank, 1996).
28. G. T. McDonald and L. Brown, "Going beyond Environmental Impact Assessment: Environmental Input to Planning and Design," *Environmental Impact Assessment Review* 15 (1995): 483–95.
29. See note 23 above.
30. See note 28 above.
31. See note 12 above.
32. B. Eggiman, "Physical planning with strategic environment assessment pursues (SMB) durability. A theoretical discussion and proposals to SMB-process with Stockholm's town sommodell." (Stockholm: Swedish Board of Housing, Building and Planning and Swedish Environmental Protection Agency, 2000).
33. R. Therivel, E. Wilson, S. Thompson, D. Heaney and D. Pritchard, *Strategic Environmental Assessment* (1992).
34. Ibid.
35. Wikipedia, "Planning Fallacy," http://en.wikipedia.org/wiki/Planning_fallacy (accessed January 10, 2008).

36. D. Lovallo and D. Kahneman, "Delusions of Success: How Optimism Undermines Executives' Decisions," Harvard Business Review (July 2003): 56–63.
37. A running debate between D. Kahneman, D. Lovallo, and B. Flyvbjerg in *Harvard Business Review* (2003).
38. See note 36 above.
39. D. A. Armor and S. E. Taylor, "When Predictions Fail: The Dilemma of Unrealistic Optimism," in *Heuristics and Biases: The Psychology of Intuitive Judgment*, ed. G. Thomas (Cambridge, UK: Cambridge University Press, 2002).
40. B. Sadler, *Towards the Improved Effectiveness of Environmental Assessment*, Executive Summary of Interim Report Prepared for IAIA'95 (Durban, South Africa, 1995).
41. R. B. Symthe, (former member of the President's Council on Environmental Quality), personal communication, 2007.
42. See note 33 above.
43. O. H. Pilkey and L. Pilkey-Jarvis, "Useless Arithmetic: Ten Points to Ponder When Using Mathematical Models in Environmental Decision Making," *Public Administration Review* (2008), 68: 470–479.

chapter six

Environmental decision-making theory and practice

> When I was younger, I could remember anything, whether it had happened or not; but my faculties are decaying now and soon I shall be so I cannot remember any but the things that never happened.

> **Mark Twain**

Case study

The Bhopal plant opened in 1969 and produced the pesticide carbaryl. Ten years later, the plant began manufacturing methyl isocyanate (MIC). This was a cheaper, but more toxic, substance used in the manufacture of pesticides. Union Carbide chose Bhopal, a city of 900,000 people in India, because of its central location and its proximity to a lake and a good transportation system.

December 2, 1984, started out as a rather ordinary day. But there would be nothing ordinary about the following week. Water inadvertently entered the MIC storage tank, which held over 40 metric tons of MIC. The water sparked a runaway chemical reaction, resulting in a rapid rise in pressure and temperature within the tank. The heat generated by the reaction resulted in a reaction of such momentum that the internal pressure could not be contained by the tank's safety systems. The MIC and other reaction products escaped from the tank and lofted into the surrounding area. There was no warning system, as the primary emergency sirens had been switched off. A cloud of deadly gas floated over the fence to nearby homes. Many died in their beds. Others awakened blinded, vomiting, struggling to breathe in a lung-smothering hell. Panic ensued. In the chaotic stampede that followed, hundreds fell dead on the streets.

An accurate death toll has never been established. Union Carbide claimed a toll on the low end of 3800. Municipal workers claimed to have cleared at least 15,000 bodies. Thousands more died months or years later. An estimated 50,000 people became invalids or developed chronic respiratory conditions. Some Indian officials estimate that nearly 600,000 became ill or had babies born with congenital defects over the 20-year period following the disaster. In the subsequent investigations and legal proceedings, the following were determined, among other things:

- The MIC tank had been filled well above the prescribed capacity.
- No emergency plans had been developed for responding to a disaster of this magnitude.
- Temperature- and pressure-monitoring gauges, including the crucial MIC storage tanks, were so notoriously unreliable that workers ignored early signs of trouble.
- Plant staff had been cut to save money.
- Workers who complained about safety violations were reprimanded or even fired.
- Tank alarms, the first line of defense for alerting personnel to a leak, had not functioned properly for at least 4 years.
- Backup systems were either not functioning or had not even been installed.
- The plant was equipped with a single backup system that lacked redundancy.
- Damaged piping and valves and other systems had not been repaired or replaced because the cost was considered too high.
- Warnings from U.S. and Indian safety inspectors about plant shortcomings had been ignored.

Some researchers back Union Carbide's claims that the accident was an act of sabotage—someone had deliberately added water to the tank. Regardless of how the water made contact with the MIC, the runaway reaction should have been contained by adequate safety and mitigation equipment.

> **Problem:** The Bhopal disaster was the result of a combination of technological, organizational, and human errors. Some observers have indicated that the accident was more the product of lack of human forethought, lax or casual attitudes, reluctance to stand up to supervisors and management, and poor leadership than flaws in the system or technology. They charge that had it not been for these psychological and behavior patterns, this accident could have been prevented. After reading this chapter, return to this problem. Consider this example in light of what you learned from this chapter, that is, how people rationalize irrational behavior, how authority figures can misdirect rational behavior, and how personal perceptions can shape one's behavior and actions. Using the concepts described in this chapter, list some of the potential human factors that may

help explain how this disaster happened. In addition, list what steps could be taken to ensure that such factors would not contribute to a future disaster.

Learning objectives

- Understand some of the common biases and factors that can affect decision making.
- Understand how real-world decision factors and methods affect the ability to implement policies, programs, or projects designed to improve environmental quality.
- Understand "groupthink" and how it can lead to irrationally based decisions and policies.
- Identify the interdisciplinary elements that must be managed within the decision process.
- Appreciate how public opinion can influence environmental decisions and how *risk communication* can be used to frame environmental decisions in their proper context.
- Learn some basic policy- and decision-making tools that can be used in reaching rationally based environmental policy decisions.

6.1   Introduction

This chapter begins by introducing some of the basic concepts essential to understanding the science of environmental decision making.

6.1.1   Nature of environmental decision making

Formal environmental decision making involves taking deliberate decisions whose intent is to prevent harm to the environment or to otherwise improve its quality and utility for sustained human uses. Such formality is embodied in environmental policies, laws, and regulations, and in the host of systems analysis methods (e.g., ecological, engineering, and economic) aimed at achieving stated objectives. Such formal analysis methods include considerations of risk and uncertainty because nature does not always respond in ways predicted by our analytic methods. Nevertheless, such methods, when properly applied, are the most effective tools available for environmental decision making.

The Bhopal disaster is a classic example of how failure to apply the available formal decision-making tools can have dire consequences. Such failures in applying well-established prevention measures are typically motivated by cost pressures that affect the profits of a business or the budgets of a government. This case study is but one example of a systemic set of problems that plague infrastructure and economic development in both industrialized and preindustrial regions. The decisions that set the stage for the Bhopal tragedy predated the widespread understanding and application of high-consequence risk planning and management tools that have evolved over the past 30 years. Yet, it is clear that the Bhopal disaster could have been averted if the tools that were available some 20–30 years ago had been adequately applied.

Table 6.1 is an executive summary of an "event assessment" of the Bhopal disaster that draws on the case study narrative. It provides an overall systems perspective on the many decisions that contributed to the disaster. Such perspectives are inherent to decision processes presented in Chapters 5 and 7 through 9. Each of these disciplines contributes to decisions integrated into policies, programs, regulations, project designs, and operations. Effective monitoring of and improvement to programs and

Table 6.1 Executive Summary—Bhopal Disaster Event Assessment

Sequence of plant failures	Failure category
Water inadvertently entered the methyl isocyanate (MIC) storage tank.	Design and operation
The water sparked a runaway chemical reaction.	Design and operation
There was a rapid rise in pressure and temperature.	Design and operation
Internal pressure could not be contained by the tank's safety systems.	Design and operation
MIC and other reaction products escaped from the tank.	Design and operation
The primary plant emergency warning system sirens had been switched off.	Operation
Root cause assessment: Preventive operational factors	
The MIC tank had been filled well above the prescribed capacity.	Plant management
Temperature- and pressure-monitoring gauges and the MIC storage tanks were notoriously unreliable.	Plant management
There was a gas-leakage single backup system that lacked redundancy.	Design plant management
Other backup systems were either not functioning or had not even been installed.	Design plant management
Workers ignored early signs of trouble.	Plant management
Damaged piping and valves and other systems had not been repaired or replaced because the cost was considered too high.	Plant management

Table 6.1 (*Continued*)

Root cause assessment: Preventive management factors	
The plant was designed and equipped with inadequate preventive safety and mitigation equipment.	Corporate neglect Regulatory neglect
Warnings from U.S. and Indian safety inspectors that plant shortcomings had been ignored.	Corporate neglect Regulatory neglect
Plant staff had been cut to save money.	Corporate neglect
Workers who complained about safety violations were reprimanded or fired.	Corporate neglect Regulatory neglect
No emergency plans had been developed for responding to a disaster of this magnitude.	Corporate neglect Regulatory neglect

Consequences of failures

Deadly gas floated over the fence to nearby homes.

Panic ensued.

Union Carbide estimated 3800 deaths.

Municipal workers cleared 50,000 bodies.

An estimated 50,000 people became invalids or developed chronic respiratory conditions.

From 1969 to 1989, nearly 600,000 people around Bhopal became ill or had babies born with congenital defects.

System and cultural factors

Regulatory system failures

Corporate leadership and management failures

Ethical failures—denial, negligence, other harmful attitudes

Lack of applied environmental impact assessments (Chapter 5), environmental management systems (Chapter 7), environmental ethics (Chapter 8), and environmental economics (Chapter 9) as essential decision tools for policy, regulation, planning, and operations.

project operations, as well as root-cause assessments of disastrous events, require the same discipline.

6.1.1.1 De facto decision making

The term "de facto environmental decision making" refers to situations in which a decision maker is misinformed about potential environmental effects, or deliberately chooses to ignore potential impacts. There are many reasons this situation can occur. In many cases, such as the Bhopal disaster, questions of criminal negligence can be raised. More often, there are complex economic, political, and cultural factors that come into play.

For example, business and government decision makers are usually aware of the inherent risks, in that every action has an associated probability of causing harm of a given magnitude. Whether or not such risks are formally assessed, the decision maker implicitly decides whether the risks are justifiable. When an investor or a politician who reaps the benefits of the risk is the principal decision maker (while the risks are borne by the public and their environment), there is an unfair bias in the decision-making system.

The remainder of this chapter explores a variety of approaches, some of which have been applied in policies, laws, and regulations. There are many approaches advocated by social science researchers concerned with human cognitive mechanisms at the core of human nature. Most, if not all, of these mechanisms were probably involved in human and environmental disasters since the onset of industrialization. However, there is no linear or one-to-one linkage between these deeper mechanisms and the issues raised by disasters such as the Bhopal tragedy. Nevertheless, these considerations, many of them the products of academic research, can provide important insight into the ethics of public officials, executives, and managers in preventing harm to the public and to the environment. Neil Carter, a political scientist who advocates more far-reaching political and cultural approaches for achieving equitable public outcomes, has analyzed how environmental decision mechanisms operate in the political environment, especially in Europe where "green parties" have become a significant factor.[1]

6.1.2   Human nature of policymaking

Good information results in rational choices, right? Well, not necessarily. As the Case Study section illustrates, environmental policymaking and decision making are often a messy process with many paradoxes, dilemmas, and fallacies quite contrary to a straightforward "rational" decision-making process. As we shall see, virtually all such decision making is driven by subjective factors that stem from peoples' diversity of survival needs, social positions, relative wealth, and personal sense of fairness and entitlement, and a whole variety of attitudes, including prejudices and biases. These are typically expressed in ideological beliefs and principles, and their means of resolution may range from political mediation processes to violent conflicts.

We are concerned with policymaking that ignores or denies relevant and available knowledge, especially policies that are likely to cause damage to the environment and thereby damage to the societies that they are presumably designed to benefit. Our intent is to help the reader assess whether the causal realities and the quality of data and information applied in designing and implementing policy are likely to produce helpful or harmful outcomes.

6.1.3 Cognitive process decision model

Decision making can be viewed as an outcome of a cognitive (mental) process leading to the selection of a course of action among multiple alternatives. Decision making is a process for choosing a course of action among alternatives.

Traditional engineering decision theory employs systems analysis to fully characterize the range of anticipated consequences for alternative decisions and causal factors. While these methods are essential tools of environmental policy, their technical description is highly specialized and outside the scope of this work.[2] Many unconscious factors, rational and irrational, influence decision making in the end, regardless of what formal analytical tools are used in reaching the decision. This is because systems models cannot capture all the real-world information and subjective decision-making factors. The upshot is that many irrational factors, particularly unconscious ones, can lead to flawed decisions that are explained away as rational ones. With respect to environmental policy, irrational decisions can lead to misdirected efforts at low-priority problems with correspondingly little attention directed at larger ones, misallocation of scarce resources, selection of unfounded or irrational courses of action based on biases, and unfounded or irrationally driven public opinion. Figure 6.1 presents one of several general decision models that have been proposed from cognitive science.[3]

In the cognitive process decision model presented in Figure 6.1,

- *Resources* employ *sensors* to gather environmental and social experiential data to the extent enabled by *intelligence*.
- *Drives* involve emotions that bias the data with purpose and meaning according to our *moods, attitudes,* and *feelings*.

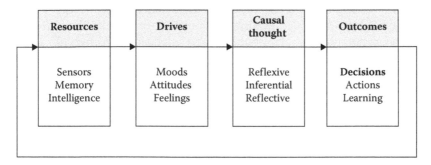

Resources	Drives	Causal thought	Outcomes
Sensors	Moods	Reflexive	**Decisions**
Memory	Attitudes	Inferential	Actions
Intelligence	Feelings	Reflective	Learning

Figure 6.1 A cognitive model of the decision-making process.*

* Frederic March, "Religion and Science as Systems of Causal Thought," *Essays in the Philosophy of Humanism*, Volume 18 (10 Spring-Summer 2020).

- *Causal thought* has the following aspects:
  - *Reflexive thought* is instinctive and immediate.
  - *Inferential thought* is rapidly processed according to memory patterns conditioned by past experience.
  - *Reflective thought* draws on all the information resources accessible to the conscious mind and can take minutes to years to plan and execute.
- *Outcomes* are the *decisions, actions,* and *learning* that are fed back to enhance the mind's resources.

The model illustrates the following intuitive understanding of how people make decisions:

- Logic and rational thought are themselves categories of inherited drives that assist human survival.
- Drives as expressed by emotional impulses determine how causal thought is applied.
- Decision outcomes result from largely emotional responses mediated by reflective thought, or rational logical faculty.

Many of the concepts described in this chapter have been studied in terms of the marketplace or in an economic context. They provide an important starting point for gauging how various factors may affect environmental policy and decision making. In the end, human intuition often becomes the final arbiter of decisions, especially when political processes are involved. Thus, it helps us know what others are thinking, including their hidden intentions. According to Nobel Laureate James Watson, human intuition is a "sort of background sense of how things should work ... intuition is logic."[4] In many cases, people use intuition throughout the day to guide them through the processes of distinguishing between difficult choices and making correct decisions. Much of the remainder of this chapter focuses on understanding factors that can make the difference between rationally and irrationally based decisions. Above all, ethical decision making in the face of uncertainty in the outcome should consider who pays the price of the risk imposed on the public if the decision turns out badly and who reaps the rewards if it turns out well.

In addition to Section 6.1, Sections 6.2 through 6.4 convey a minimum set of concepts, principles, and practices that we have found helpful in our respective professional involvements in many environmental decision contexts. The rest of the chapter, starting with Section 6.5, provides a wide sampling of additional concepts, principles, and practices from the literature—some fairly limited in their application and some advocated by academic and research workers. Some of these topics are advanced subject matter and may be of more interest to graduate students than to undergraduates.

6.2   Biases and mindsets

Nicholas Taleb's popular book, *The Black Swan*, illustrates several ways in which we irrationally respond to natural risks, including the certainty of death, with denial. As used by Taleb, a *black swan* is an extreme event that can occur at any time, such as an earthquake, flood, or volcanic eruption. But people tend to ignore such threats because they are believed to be too far in the future to require the investment of much thought or action into preparing for them; or worse, people deny the very reality of the threat. Taleb explains as follows[5]:

- We behave as if the black swan does not exist: human nature is not programmed for black swans.
- What we see is not necessarily all that is there. History hides black swans from us and gives us a mistaken idea about the odds of these events; this is the distortion of evidence.
- We "tunnel," that is, we focus on a few well-defined sources of uncertainty at the expense of the others.

Table 6.2 expands on Taleb's observations.[6] The topics discussed in Sections 6.2.1 through 6.2.3 describe issues commonly involved in individual and group decision making, no matter how much study or formal decision analysis has been presented to them.

Table 6.2 Common Decision-Making Biases

Base rate fallacy: Ignoring statistical data in favor of particular data or events.

Bandwagon effect: The tendency to believe or do things because many other people believe or do the same.

Choice-supportive bias: Tendency to remember one's own choices as better than they actually are.

Confirmation bias: Tendency to interpret information in a way that confirms one's preconceptions.

Conservatism bias: Tendency to ignore the consequence of new evidence.

Deformation professional: Tendency to consider things according to the conventions of one's own profession, rather than any broader point of view.

Distinction bias: Tendency to view two options as more dissimilar when evaluating them simultaneously than when evaluating them independently or separately.

Expectation bias: Tendency for experimenters to believe, certify, and publish data that agree with their expectations, and to discard or downgrade corresponding weightings for data that appear to conflict with those expectations.

(Continued)

Table 6.2 Common Decision-Making Biases (*Continued*)

Extraordinarity bias: Tendency to value an object more than others in the same category because of an extraordinarity of that object, which does not, by itself, change the value.

Extreme aversion: Tendency to avoid extremes, being more likely to choose an option if it is the intermediate choice.

Hyperbolic discounting: Tendency of people to have a stronger preference for more immediate payoffs relative to later payoffs.

Illusion of control: Tendency to believe one can control or at least influence outcomes that one clearly cannot.

Impact bias: Tendency for people to overestimate the length or the intensity of the impact of future feeling states.

Information bias: Tendency to seek information even when it cannot affect an action.

Irrational escalation: Tendency to make irrational decisions based on rational decisions in the past or to justify actions already taken.

Loss aversion: The disutility of giving up an object is greater than the utility associated with acquiring it.

Mere exposure effect: Tendency for people to express undue liking for things merely because they are familiar with them.

Need for closure: A need to reach a verdict in important matters, to have an answer, and to escape feelings of doubt and uncertainty.

Neglect of probability: Tendency to completely disregard probability when making a decision under uncertainty.

Not invented here: Tendency to ignore that a product or solution already exists because its source is seen as an "enemy" or as "inferior."

Omission bias: Tendency to judge harmful actions as worse or less moral than equally harmful omissions (inactions).

Outcome bias: Tendency to judge a decision by its eventual outcome rather than by the quality of the decision at the time it was made.

Planning fallacy: Tendency to underestimate task-completion times.

Pseudocertainty effect: Tendency to make risk-averse choices if the expected outcome is positive but make risk-seeking choices to avoid negative outcomes.

Reactance: An urge to do the opposite of what someone wants you to do out of a need to resist a perceived attempt to constrain your freedom of choice.

Selective perception: Tendency for expectations to affect perception.

Status quo bias: Tendency for people to like things to stay relatively the same.

Von Restorff effect: Tendency for an item that "stands out like a sore thumb" to be more likely to be remembered than other items.

Wishful thinking: Formation of beliefs and the making of decisions according to what is pleasing to imagine instead of by appeal to evidence or rationality.

Zero-risk bias: A preference for reducing a small risk to zero over a greater reduction in a larger risk.

6.2.1   Mindset

In cognitive decision theory, a mindset refers to a set of assumptions or methods held by one or more people (group), which is so well established that it creates a momentum to continue accepting prior choices or behaviors, even when they are flawed. A mindset can adversely affect the decision-making process, particularly when a decision involves new circumstances. For this reason, more optimal decisions can often be identified by actively challenging "standard" thinking or the assumptions that influence one's mindset. Mindsets are evident in the often polar opposite views of political players in the policy process.

6.2.2   Attitude polarization

The term "attitude polarization" refers to the tendency of people who hold a particular belief or attitude to interpret evidence for or against that belief or attitude selectively, in a manner that shows a bias in favor of their prejudices and views.[7] When people are given evidence that agrees with their belief, they tend to accept it as supporting their position. When given evidence that contradicts their belief, they tend to ignore, criticize, or reinterpret it in a manner that supports their prevailing views. This may explain why people often do not seek out views that challenge their prevailing mindsets.[8]

6.2.2.1   Group polarization

Some studies suggest that decisions made by a group tend to display more experimentation and less conservatism and advocate riskier positions than those made by individuals acting alone; a group's position will frequently be an extreme version of each individual's preferred action.

The term "group polarization" refers to the tendency of people to make decisions that are extreme when they are in a group, as opposed to those made alone or independently; consequently, people on both sides of an issue can be pushed farther apart, or polarized, when they are presented with the same mixed evidence. However, depending on the initial tendencies of group members, "cautious shift" outcomes are also possible.

6.2.3   Confirmation bias

Confirmation bias is a tendency to interpret new information in a way that confirms one's preconceptions. Conversely, it also includes the tendency to avoid information that contradicts one's personal beliefs.[9] With respect to critical thinking, this bias is important because it can lead people to only critically examine evidence challenging one's preconceived idea but not to scrutinize evidence to the contrary.[10]

For instance, Evans et al. performed experiments in which subjects were presented with deductive arguments where a series of premises and a conclusion were provided. They were asked to logically deduce if each conclusion followed from the given premises. Many subjects exhibited confirmation bias, rejecting valid arguments with believable conclusions and accepting invalid arguments with unbelievable conclusions. The subjects appear to have based their assessments on personal beliefs.[11]

With respect to environmental policy, those who choose not to believe in a particular environmental concern such as global warming may be more willing to dismiss this issue regardless of the facts. Conversely, those who choose to believe in global warming may tend to look for evidence that supports this theory, even if the evidence is questionable, and reject information to the contrary. These issues are especially relevant when communicating risk and seeking to inform public opinion, as discussed in Section 6.3.

6.3   Public opinion and risk communications

Public policy can affect public opinion. Likewise, public opinion can shape, and has profoundly shaped, environmental policy. There are also factors, overt and covert, that influence both public policy and public opinion. Sections 6.3.1 and 6.3.2 examine factors that influence public opinion and, indirectly, environmental policy.

6.3.1   Public opinion and herd behavior

The *bandwagon effect* or *herd behavior* is a well-documented observation that people frequently believe things simply because many others believe the same thing.[12] People tend to follow a crowd without carefully examining the merits of an issue. As more people come to believe something, others also "come on board," regardless of the underlying evidence. Such phenomena can spread rapidly among people, as fads and trends. The probability of an individual adopting something frequently increases with the proportion of those who have already followed suit. The term is attributed to a "bandwagon"—a wagon that was used to carry a band in a parade or circus. The bandwagon effect has been applied to situations involving majority opinion, such as political outcomes, where people alter their opinions to conform to the majority view.

The observed tendency to follow actions or beliefs of others occurs because many people prefer to conform; it can also occur as a result of many people attempting to derive information from others. Both of these explanations have been used for evidence of conformity in psychological experiments. For instance, peer pressure has been used to explain Asch's conformity experiments[13] (described later in Section 6.7.1).

Collective effervescence is a perceived energy formed by a gathering of people, such as a sporting event, riot, or other emotionally charged arena. This energy can cause people to act differently from their actions in their everyday life.

6.3.2   Communicating risk to the public and policymakers

How a risk is explained and framed can have a significant effect on public opinion. Some common factors influencing risk perception are depicted in Table 6.3.[14] The field of risk communication (RC) involves a multidisciplinary process designed to convey risk in terms that nontechnical people can more readily digest. Its applicability spans a spectrum of environmental policy and decision-making issues such as global warming, air pollution, hazardous waste sites, lead, pesticides, drinking water, and asbestos.

As expressed in Table 6.4, the U.S. Environmental Protection Agency (EPA) and several of the field's founders have identified seven cardinal rules for the practice of RC (Table 6.4)[15]. According to the EPA:

> Parties can communicate better if they remind each other that most technologies are neither "safe" nor "unsafe" in absolute terms. Therefore, the debate should be framed not in terms of whether something is "safe" but in terms of whether it is "safe enough." Defining the key safety question "Safe enough?" forces parties on both sides to discuss the value components of the decision, as well as what

Table 6.3 Some Common Factors Influencing Risk Perception

Risks that are under an individual's control tend to be more acceptable than those that are not.

Voluntary risks are more acceptable than those that are imposed on an individual.

Risks that have clear benefits are more acceptable than those with little or no benefit.

Unfairly distributed risks are less acceptable than those deemed to be fairly distributed among a population.

Familiar risks tend to be more acceptable than exotic ones.

Risks that affect adults are viewed as more acceptable than those that affect children.

Natural risks tend to be more acceptable than man-made ones.

Risks generated by a trusted source tend to be more acceptable than those generated by a mistrusted source.

Table 6.4 Rules for Communicating Risks

Accept and embrace the public as a legitimate partner in planning and
assessing the risk.

Exercise prudence in planning and evaluating your communication efforts.

Listen to the public's underlying concerns.

Be open, frank, and honest.

Coordinate and collaborate with other credible sources.

Meet the needs of the communication media.

Speak clearly and with empathy.

is known and not known on the scientific side. It
is a major step toward more effective and clear
communication.

6.4   Policy- and decision-making tools

Funtowicz and Ravetz have written that

Procrastination is as real a policy option as any
other, and indeed one that is traditionally favored
in bureaucracies; and inadequate information is the
best excuse for delay.[16]

As witnessed, failure to deal appropriately with uncertainty can lead to
misleading or false conclusions. It should come as no surprise that critics
have seized on the element of uncertainty as a tool for delaying or even
distorting public debate. Thus, the science of uncertainty is not simply a
subject of academic interest but a pressing problem in search of practical
solutions.

An excellent case involving uncertainty concerns the decision made
in 1922 by chemical and automobile corporations to introduce tetraethyl
lead into gasoline. When the decision was announced, a number of health
experts warned that it was a dangerous idea and urged a delay in order to
allow time for scientific study. The chemical and automotive corporations
countered that there was no scientific agreement concerning the threat,
and in the absence of solid scientific evidence to the contrary, they had the
right to proceed. Tetraethyl lead became a standard gasoline additive. As
a result, many medical scientists today believe that millions of children
were adversely affected in a range of ways that may have included brain
damage or even permanent impairment of their IQs. This example illustrates that prudence should be exercised where a major decision involving scientific uncertainty exists. The *precautionary principle*, introduced

in Chapter 2, presents one approach for dealing with scientific uncertainty where the consequences could involve significant environmental risks. An alternative approach, cost–benefit analysis, is described in Chapter 9.

6.4.1   Dealing with uncertainty in environmental impact assessment documents

Environmental impact assessments (EIAs), such as the U.S.'s environmental impact statements (EISs), provide a rigorous procedure for studying proposals that may involve a considerable degree of scientific uncertainty (see Chapter 5). The U.S. National Environmental Policy Act (NEPA) regulations for preparing EISs provide detailed procedures for evaluating proposals that involve uncertainties or unknown information. Against this backdrop, the NEPA regulations prescribe specific procedures requiring federal agencies to publicly disclose environmental impacts involving uncertainty and evaluate the possible impacts based on theoretical approaches or research methods generally accepted by the scientific community.[17]

6.4.2   Decision making and the Delphi method

In group settings, participants frequently tend to hold irrational and steadfast opinions; weaker personalities may be swayed by more dominant ones. The Delphi method provides a systematic, interactive, and highly structured technique in which a "blind" panel of independent experts provides its assessment of likely future outcomes by responding to several rounds of questions. When properly implemented, this technique has frequently been very effective in generating an accurate consensus, particularly with respect to making forecasts. Because all panel members maintain anonymity, the Delphi method avoids the negative effects of face-to-face discussions and personality conflicts, and resolves many problems associated with traditional group dynamic. Its disadvantages are that it can involve a lengthy and relatively expensive process, and has mixed history of success.

6.4.3   Rogerian argumentation

Sometimes, people can feel intimidated or even threatened by views different from their own, and in such cases persuasion may be all but ineffective. One of the greatest challenges to a policymaker is to keep the audience from becoming so defensive or annoyed that they will not listen to anything that is said or written.

In response to this dilemma, Carl Rogers developed a negotiating strategy called *empathic listening* or *Rogerian argument* for avoiding such clashes. Under a Rogerian strategy, a policymaker (or speaker, writer, etc.) refrains from passing judgment on the audience's views until the audience's position has been heard. The policymaker attempts to seriously follow the audience's reasoning and acknowledges the validity of the audience's viewpoint. The policymaker shows empathy for the audience's viewpoint by making a serious effort to understand their perspective. This can open the door to dialog, mutual understanding, and respect. This strategy attempts to avoid an "I win—you lose" scenario; instead, it emphasizes a "you win and I also win" solution, under which negotiation and mutual respect are valued. Ultimately, this strategy is focused on encouraging people to listen to each other rather than attack each other.

Consider the divisive issue of global warming. Under a Rogerian approach, the policymaker might attempt to establish common ground about the future of our children's environment and factors raised by critics such as the cost of controlling greenhouse gases and its effect on consumers. The policymaker might begin by pronouncing respect for the scientific uncertainty and acknowledging the monetary price tag; he or she would acknowledge the audience's side of the issue fairly and objectively, so as to show that their concerns are understood and appreciated. The policymaker might then provide an objective statement of his or her position, again trying to avoid attacking language and trying not to imply that this position is somehow morally superior. Finally, the policymaker might phrase this view in such a way as to show that some concessions have been made toward the audience's position. By conceding some ground, the policymaker invites the audience to also concede the same in an effort to reach common agreement.

6.4.4   Computer modeling and environmental decision making

Computer modeling is another important method for assessing the viability of environmental plans and policies. Coauthor Frederic March led a team that applied simulation as a tool to help formulate federal solar energy policy. The objective was to assess the prospects of a utility company in producing economical solar and wind power. A monthly production simulation program gauged the cost of generating a given amount of electricity, and a financial simulation program measured the financial impact of the investment on the utility company. The methodology considered an array of wind energy scenarios and assumptions of future fossil-fuel prices, which were tested by the model for their economic feasibility. Interviews with policymakers and utility executives were conducted to establish scenario outcomes that would motivate investment in wind energy. Simulation results were analyzed and shown to the interviewees. This resulted in a set of policy conditions such as taxes on fossil-fuel emissions, and tax and financing incentives over a range of future fossil-fuel prices that would motivate utility investment.[18] These results were useful in making decisions about the economic viability of wind and solar energy projects. Unfortunately, many modeling studies are less successful.

Edward A. Parson of Harvard's Kennedy School of Government summarizes the potential benefits and risks of using computer models[19]:

> Practice and research in assessment of global environmental changes are dominated by two conventional assessment methods: formal models and expert panels. Models construct a representation of biophysical and socio-economic components of a policy issue, to project future trends or consequences of interventions. Panels articulate consensus views of policy-relevant knowledge through deliberations among selected experts. These methods make valuable contributions, but are weak in addressing certain kinds of knowledge needs that are typical of global change issues.

P. B. Hammond is concerned that model builders overstep their expertise when they try to advocate policy. He writes, "Models are used as symbols, as claims to authority, whether or not the underlying knowledge is technically up to snuff."[20]

6.5   Environmental decision making and choices

A "choice" can be defined as a mental process of considering the merits of multiple options and selecting one of them. Sections 6.5.1 through 6.5.8 consider several theoretical aspects of societal decision making, ranging from individual choices to the collective choices of a society as a whole.

6.5.1   Size of the choice set

A severely limited set of choices can lead to the dilemma of having to choose an unsatisfactory outcome; thus, most people tend to consider the freedom to explore a wide range of choices a good thing. However, a large or unlimited set of choices may lead to a feeling of being overwhelmed and regret for the alternatives not taken.

In his 2004 book, *The Paradox of Choice—Why More Is Less*, Barry Schwartz argues that eliminating consumer choices can greatly reduce anxiety for shoppers.[21] Autonomy and freedom of choice are critical to our well-being, and a well-developed set of choices is critical to our freedom and autonomy. Nevertheless, in his view, there is a problem: while

Westerners have more choice than any group of people in history, and presumably more autonomy and freedom, we do not appear to be benefiting from it psychologically. If his analysis is correct, the question arises, "Could dissatisfaction also extend to environmental policy problems?" As described in Sections 6.5.1.1. and 6.5.1.2, personality types appear to play an important role in how individuals deal with a large set of choices.

6.5.1.1   Maximizers–satisfiers

According to Simon's theory, *maximizers* are those who tend to seek an optimum option from among a set of choices. A maximizer is akin to a perfectionist who needs to be continually assured that his or her every decision has been the optimum one. Maximizers do this by considering all the alternatives that can be imagined. This, in itself, can be a stressful and daunting task. They may experience anguish in terms of second-guessing their choice and whether they made the best possible choice. Not surprisingly, maximizers are more likely to avoid making a choice when the choice set size is large.[22]

In contrast, *satisfiers* also have standards to be met, but they are much less concerned about the possibility that there might be a better option that is not taken. Although a satisfier may place high standards on themselves, they are content with a good choice and place less priority on making an optimal choice. Environmental policymaking is a complex science, often with a large array of potential options. Could decision makers, and perhaps society at large, be overwhelmed by such a complex array of potential environmental and energy paths and options?

6.5.1.2   How the size of the choice set may affect environmental policymaking

In some circumstances, choosing an option or a course of action may also lead to the illusion of having responsibility or control over the course of action, which may not always be the case. Some studies show that when individuals are forced to choose from among a large set of options, they are often unable to make a choice.[23] One example involves choices that a policymaker is confronted with in deciding how to develop water resources for a region. Development schemes often require multicommunity allocation of scarce water resources (for agriculture, urban centers, power generation, recreation, conservation, and flood control). Hence, there can be many stakeholders with conflicting interests. Outcomes can include gridlock, negotiated decisions, or decisions imposed on the stakeholder community by investors and governments.

It is important to note that assertions are sometimes made by politicians or public interest groups that there is no reasonable option or alternative to a particular problem. Although the proponents of such a statement may believe what they are alleging, in fact, most problems have options and alternatives. Sometimes the ones making such a statement do not believe there are no alternatives, as otherwise they would not bother to argue the point; instead, their strategy is to oppose other alternatives and thereby seek to dismiss them by denying their existence.

6.5.2 Invalid or illogical choice-set premises

What has been referred to as the *false dilemma* or *informal fallacy* involves a situation in which only *two* alternatives are seriously considered, despite the fact that other options exist. When only two alternatives are presented, they are often, although not always, the two extreme cases of a spectrum of possibilities.

Moreover, the two options are often presented as being exhaustive. Consider the current debate on greenhouse emissions. The issue is often framed in terms of a large and costly reduction in emissions or taking no action at all to reduce emissions. Consider for a moment the vast middle ground of policy options that lie somewhere between these two extremes. One obvious way of countering such a dilemma is by simply suggesting, arguing, or publicizing other possibilities.

Closely related to the problem of false dilemma is the failure to consider a range of options or to only think in terms of extremes, also referred to as *black-and-white thinking*. For example, an action may be automatically labeled as environmentally good or bad, when in reality it may have both good and bad attributes.

6.5.3 Falsus omnibus

The Latin phrase *"falsus omnibus"* can be roughly translated to mean "false in one thing, false in everything." It is often witnessed in the following manner: if someone is found to be wrong in one issue, a critic concludes he or she must also be wrong in other (or all other) issues. The fallacy is that it is not necessarily correct to conclude if someone is incorrect in one issue, he or she is also incorrect in all other respects.

6.5.4 Hobson's choice

Hobson's choice is a situation in which only one option is offered, and a person may refuse to take that option. The term is attributed to Thomas Hobson (1544–1630), a livery stable owner who worked in England. He rotated his horses by offering customers the choice of either renting the horse in a particular stall near the door or having no horse at all. The term "Hobbesian choice" is often used in referring to this dilemma. In marketing his Model T automobile, Henry Ford presented his customers with a famous example of Hobson's choice; he said, "You can have the car in any color, so long as it is black." Variations of Hobson's choice are occasionally encountered in environmental policy and decision making.

6.5.5   Morton's fork

Sometimes, a choice between two equally unpleasant options is actually a false choice set. This dilemma has been referred to as *Morton's fork*, which originated from an argument over taxing English nobles. The argument was presented as such: "Either the country's nobles appear wealthy, in which case should be taxed; or they appear poor, in which case they are living frugally and therefore must have money tucked away that can be taxed." The assertion was not necessarily correct because some members of the nobility lacked liquid assets that could be readily taxed.[24]

A policymaking example for this situation would be limiting global warming policy to the following false choice set: (1) taxing profitable corporations for damage done to the environment or (2) if they say they cannot afford such a tax rate, it must mean that they are paying their executives and shareholders too much. Obviously, this is a false conclusion. The profit margin of some companies is very narrow, if not in the red, and a greenhouse emissions tax would mean doom for some weak companies, not to mention the employees who work for them; this would also lead to less competition and potentially higher prices for consumers.

6.5.6   Slippery slope

A *slippery slope* argument states that a relatively small first step may lead to a chain of subsequently related events, eventually culminating in a bad result. For example, a judiciary decision not to require an EIA may lead to other such decisions resulting in an erosion of otherwise effective policies.

In another example, some environmentalists have proposed that a global power grid be developed. Despite the dubious and unsubstantiated benefits, they claim that it would promote a single global community. Critics counter that such a grid would constitute a slippery slope, where nations would begin to sacrifice their security and sovereignty for benefits, which principally undeveloped countries would reap. In either case, such claims are difficult to unequivocally prove. But regardless of the case, logical premises should be carefully scrutinized.

6.5.7   Prospect theory and risk-based decisions

The second-most cited paper ever to appear in the *Journal of Econometrica,* a prestigious journal of economics, was written in 1979 by two psychologists, Daniel Kahneman and Amos Tversky.[25] Their paper considered decision making in risky economic situations and proposed the concept of *prospect theory*, which led to a Nobel Prize and its subsequent application in marketing campaigns.

Prospect theory describes how we tend to assess choices between alternatives that involve risk. That is, people behave differently depending on how a situation is presented to them: they tend to value a gain that is certain more than one that is less than certain, even when the expected value or probability of each is the same. The opposite is even more evident for losses: We will "clutch at straws" to avoid a *certain* loss, even if it means assuming a greater risk to avoid a certain loss. When we perceive higher risk we focus on loss; when risk is viewed to be lower, we focus on gains. If you want someone to focus on loss or gain, advertise the perceived risk accordingly.

Thus, prospect theory can be used to one's advantage to achieve certain desired results: If you wish to get people to adopt something, package your message in terms of what they stand to gain. To get people to reject something, focus your message in terms of what they stand to lose. If the perceived risk is high, focus on loss. If the perceived risk is low, focus on gain.

Although this theory is largely focused on economic choices, it is also applicable in how people assess environmental risks and make choices. For instance, how will the public respond if policymakers frame climate change in terms of economic benefits created by pursuing green energy? Or, will people accept tough choices if they perceive that a grave environmental risk will almost certainly result if no action is taken to mitigate the effects? The authors are unsure, but how environmental policy choices are framed may have a significant effect on how the public accepts them.

6.5.8 Beck's theory of a risk society

Scott Lash and Brian Wynne characterize Ulrich Beck's "risk society" as follows[26]:

> One of the most influential works of social analysis in the late 20th century ... Beck's book has had an enormous influence. First it had little short of a meteoric impact on institutional social science. Risk Society further played a leading role in the recasting of public debates in German ecological politics.

Hooman characterizes Beck's concept of a risk society as follows:

> In the advanced modern world, the social production of wealth systematically goes hand in hand with

social production of risks. Accordingly, the problems and conflicts of distribution in a society of shortages are over layered by problems and conflicts that arise from over-production, definition and distribution of scientifically and technologically produced risks, says Beck. ... To sum up, Beck gives a new identity to risk which long has been dominated by rational doctrine. Manufactured Risks, which are at the heart of the modern society have become a taboo. The modern corporations have build a "Family of Myths" among which the most important is the "Myth of Rationality."[27]

William Leiss summarizes Beck's core concept as follows:

Beck writes: "Risk may be defined as a systematic way of dealing with hazards and insecurities induced and introduced by modernization itself" (21). And: "In contrast to all earlier epochs (including industrial society), the risk society is characterized essentially by a lack: the impossibility of an external attribution of hazards. In other words, risks depend on decisions, they are industrially produced and in this sense politically reflexive" (183). Now, the very important point here, where we can agree with Beck, is that industrial society marks a transition—a watershed in human history, in fact—from a human condition where naturally occurring hazards (disease, flood, famine, and the like)—along with socially determined hazards such as invasion and conquest, regressive forms of thought and culture, and rigid class structures—molded the fate of individuals and groups, to one where increasingly our fate is bound up with risks that are deliberately undertaken—for the sake of benefits conceived in advance—by means of our technological mastery over nature.[28]

6.6   Cognitive dissonance and effort justification

Cognitive dissonance refers to an uncomfortable feeling caused by simultaneously holding two conflicting or contradictory ideas, attitudes, or beliefs.[29] For example, I may wish to make environmentally responsible decisions such as voting for a tax that would reduce carbon emissions.

But I may also decide to buy a gas-guzzling sport utility vehicle (SUV). Such conflicts in one's value system may produce uncomfortable conflicts in their personal values, that is, cognitive dissonance. Since I can afford an SUV, I might, therefore, attempt to justify this conflict by driving my hybrid Prius all over town and using the SUV only to travel to my country hideaway with my big family, loaded with groceries for the long weekend. This would also be an example of *effort justification*.

Effort justification is a dissonance between the amounts of effort exerted into completing a task. More specifically, it is a person's tendency to attribute a greater value (greater than the objective value) to an outcome that they expended considerable effort to achieve.[30] Here, a high degree of effort is acquainted with a high cost, and the reward for that effort is lower than that expected for such an effort.

6.6.1 Doublethink

The term "doublethink" refers to the act of simultaneously accepting as correct two mutually contradictory beliefs. The original concept is derived from George Orwell's 1949 novel, *Nineteen Eighty-Four*.[31] The term has since become synonymous with the concept of cognitive dissonance by ignoring the contradiction between two worldviews, or of even deliberately seeking cognitive dissonance. Recall from Chapter 2, Section 2.2.2, what Mark Sagoff wrote of his environmental schizophrenia. Stop for a moment and consider how many instances you can think of in which doublethink can explain the values our society appears to place on a healthy environment versus the reality of how people live in the twenty-first century.

6.7 Groupthink

When the group's expertise, attitude, and beliefs are overly narrow, it can lead to harmful and even disastrous decisions. When policies are made by a small group of counselors to a charismatic or powerful leader, members are often prone to "groupthink." Loyalty to the leader can trump a full hearing of diverse views. Such leaders often have preconceived notions of a policy and do not want them questioned. For example, we can speculate that the government leaders cited in the two case studies mentioned in Chapter 13 did not entertain any meaningful discussion linking a policy to invite oil companies to develop petroleum resources to meaningful policies that would protect the environment and the livelihoods of people from the readily predicted impacts of an unconstrained oil policy.

A group of people may act simultaneously to achieve a goal in a manner that differs from what individuals would do acting alone. This phenomenon, also known as the bandwagon effect or herd behavior, is described in more detail in Section 6.3.1.

6.7.1   Asch's conformity

Asch performed a set of experiments in which all but one of the participants were actually confederates (experimental cohorts united with the experimenter to deceive the single student being studied) of the experimenter.[32] In reality, the study was directed at discerning how the remaining student would react to the confederates' behavior.

Participants (the deceived subject and the confederates) were asked a variety of questions about comparing the length of lines (e.g., which line was longer than the other). Each participant was told to announce his or her answers out loud. However, the confederates were instructed to quickly announce their answers before the study participant; they were also tutored to always provide the same answer. The confederates answered a few questions correctly in the beginning of the experiment; but then they eventually switched, providing incorrect responses.[33]

In the control group (with no pressure to conform to an erroneous view), only 1 out of 35 subjects provided an incorrect answer. When surrounded by individuals all voicing an incorrect answer, however, participants provided incorrect responses on a high proportion of the questions. This experiment shows the effect that even a small dissenting minority can have. It also demonstrates how a small minority (be it politicians or special interests) may shape public environmental policy.

6.7.2   Abilene paradox

The *Abilene paradox* describes a situation in which a group of people collectively decides on a course of action that runs counter to the preferences of individual group members.[34] It can lead to a breakdown of communication in which each individual member mistakenly believes that his or her own preferences run counter to that of the group's, and therefore the individuals do not raise objections.

This paradox was first noted by Jerry Harvey in his article, "The Abilene Paradox and Other Meditations on Management."[35] According to Harvey, one hot afternoon his family was playing dominoes. The father-in-law suggested that they make a trip to Abilene, a 50-mile drive, for dinner; Harvey's wife replied that it was a great idea. Despite the long and hot drive that would be involved, the husband believes his negative preferences must be out of step with the group and replies that it "Sounds good to me. I just hope your mother wants to go." The mother-in-law replies, "Of course I want to go. I haven't been to Abilene in a long time."

The drive was indeed hot and dusty. To make things worse, the food was as bad as the drive. They return home many hours later, hot and tired. One of the family members dishonestly states, "It was a great trip, wasn't it?" To which the mother-in-law replies that actually she would have preferred to have stayed home but wanted to please the other three. The husband says he was not that excited about heading off in the first place, but did so to satisfy the rest of the family. The wife says, "I just went along to keep you happy. I would have had to be crazy to want to go out in the heat like that." The father-in-law replies that he only made the suggestion because he felt the others might be bored. The family sits back, perplexed that they had collectively decided to make a trip that none of them really wanted to make in the first place. Similarly, the Abilene paradox might lead a worker to oppose an environmental remediation tax because he believes that everyone opposes it, while in reality, most others wish to clean the junk yard up, but no one expresses this desire due to the fear of being ostracized for supporting a move to raise taxes.[36]

6.7.3 Spiral of silence

The *spiral of silence* asserts that a person holding a minority view is less likely to voice an opinion for fear of reprisal or isolation from the majority, which may then escalate.[37] This often occurs as the mainstream media's coverage of the majority opinion becomes accepted as status quo such that the minority becomes less likely to speak out. In general, this theory applies only to opinion; it is much less witnessed with issues that can be proved right or wrong from a factual basis. However, a vocal minority of influential or cavalier individuals who do not fear isolation is more likely to speak out regardless of public opinion. Furthermore, this minority can be an essential factor of change, while the complacent majority is a factor of stability.

This theory has important applications to environmental policy, as much of the environmental movement was sparked by a few isolated individuals who had sufficient resolve to take on large or entrenched corporate and government interests. In particular, some scientists who do not accept the theory of global warming claim that they are unfairly attacked and criticized for their views by the mainstream media and advocates of the theory.

6.7.3.1 Internet and promoting equality

The Internet brings empowerment, voluminous amounts of easily accessible information, and audiences that can be specifically targeted. The Internet has the capability of freeing people from the spiral of silence.

Stromer-Galley believes that the characteristics of online conversation free people from the psychological walls that have traditionally kept them from engaging in face-to-face deliberations. Wallace states that when people believe their thoughts or actions cannot be attributed to them personally, they tend to become less inhibited by social conventions and restraints.[38] Over time, this could lead to a more equally informed citizenry, with a more level playing field in terms of being able to effect environmental policies.

6.8   Power and limits of collective intelligence

In contrast to groupthink, *collective intelligence* encourages inquiry, investigation, troubleshooting, and debate. Collective intelligence refers to a shared or group intelligence that frequently emerges from competition or collaboration among many individuals. In contrast, groupthink often obstructs collective intelligence by reducing input to a few select individuals or filtering potential suggestions without fully developing them into tangible options.

Decision making plays a major role in crowd behavior. Sigmund Freud's theory of *crowd behavior* argues that people in a crowd tend to act differently from those who are thinking or acting individually. The minds of the group tend to merge to form a single and collective way of thinking. Individual enthusiasm begins to take over, and one becomes less aware of the true nature of their actions. Le Bon argued that crowds foster anonymity, which can foster new and different emotions. However, some studies indicate that "the madding crowd" does not take on a life of its own beyond the intentions of its members.

Yet another school of thought holds that crowd behavior is not actually a product of the crowd itself, but is the product of a few, particularly influential individuals. We leave it to the reader to determine how such theories might explain the behavior of organizations such as Greenpeace, antiglobalization demonstrations, or antienvironmental rallies, particularly in Europe. We will now turn our attention to what has aptly been referred to as the "wisdom of crowds."

6.8.1   Wisdom of crowds

In 1906, Francis Galton, a statistician, visited a livestock fair, where he stumbled upon a most interesting contest. An ox was on public display. Local citizens were invited to guess the animal's weight. Nearly 800 citizens gave it a go. As you might expect, not one guess hit the exact mark of 1198 pounds. But what Galton later learned bordered on the remarkable. The mean (average) of those 800 guesses, 1197 pounds, was in error only by one pound![39] What Galton stumbled upon was that in the "right circumstances," crowds are remarkably intelligent and are often smarter than the smartest member.

6.8.1.1   Mystique of collective intelligence

James Surowiecki recounts a story involving the 1968 disappearance of the American submarine *Scorpion* somewhere in the north Atlantic.[40] The navy had no idea what had happened and had only scanty data as to where the vessel might have gone down. When they started the search, they used a circle of 20 miles radius in an ocean thousands of feet deep. After an initial search, they failed to find any trace of the submarine. Fortunately, one naval expert, John Craven, had a unique idea. He assembled a team, but he did not simply look for navigation or submarine experts. Instead, he assembled a very diverse team, all knowledgeable people, who came from different parts of the navy. The team included mathematicians, submariners, salvage experts, and navy sailors. Instead of asking them to consult with each other, he asked them to bet on a series of different scenarios as to the various possibilities of what had happened. He also had them bet on things like, "How fast do you think it was going?" and "How steeply was it falling toward the ocean floor?"

By soliciting bets, he was essentially encouraging them to express how confident they were in their guesses. He took all the guesses and ran them through a formula for aggregating the probabilities and guesses. The result was their best guess as to the submarine's location. He told his naval superiors, and they were surprised because it was a place no one had anticipated. What happened next was nothing short of amazing. They searched the coordinates Craven had given them and found the submarine. It was a mere 220 yards from where Craven's team had predicted.

6.8.1.2 Wisdom of crowds vs. independent experts

Noted here are some popularly cited successes and failures of crowd intelligence:

- The algorithm of the world's most popular search engine, Google, relies on how many hits a page receives to determine the best Web page for each topic (i.e., the more hits a Web page has received from a random audience, the higher it will fall in Google's output).
- The ever popular television show *Who Wants to Be a Millionaire?* offers contestants the option of polling the audience for the answer or asking an expert that the contestant has chosen for his or her opinion. An expert should win out over a "dumb" audience, right? The results may be surprising:
  - Audience poll: 91% accurate
  - Soliciting the opinion of a preselected expert: 64% accurate

Now, consider a couple of findings with respect to "independent experts" working alone:

- Nearly 90% of independent mutual fund managers underperformed the market.
- Studies have shown that groups or panels of nonpsychologists are better at predicting people's behavior than psychologists.
- Independent medical pathologists presented with the same evidence disagree on a diagnosis 50% of the time.

6.8.1.3   When the system fails

Many failures have been cited in which organizations had the collective diversity and knowledge to succeed in their missions, yet failed because some members were not allowed to participate in the final decision. One of the famous examples involves the space shuttle *Challenger*. The Thiokol team (responsible for shuttle booster rockets) reported to the National Aeronautics and Space Administration (NASA) leadership; this team did not suffer from groupthink and was a fully functional collective intelligence operation well-grounded in shuttle engineering safety requirements. Yet, the night before the *Challenger* was launched, the following events took place[41]:

- High-level NASA and Thiokol officials discussed whether to launch.
- The Thiokol engineers who designed the booster rocket opposed the launching because of potential problems involving performance of the O-rings in cold weather.
- A high-level NASA manager was "appalled" by their opposition.
- Due to pressure from the NASA, the Thiokol managers changed their minds and supported the launch.

Such experience is shared by similar environmental disasters such as the Bhopal chemical plant accident (see the Case Study at the beginning of this chapter) and coastal oil spills.

6.8.1.4   Expert collective intelligence teams

In contrast to the wisdom of crowds, expert intelligence teams can be assembled to specifically recruit members with a variety of individual expertise related to the task at hand. Surowiecki suggests four key criteria that separate a "wise crowd" (such as expert teams) from irrational ones:

1. *Diversity of opinion*: Individuals display diversity in knowledge, skills, abilities, and experience.
2. *Independence*: The opinions of individuals are not determined by the opinions of those around them.
3. *Decentralization*: Individuals are able to specialize and draw on their own knowledge base.
4. *Aggregation*: A mechanism exists for turning private judgments into a collective decision, such as voting.

6.8.1.5   Criticism

Surowiecki has studied less successful situations (i.e., stock market bubbles) in which a crowd exercises very bad judgment and makes poor decisions. He argues that in such situations, the crowd's cognition and cooperation failed because the members of the crowd were actually too conscious of the opinions of others; they began to conform to the crowd and emulate each other rather than think independently (i.e., groupthink). While he offers experimental evidence that crowds can be collectively swayed by an irrationally persuasive speaker, he argues that the principal reason for such conformity is that the decision-making system is flawed.

Collective intelligence appears to be effective in restricted circumstances, such as when all participants have relevant knowledge and experience to share, their judgment is not clouded by ideology or preconceived solutions, and they are able to refrain from impulsive behaviors and actions. In his article "The Wisdom of Checking Sources," Meyrick severely criticizes Surowiecki's examples. Meyrick argued that too many of the cited examples invoke "mystical crowd" wisdom and hence lack scientific rigor.[42]

6.9 Uncertainty in decision making

Many policies, particularly environmental ones such as global warming, involve some degree of uncertainty. Sections 6.9.1 and 6.9.2 explore the effects this uncertainty can have on the decision-making process.

6.9.1 Ellsberg paradox

In spite of the fact that even his own biographer has questioned Daniel Ellsberg's true intentions and ethical motives for releasing the Pentagon Papers, few question that he made an important contribution to decision-making theory.[43]

The Ellsberg paradox arises from a series of experimental games.[44] Consider two urns, each containing 100 poker chips. The poker chips are colored either red or black. The first urn contains 50 red and 50 black chips. The second urn also contains 100 chips, but the proportion of red to black in this urn is unknown: it might contain 100 red chips, or 100 black chips, or any proportion thereof. A blind is placed over your eyes. A facilitator asks you to draw one chip from either one of the urns. If you draw a red poker chip, you win $10,000. Which urn do you choose to draw the chip from?

Now, stop here and make your own mental choice before reading on. Which urn would you choose from? There is no rational reason to believe that your chance of picking up a red chip is any higher from one urn than the other. Yet, most people choose the urn containing the 50 red and 50 black chips over the urn in which the proportion is unknown. If your answer was the first urn containing a 50-50 proportion of poker chips, it is logical to assume you had a hunch that the other urn contained more black chips. So let us test this hypothesis.

Consider a variation on this experiment. The facilitator makes a second wager: "It's clear that you must believe that the first urn is more likely to let you draw a red chip. Now, I'm offering you $10,000 if this time you draw a black chip." Which urn would you draw from this time, the same urn or the other urn?

Given the facilitator's rational, you should logically draw your black chip from the second urn. But in experiments that have been performed, this does not appear to be the case. Again, subjects overwhelmingly chose the 50-50 urn. This is true despite the fact that the chance of picking either color is identical in both gambles.

This raises unsettling questions in terms of how people reach decisions when confronted with uncertainty. This paradox suggests that people strongly prefer definite information over ambiguity and will make their decisions accordingly (i.e., ambiguity aversion). The value of such experiments resides in the fact that they illuminate the preference people tend to have for choices that seem to involve least risk or uncertainty.

Perhaps the best explanation for this paradox is that people tend to instinctively avoid circumstances that may result in the worst-possible outcome as opposed to making a rational assessment of the choices or optimum course of action to pursue. Additional research has found that uncertainty about technological and other risks tends to make less ambiguous technologies more acceptable to the public.

6.9.1.1 Application of the Ellsberg paradox to environmental policy and decision making

Environmental decision making often involves a great deal of risk, ambiguity, and uncertainty. While the aforementioned Ellsberg paradox is widely discussed in fields such as economics, politics, and defense strategy, a review of the literature indicates that as of this writing no serious attention has been devoted to its implications with respect to environmental policy and decision making.[45]

The paradox suggests that many environmental policy decisions are probably reached by a process in which one option is chosen over another simply because decision makers have a natural desire to avoid risk or uncertainty. Some superior alternatives and courses of action have probably even been rejected simply because they involved a greater degree of ambiguous information or circumstances. This implies that many bad choices have probably been made simply because they involved decisions that did not have a significant degree of risk or uncertainty. Thus, the way in which decisions are made in relation to this paradox may actually endanger society.

6.9.2 Risk homeostasis hypothesis

The *risk homeostasis* hypothesis proposed by Gerald Wilde says that everyone has a constant or fixed level of acceptable risk. As the level of risk in one part of an individual's life increases, it tends to trigger a corresponding decrease of risk elsewhere, bringing one's overall risk back into equilibrium.[46]

An often-cited study in this context involves taxi drivers in Germany. Half the fleet of taxicabs was equipped with antilock brakes, while the other half had conventional brake systems. Yet, surprisingly, the crash rate remained about the same for both groups. Wilde asserts that this was due to the fact that antilock-equipped drivers assumed that their braking system would protect them, and consequently they took more risks; the drivers without antilock brakes drove more carefully since they lacked this additional protection. How do you think such findings might apply to how decisions are made by a society when additional mitigation and safety features are added to a potentially risky technology?

6.10 Some food for thought

Although there are many books and courses that illustrate good decision-making practices, there is no single foolproof method for reaching wise decisions. The problem is as old as civilization itself; approximately 5000 years ago, the first rulers of Sumer and Egypt recruited "wise men" and various experts to counsel them on the basic policy matters of their day. Although modern human culture enjoys the benefits of a scientific outlook and technology that would be considered magical to just about everyone who lived prior to the twentieth century, the ability of human communities and their governance to discover, embrace, and effectively apply wise policies on a consistent basis has hardly improved.

This does not mean we should surrender humanity to the fate of its own environmentally destructive trends or that we should shrink from efforts to instill wisdom into the environmental policymaking process. The most important lessons learned can be summarized as follows:

- Be humble in your ability to radically change human nature.
- Recognize when people's attitudes, knowledge, and commitment are likely to make them good or poor policy team members. Members holding contrary views are essential to illuminate all sides of an issue, but ideologically rigid thinkers can be problematic.
- Cultivate the skills of open and respectful communication that fosters effectiveness of collective intelligence.
- Learn to recognize wise leadership and support it.
- By all means cultivate reasonable analytical and modeling tools, but be open to the human factors and to independent critiques of causal relationships, data reliability, and your own personal inferences of results.

Discussions, problems, and exercises

1. Briefly describe the principles and dilemmas of environmental policy decision making.
2. Make a list of all the varieties of expertise that can affect the decision-making process.
3. Imagine a policy that would free our nation from imported fossil energy in the next 10 years, and a scenario for how that would be done. List the questions you would want to include in an environmental–economic study of your approach to implementing the policy.
4. An EIS has been prepared for a flood control project. The proposed action involves constructing a levy control system that has a 50/50 chance over the next 20 years of being breached, which would result in limited flooding with some deaths and damage amounting to $100,000,000. The alternative involves constructing a flood control dam that would entirely eliminate the risk of any future flooding but could catastrophically fail under certain extreme seismic conditions. The chance of a dam failure is uncertain but experts believe it is very remote; the consequences of a dam failure are likewise uncertain but might be potentially catastrophic. What decision-making factors should policymaking officials be aware of before reaching a final decision to pursue either the levy or the dam?
5. Describe the term doublethink.
6. How might the spiral of silence affect scientific debate (for and against) in terms of the greenhouse effect?
7. Suppose that a pilot plant is constructed to manufacture a new but extremely toxic pesticide. The plant management prepares an environmental policy review of the plant. To increase plant safety, they decide to automate several safety features that are currently conducted manually by employees. Consider the problem with respect to the risk homeostasis hypothesis. How might this affect other essential plant employees such as the inspectors, safety officer, and plant operators? Can we be certain that such changes would actually increase the overall safety level of the plant? If not, what other measures could be taken to counter potential effects from risk homeostasis?
8. Consider the following problem in terms of the Ellsberg paradox: Is the public or a policymaker more likely to accept a national coal-fired energy policy that will definitely result in 5000 early deaths from cancers and lung ailments per year but is well understood over a second technological alternative that results in far fewer direct ailments but in which the risk of a catastrophic accident is much more ambiguous? What can be done to reduce the influence of the paradox in terms of decision making?

Notes

1. N. Carter, *The Politics of the Environment: Ideas, Activism, Policy* (Cambridge: Cambridge University Press, 2001).
2. There are numerous books and actual examples of engineering systems analysis. An excellent application is the text for a course taught by Sandia National Laboratories, *System Safety for High Consequence Operations and Organizations* by Richard L. Perry (Albuquerque, NM: Sandia National Laboratories, April 2001).
3. The figure is based on narrative in Todd Tremlin's *Minds and God: The Cognitive Foundation of Religion* (Oxford, UK: Oxford University Press, 2006).
4. P. Cockerell, "Learning to Trust Your Intuition: 'Gut Feeling' Sometimes Right," *Washington Examiner,* June 2, 2009, http://washingtonexaminer.com (accessed June 6, 2009).
5. N. Taleb, *The Black Swan* (New York: Random House, 2007), 50.
6. Wikipedia, "List of Cognitive Biases," http://en.wikipedia.org/wiki/List_of_cognitive_biases (accessed May 22, 2009).
7. B. Delia and P. Bearman, "Dynamics of Political Polarization,"*American Sociological Review,* 72 (October 2007): 784-811.
8. D. M. Mackie and J. Cooper, "Group Polarization: The Effects of Group Membership," *Journal of Personality and Social Psychology* 46 (1984): 575–85.
9. R. S. Nickerson, "Confirmation Bias: A Ubiquitous Phenomenon in Many Guises," *Review of General Psychology* 2 (1998): 175–220.
10. T. van Gelder, "Heads I Win, Tails You Lose: A Foray into the Psychology of Philosophy," http://www.philosophy.unimelb.edu.au/tgelder/papers/HeadsIWin.pdf (accessed May 20, 2009).
11. J. Evans, J. L. Barston and P. Pollard, "On the Conflict between Logic and Belief in Syllogistic Reasoning," *Memory and Cognition* 11 (1983): 295–306.
12. A. Colman, *Oxford Dictionary of Psychology* (Oxford, UK: Oxford University Press, 2003), 77.
13. S. E. Asch, "Opinions and Social Pressure," *Scientific American* 193 (1955): 31–5.
14. Prepared by Max R. Lum and Tim L. Tinker, *A Primer on Health Risk Communication Principles and Practices* (Washington, DC: U.S. Department of Health and Human Services, Public Health Service, Agency for Toxic Substances and Disease Registry, 1994); P. Slovic, "Perception of Risk," *Science* 236 (1987): 280–5.
15. Pamphlet drafted by V. T. Covello and F. H. Allen, *Seven Cardinal Rules of Risk Communication* (Washington, DC: U.S. Environmental Protection Agency, April 1988).
16. Funtowicz and Ravetz, *Uncertainty and Quality in Science for Policy* (Dordrech: Kluwer Academic Press, 1990).
17. 40 Code of Federal Regulations (CFR) pt 1502.22.
18. F. March and others, *Wind Power for the Electric-Utility Industry?—Policy Incentives for Fuel Conservation* (Lexington, MA: Lexington Books, 1982).
19. E. A. Parson, "Informing Global Environmental Policy: A Plea for New Methods of Assessment And Synthesis," *Environmental Modeling and Assessment* 2 (1997): 267–79. Available at http://www.springerlink.com/content/h451671m48u67830/ (accessed July 8, 2010).
20. P. Brett Hammond, "The Energy Model Muddle," *Policy Sciences* 16, no. 3 (1984), 227–243.

21. B. Schwartz, *The Paradox of Choice—Why More Is Less* (New York: Harper Perennial, 2005).
22. B. Schwartz, *The Paradox of Choice: Why More Is Less* (New York: Harper Perennial, 2005).
23. S. S. Iyengar and M. R. Lepper, "When Choice is Demotivating: Can One Desire Too Much of a Good Thing?" *Journal of Personality and Social Psychology* 70, no. 6 (2000): 996–1006; F. Norwood, "Less Choice Is Better, Sometimes," *Journal of Agricultural and Food Industrial Organization* 4, no. 1 (2006), article 3; F. B. Norwood and others, "An empirical investigation into the excessive-choice effect," *American Journal of Agricultural Economics*, 91, no. 3 (August 2009): 810–825.
24. I. H. Evans, ed., *Brewer's Dictionary of Phrase & Fable*, 14th ed. (New York: Harper & Row, 1989).
25. D. Kahneman and A. Tversky, "Prospect Theory: An Analysis of Decision under Risk," *Econometrica* 47 (1979): 263–91.
26. S. Lash and B. Wynne, *Introduction to Risk Society, Towards a New Modernity* (New York: Sage Publications, 1992) [originally publ. 1986]. Available at http://books.google.com/books?hl=en&lr=&id=QUDMaGlCuEQC&oi=fn d&pg=PA1&dq=ulrich+beck+risk+society&ots=8DsOrCbyUV&sig=pOJCv XGvco5l5YS2IhKq6uu40O0#v=onepage&q=&f=false (accessed July 8, 2010).
27. Amazon book review by Hoomans of Beck's *Risk Society*. Available at http://www.amazon.com/review/RJVZS1BQEC9TU (accessed July 8, 2010).
28. Review by William Leis of Simon Fraser University of Ulrich Beck, *Risk Society*. http://www.cjsonline.ca/articles/leiss.html (accessed July 8, 2010).
29. L. Festinger, *A Theory of Cognitive Dissonance* (Stanford, CA: Stanford University Press, 1957).
30. Ibid.
31. G. Orwell, *Nineteen Eighty-Four* (London: Martin Secker & Warburg Ltd, 1949).
32. S. E. Asch, "Effects of Group Pressure upon the Modification and Distortion of Judgment," in *Groups, Leadership and Men*, ed. H. Guetzkow (Pittsburgh, PA: Carnegie Press, 1951).
33. S. E. Asch, "Studies of Independence and Conformity: A Minority of One against a Unanimous Majority," *Psychological Monographs* 70, no. 416 (1956).
34. J. McAvoy and T. Butler, "Resisting the Change to User Stories: A Trip to Abilene," *International Journal of Information Systems and Change Management* 1, no. 1 (2006): 48–61.
35. J. B. Harvey, "The Abilene Paradox and other Meditations on Management," *Organizational Dynamics* 3, no. 1 (summer 1974): 63.
36. J. A. Kitts, "Egocentric Bias or Information Management? Selective Disclosure and the Social Roots of Norm Misperception," *Social Psychology Quarterly* 66, no. 3 (2003): 222–37.
37. J. A. Anderson, *Communication Theory: Epistemological Foundations* (New York: The Guilford Press, 1996).
38. P. Wallace, *The Psychology of the Internet* (Cambridge, UK: Cambridge University Press, 1999).
39. See note 1 above.
40. J. Surowiecki, *The Wisdom of Crowds: Why the Many Are Smarter Than the Few and How Collective Wisdom Shapes Business, Economies, Societies and Nations* (New York, NY: Random House, Inc., 2004).

41. Sandia National Laboratories, Manual for course: *System Safety for High Consequence Operations and Organizations*, April 2001.
42. C. Meyrick, "The Wisdom of Checking Crowds," *Competitive Intelligence Magazine* 10, no. 6 (November–December 2007): 55–7.
43. T. Wells, *Wild Man: The Life and Times of Daniel Ellsberg* (New York: Palgrave Macmillan, 2001).
44. D. Ellsberg, "Risk, Ambiguity, and the Savage Axioms," *Quarterly Journal of Economics* 75 (1961): 643–69.
45. C. H. Eccleston, *NEPA and Environmental Planning: Tools, Techniques, and Approaches for Practitioners* (Boca Raton, FL: CRC Press, 2001).
46. Gerald J. S. Wilde, *Target Risk 2: A New Psychology of Safety and Health* (Toronto: PDE Publications, 2001).

chapter seven

Environmental management systems

> Man is a complex being; he makes the deserts bloom
> and lakes die.
>
> **Author unknown**

Case study

Samsung Austin Semiconductor employs 1000 people and provides services and products to the electronics industry. Prior to receiving its ISO 14001 registration, the company had an ill-defined waste reduction or minimization program. Moreover, there was no formalized environmental management system (EMS) to implement its existing and planned environmental processes.[1]

The company began an effort to install an ISO 14001 EMS that would allow more of the employees to become involved in managing the company's environmental aspects and impacts. The EMS would assist managers in more clearly determining what they should focus on.

A core team was assembled, consisting of about 20 people from different areas of the company. The diverse group allowed managers to look at all potential environmental aspects that could be environmentally significant. This team was able to tap into areas that the environmental health and safety (EHS) group might have missed for lack of expertise. It was decided that the company would focus on three main targets and objectives. All three dealt with waste reduction or minimization. The company was also able to set ambitious yearly targets that included the following:

- Reduction of 4% of the company's total landfill waste
- 10% reduction of arsenic waste in the company's implant area
- Reduction of 5% total (or 1.5 tons per month) of landfill waste

The company reported that staff morale increased after it received its ISO 14001 registration. There is more focus on environmental performance and awareness of the company's environmental efforts. Everyone is involved with meeting the company's environmental goals. The company management reported that it was enthusiastically looking forward to combining its quality and EHS management systems with its EMS. The company reported that it has reduced the hazardous-waste management expenses associated with its arsenic waste management operations. It has also saved money with its sulfuric acid recycling project.

> **Problem:**   After reading this chapter, return to this problem. Consider this case study in light of what you have read. Develop an EMS plan for your university, your local government, or a local business. How would you integrate this plan with an environmental impact assessment (EIA) process and the goal of sustainability? This problem may be pursued either as an individual assignment or as a class project.

Learning objectives

After you have read the chapter, you should understand the following:

- The ISO 14000 series of standards
- The basic elements of an ISO 14001 EMS
- How an EMS can be used to help ensure that environmental decisions and policies are correctly implemented
- How an EMS and an EIA process can be integrated to help policy-makers develop and implement sustainable development decisions and plans

Modern businesses, organizations, and government agencies must comply with many environmental laws, rules, and regulations. It is vital that an organization safeguard itself against shortcuts and noncompliances that can leave it vulnerable to legal or even criminal violations of the law.

Regardless of what the phrase might suggest, environmental management is *not* the management of the environment as such. Rather, it deals with the effective management of an organization's structures, functions, and activities that can affect the environment. The presumption is that if the organization's components are managed with an eye to the environment, such management will reduce the biophysical impact on the environment. In this context, the environment also involves the relationships of the "human environment," such as socioeconomic and cultural impacts and their interaction with the biophysical environment.

For any organizational management process, effective management standards and tools are required. In this respect, environmental management is no different. An EMS provides a systematic procedure for managing the functions of an organization that can affect the environment. That is, it is designed to assist organizations in managing the environmental effects of their functions, operations, and practices. As you will soon see, an EMS provides a particularly efficient mechanism for developing and implementing a plan that executes an established environmental policy.

7.1 International Organization for Standardization

Incongruent standards can lead to barriers in international trade, giving some organizations advantages while others are placed at a disadvantage. Internationally accepted standards and protocols provide a common functionality that is recognized and accepted worldwide.

The International Organization for Standardization (ISO) is a body composed of representatives from various nations around the world, whose mission is to develop common standards for products and services. It is important to note that ISO is not only an acronym. This term is in fact derived from the Greek word *iso* meaning "equal," or in this case, "equivalent standards." The inference is that if two objects meet the same (equal) standard, they should be equivalent in terms of their results or functions. This name also eliminates the confusion that could result from the translation of International Organization for Standardization into different languages, each of which would be translated into different acronyms.

Headquartered in Geneva, Switzerland, ISO is a voluntary organization whose purpose is to promulgate internationally accepted standards. Its members are recognized authorities, each member representing one country. Most of the work done by ISO is performed by one of its 2700 technical committees, subcommittees, and working groups. ISO's standardization documents are copyrighted. It charges for copies of most standards it publishes, which provides funding that supports the organization. Each proposed standard goes through a six-stage review and adoption process, which is detailed in Table 7.1.

Table 7.1 ISO's Six-Stage Review and Adoption Process

1. Proposed standard.
2. If it makes it through the proposed standard stage, the standard enters the preparatory stage where a working draft of the standard is developed.
3. Next it enters the committee stage where it is distributed for review and comment.
4. In the fourth stage, a DIS is prepared.
5. The DIS is circulated among all ISO bodies for a vote. If the DIS does not receive 75% of the vote, it returns to lower stages for refinement. If it passes, it becomes a Final Draft International Standard.
6. The last stage is the approval stage. The Final Draft International Standard is circulated for a final vote, which requires a 75% majority for passage. If it passes, it is published as an ISO international standard.

DIS: draft international standard.

Before introducing the ISO 14000 environmental standard, it is worth taking a moment to review the ISO 9000 quality (business) management standard on which the ISO 14000 environmental management standard was modeled. The ISO 9000 series was designed to provide organizations with a rigorous, yet flexible, quality assurance management process; specifically, it establishes an internationally accepted management process to help assure the quality of manufactured products and services. It is important to note that certification to the ISO 9000 standard does not guarantee a superior or even adequate quality of products and services; rather, it certifies that a formalized management process is in place to ensure that all business processes affecting quality are correctly implemented.

While ISO 9000 deals with effective implementation of business management processes that can affect the quality of a product or service, ISO 14000 deals with the management of an organization's processes and functions that can impact environmental quality. Both standards are so similar in their procedures and goals that there has been talk of combining the two series into one.

7.1.1   Development of the ISO 14000 standard

Building on the footsteps of the rapid and very successful acceptance of its 9000 series, ISO began evaluating the advantages of an international environmental management standard, which would run along a similar process. ISO established the Strategic Advisory Group on the Environment to define the basic requirements for a series of environment-related standards.

Many companies and organizations had underestimated the interest with which ISO 9000 would be accepted around the world; consequently, they had shown only limited interest in providing input to the development of the ISO 9000 series of business standards. These organizations were surprised by the overwhelming acceptance that ISO 9000 received among the international community. Based on this earlier miscalculation, many of the same businesses and organizations were wary of the proposed ISO 14000 standard and consequently took much more ownership in the development of ISO 14000.

Some organizations, such as U.S. companies that are routinely subjected to expensive civil suits, were hesitant to endorse a new stringent set of additional environmental management standards. They feared that violating stringent ISO 14000 standards could set the stage for fierce and costly legal battles, as well as frivolous lawsuits for minor violations of such standards; such concerns only enhanced participation in the development of ISO 14000.

7.2   ISO 14000 series of standards

This section describes ISO 14000 and the ISO 14001 EMS.

7.2.1 Difference between ISO 14000 and ISO 14001

The terms ISO 14000 and ISO 14001 are commonly cited in literature. Not surprisingly, this has led to significant confusion concerning the differences between these two terms. As illustrated in Table 7.2, the term ISO 14000 refers to a series or family of related environmental management standards (e.g., EMSs, environmental labeling, environmental auditing, and environmental assessment [EA] of organizations). In contrast, the term ISO 14001 (the first element in the ISO 14000 series) deals exclusively with the requirements for establishing an ISO 14001–compliant EMS.

7.2.1.1 Improving the EMS vs. improving environmental performance

With respect to an ISO 14001 EMS, the term "environmental performance" typically refers to the act of reducing an environmental impact, such that environmental quality is improved. Instead, the ISO 14001 standard focuses on improving the *management process*, which manages or administers an organization's functions and activities that can affect the environment. ISO 14001 does not actually require that an EMS improve environmental performance (i.e., reduce environmental impacts). For instance, ISO 14001 does not prescribe a particular level of pollution or environmental performance, require the use of particular technologies, or

Table 7.2 ISO 14000 Series of Standards

Series	Explanation
ISO 14001	Requirements and guidance for using EMSs.
ISO 14004	EMS (general guidelines) principles, systems and support techniques.
ISO 14015	EA of organizations and sites.
ISO 14020	Series (14020–14025), environmental labels and declarations.
ISO 14031	Environmental performance evaluation—guidelines.
ISO 14040	Series (14040–14049), lifecycle assessment; discusses preproduction planning and environmental goal setting.
ISO 14050	Definitions.
ISO 14062	Describes making improvements to environmental impact goals.
ISO 14063	Environmental communication—guidelines and examples.
ISO 19011	Specifies one audit protocol for both the ISO 14000 and 9000 series together. It replaces ISO 14011 metaevaluation. Currently, ISO 19011 is the only recommended way to determine this.

EMS = environmental management system; EA = environmental assessment.

establish regulatory standards for environmental outcomes. In fact, some organizations, engaged in similar activities, may have widely different effects on the environment, yet all comply with ISO 14001.

The focus of an ISO 14001 EMS is on improving management processes, practices, and procedures that control an organization's functions and activities, which can affect the environment. The overarching intent is that by implementing a management process, which administers an organization's functions, products, and services, and by continually improving this management system, this process will eventually lead to improved environmental performance. Although this is generally true of organizations that are truly committed to the goal of improving environmental quality, it may not be true of an organization that lacks a serious commitment; in the latter case, an EMS may amount to nothing more than "window dressing" to improve a business' or organization's image with the public and consumers.

It is important to note that adherence to ISO 14000 standards does not, by itself, release an organization from full compliance with other local or national environmental laws and regulations regarding specific environmental performance standards that must be met. In fact, it provides procedures to help ensure that all applicable laws and regulations have been identified as well as an auditing or a monitoring procedure to identify any noncompliance.

7.2.2 *Proponents and critics of ISO 9000 and ISO 14000 series*

Both ISO 9000 and ISO 14000 have proponents and critics. Critics assert that the quality management policies proposed by ISO 9000 do not necessarily result in quality products. Likewise, they argue that the ISO 14000 environmental management policies and procedures do not guarantee that an organization is not damaging the environment or that it will lessen its environmentally destructive practices in future.

Moreover, the ISO 9000 and ISO 14000 series provide for third-party certification, which is a lucrative and growing business. Questions have been raised regarding the fairness and impartiality of this certification process. Other critics state that both series are overly burdened with extensive documentation requirements, which add excessively to the cost, resources, and time required to comply with the ISO requirements.

7.2.2.1 *Proponents of ISO 9000 and 14000 series*

Proponents of both the ISO 9000 and ISO 14000 series point out the economic benefits gained by implementing these standards. These benefits include opening new markets in other nations, as a result of less opposition based on environmental deficiencies, and streamlined procedures, which can lead to increased profitability. Nontangible benefits of ISO

14000 certification can include improved employee morale and improved corporate image among the public and consumers.

7.2.2.2   Specific ISO 14001 criticisms

As with most systems, ISO 14001 also has many shortcomings. Obstacles to implementing an EMS include the following:

- High cost
- Lack of time
- Insufficient resources
- More requirements to conform to
- Lack of senior management support
- Lack of appreciation of the EMS process and its importance in managing environmental impacts

As stated earlier, one of the most frequently voiced criticisms is that this certification does not necessarily reflect any actual environmental performance or long-term reduction in the organization's environmental footprint. Although there is an ISO guideline for evaluating environmental performance (ISO 14031), there is no actual requirement to verify that an organization is achieving continuous environmental improvement in terms of environmental performance. The ISO 14001 standard simply assumes that a well-orchestrated EMS will lead to continuous improvement in environmental performance (i.e., less environmental impact). Critics argue that such a presumption is invalid and that independent external audits should be required to verify that environmental performance is actually occurring.

Compared to BS 8555, ISO 14001 places less attention on legal compliance. Moreover, ISO 14001–certified organizations only need to publish their environmental policy. Although organizations are required to perform audits and performance monitoring, the results of such activities are not required to be made public. Many stakeholders find such practices unacceptable.

7.3   ISO 14001 environmental management system process

A gap analysis is often prepared to identify "gaps" between existing management practices and the requirements for complying with the ISO 14001 standard. It is also used to identify the means of improving an organization's environmental management practices and to set priorities when actions are needed. Once the gap analysis has been performed, the stage is set for developing the EMS. The basic five-stage EMS process is described in Figure 7.1. The basic components of each stage are described in more

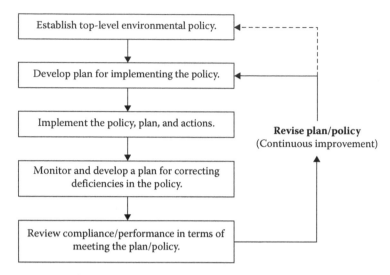

Figure 7.1 Simplified overview of a typical ISO 14001 environmental management system.

detail in Section 7.3.1. As illustrated in the figure, an EMS provides a structured system (i.e., "plan-do-check-revise") in which a set of management procedures are used to systematically identify, evaluate, manage, and address environmental issues and requirements. Table 7.3 provides a brief overview of the typical ISO 14001 development and maintenance process. Section 7.3.1 describes the process in more detail.

7.3.1 Essential EMS functions

These sections provide a more detailed description of essential EMS functions.

7.3.1.1 Environmental policy

The ISO 14001 standard requires establishing a high-level organizational *environmental policy*. This policy is a statement from top management that establishes an environmental commitment and direction for the entire organization. The policy is important because it provides the programmatic direction and goals of the organization (business, company, government agency, and so on). It also provides direction for the EMS process.

The policy is unique and should be tailored to each individual organization. It must be appropriate for the size and complexity of the organization and generally should not exceed one to two pages in length. The policy must also be written in nontechnical language so that it can be

Table 7.3 Typical ISO 14001 Development and Maintenance Process

Stage 1—Environmental policy: The EMS process begins with preparing and establishing an environmental policy.

Stage 2—Planning: The next stage of the EMS process involves developing a plan for implementing the system. Although the planning function is often performed to determine how an organization will meets its quality policy, it can also be used more comprehensively to develop detailed environmental plans. Environmental aspects are identified, environmental objectives and targets are established, and a program to achieve them is developed. This plan includes identifying the following:

Environmental aspects: Operations, activities, products, and services are reviewed to identify how they interact with and may affect the environment.

Legal and other requirements: The plan identifies legal and other requirements that apply to the organization's environmental aspects.

Objectives and targets: Environmental objectives and targets are developed and communicated throughout the organization. A program is developed for achieving objectives and targets.

Stage 3—Implementation: Once the plan has been formalized, the EMS is ready for actual implementation and integration with the organization's functions and activities. Environmental and EMS responsibilities are assigned. Employees are trained to ensure that they are aware of the plan and are able to perform required duties in compliance with the EMS policy and plan. Specific work procedures are developed, defining how specific tasks are to be conducted. These implementation requirements are summarized as follows:

Structure and responsibility:

Roles, responsibilities, and authorities are defined for personnel whose activities may directly or indirectly affect the environment.

An individual(s) is appointed by top management as the "management representative(s)." The management representative(s) is assigned responsibility and authority for ensuring that the EMS complies with the ISO 14001 standards and for reporting EMS performance to top management.

Training, awareness, and competence:

Training: The organization identifies training requirements of personnel whose work may significantly impact the environment. Personnel must receive appropriate education and training and/or must have experience in dealing with environmental requirements.

Communication: Relevant information concerning environmental aspects is required to be communicated throughout the organization.

EMS documentation: Information must be maintained that describes the basics of the EMS. The documents must be reviewed on a regular basis. This documentation must be managed and maintained through an established DCS.

(Continued)

Table 7.3 Typical ISO 14001 Development and Maintenance Process (*Continued*)

Operational control: Activities that can significantly impact the environment and are relevant to the organization's objectives and targets must be identified. The organization must ensure these operations are performed according to the EMS plan under controlled conditions. Controlled conditions can include documented procedures with specific operating criteria.

Emergency preparedness and response: The organization must identify potential accidents and emergency situations that may result in an environmental impact. Procedures must be developed for responding to such accidents and emergencies.

Stage 4—Monitoring and corrective action: This stage involves checking and audits, control of nonconformances, corrective action, and preventative action. Characteristics of operations and activities that can significantly impact the environment need to be regularly monitored and measured. Monitoring and measurement results need to be compared with legal and other requirements to assess compliance.

Stage 5—Management review: The final stage involves a review by the organization's management. This step helps ensure that the system is operating effectively and provides the opportunity to address changes that may be made to the EMS.

understood by a typical reader. It is communicated to all employees and must also be made publicly available.

The policy must include senior management's commitment to (1) pollution prevention (P2), (2) continuous improvement throughout the organization, and (3) compliance with applicable environmental regulations and standards that affect the organization. The policy provides a starting point for establishing the organization's EMS objectives and targets (described in Section 7.3.1.4).

7.3.1.2   Planning function

The planning function is often performed to determine how an organization will meet its quality policy. However, it can also be used more comprehensively to develop detailed environmental plans. The EMS team identifies legal requirements and also considers how the organization's functions and activities interact with the environment. The team develops a plan for reducing the adverse environmental aspects of the organization's operation. The plan spells out how the EMS will be implemented. The planning stage should involve employees from all levels within the organization.

7.3.1.2.1 Emergency preparedness and response Procedures must be developed, communicated, and tested to help ensure that any unexpected incidents are effectively and efficiently responded to by internal and external personnel. A process must be established for identifying the potential emergencies as well as procedures for mitigating their effects.

7.3.1.3 Environmental aspects

One of the more important functions of an EMS involves identifying an organization's significant environmental aspects. The term "aspect" is defined broadly to mean the organization's activities, products, or services that interact with the environment. In contrast, the "environmental impact" is how a given aspect actually affects or changes the environment.

A systematic and verifiable procedure must be followed to identify environmental aspects and determine which may significantly affect environmental quality. ISO 14001 does not describe what aspects are significant; nor does it specify how *significance* is to be determined (more about this in Section 7.6). The ISO 14001, 14004, and 14031 standards provide helpful guidance for identifying significant environmental aspects. The EMS policy and plans for improving environmental performance are documented and communicated to the employees.

7.3.1.4 Objectives and targets

Once environmental aspects have been identified, attention turns to developing a plan for reducing them. *Environmental objectives* and *targets* are established to meet the goals documented in the organization's environmental policy. The term "objectives" refers to general, long-term goals. In contrast, the term "targets" refers to more specific, measurable events. Targets will generally vary over time and across various organizational functions and activities.

Objectives and targets need to be defined for appropriate functions and levels of the organization, and should be measurable, where practicable. The EMS is designed to identify specific objectives or targets, and describes the means to achieve them. The EMS process must be designed to ensure that objectives and targets are consistent with the environmental policy, including a commitment to comply with legal and other requirements, continual improvement, and P2. Work procedures, instructions, and controls are developed to ensure that the policy is implemented and the targets can be achieved.

7.3.1.4.1 Defining objectives The organization identifies its principal environmental objectives. As indicated earlier, environmental objectives tend to override considerations such as the development of better

Table 7.4 Example of Environmental Objectives and Targets for
a Small Company

Objectives	Targets	Responsibility
1. Environmental compliance: Comply with all applicable environmental laws and regulations.	Zero penalties or fines per year.	Principal regulatory manager
2. Minimize waste and prevent pollution.	Recycle 75% of all paper products and 50% of aluminum waste.	Chief process engineer
3. Conserve energy.	Reduce electricity consumption by 20%.	Chief plant engineer
4. Improve the EMS.	Obtain ISO 14001 certification.	EMS program manager

employee environmental training or improved environmental communication with other interested parties. These objectives will become the primary areas of focus within the improvement process.

7.3.1.4.2   Defining targets   In contrast to objectives, environmental targets tend to be specific points or items such as the reduction of energy utilization by 20% or reduction in sulfuric acid waste by-products by 10%. Targets should quantify the organization's commitment to an environmental improvement. Table 7.4 shows the difference between objectives and targets, as well as the assignment of responsibilities for ensuring that these objectives and targets are met.

7.3.1.5   Identifying legal and other requirements
An EMS procedure provides an ideal mechanism for ensuring that applicable legal and other requirements are identified and ensures that this information is relayed to key organizational functions.

7.3.1.6   EMS documents and document control
The ISO 14001 standard requires organizations to establish procedures for controlling documents related to the implementation of the EMS. This documentation must provide guidance for effectively planning, implementing, and controlling processes, and must be sufficient to demonstrate conformance with the ISO 14001 standard.

Specific procedures must establish a document control system (DCS) to ensure that documents that can demonstrate that the organization's operations conform to the ISO 14001 standard. These procedures must ensure that documents are approved prior to use and are reviewed and updated as necessary. Procedures must ensure that such records are identifiable, retrievable, secure, and traceable. The EMS could also be used to maintain sustainability plans and National Environmental Policy Act (NEPA) documents (see Section 7.6).

7.3.1.7   Monitoring and measurement

To gauge progress, an EMS must be monitored to determine its effectiveness and provide data for improvement. Many EIA processes, such as NEPA, do not have enforceable requirements to perform postmonitoring. Thus, NEPA lacks a systematic process for ensuring that the final decision, including any adopted mitigation measures, is properly implemented. In contrast, monitoring is a basic element inherent in an EMS. A properly integrated EIA/EMS system can contribute to environmental protection as it ensures that monitoring procedures are likewise executed (see Section 7.6). These ISO 14001 requirements can also help ensure that the staff in charge of maintaining sustainability measures are properly trained and that sustainability commitments are monitored.

7.3.1.7.1   Internal audit   A procedure must be developed for periodically monitoring conformance with the 14001 standard and EMS plan and for assessing how well the organization is doing in terms of managing its environmental functions and operations, including compliance with applicable environmental requirements. It can also be used to assess the performance of the EMS in terms of achieving environmental objectives and targets. The EMS procedures also must specify who is responsible for performing the audit and the means of reporting the results.

Implementing mitigation and other EIA-related commitments can be included as part of the EMS audit function (see Section 7.6). Thus, the EMS audit provides another mechanism for ensuring that an agency's EIA commitments are appropriately implemented. Similarly, these audits can help ensure that sustainability commitments are performed correctly.

7.3.1.8   Continuous improvement

The ISO continual-improvement process is based on the following: (1) monitoring; (2) a nonconformance, and corrective and preventive actions; and (3) management review with a commitment to improve the EMS process. It can also help ensure that sustainable management practices are improved over time.

7.3.1.8.1   Management review   The organization's top management is required to periodically review the EMS in order to ensure the system is operating as planned and is effectively performing its intended goals and objectives. The management review provides the ideal forum for determining how to improve environmental practices in the future.

7.3.1.8.2   Nonconformance and corrective and preventive actions   An independent internal or third-party audit monitors conformance with EMS requirements, as well as applicable environmental regulations and requirements. A corrective-action procedure is implemented when there is an environmental incident or a nonconformity; for instance, nonconformance may include a breach of an EMS procedure or a violation of an applicable environmental regulation. The benefit of this step is that under ISO 14001, corrective actions can be viewed as good things—unlike traditional noncompliance situations; in other words, an EMS can actually be used to put a positive spin on a less-than-optimal result.

As applicable, a root-cause analysis may be conducted to determine the underlying cause of an incident or noncompliance; corrective actions are then taken to ensure that this problem does not happen again.[2] Findings and/or recommendations resulting from EMS monitoring and auditing phase provide the basis for identifying and managing preventive or corrective actions. Such preventative or corrective actions can also be designed to promote and maintain sustainability goals and objectives (see Section 7.6).

A preventative-action program is performed in the same manner as corrective actions, except there will be no actual incident or nonconformity to address. Instead, emphasis is placed on identifying potential future problems and taking measures to prevent them before they occur. Sometimes, managers encounter opposition while attempting to justify a preventative-action program (when no actual incident or nonconformance has actually occurred); this is because it is often difficult to determine the effectiveness of a resulting initiative that prevents problems from occurring in the first place.

7.3.2   Implementation requirements

These sections outline key requirements for effectively implementing the EMS.

7.3.2.1   Responsibilities

Top management is responsible for ensuring resources are available so that the EMS can be effectively implemented. EMS roles, responsibilities, and authorities must be defined and communicated.

7.3.2.2 Competence, training, and awareness

Persons performing tasks that have or can result in significant environmental impact or that relate to legal and other compliance requirements must receive appropriate training to ensure they are competent in performing their tasks. An EMS procedure must be in place to ensure that such persons are aware of the need to comply with all EMS requirements and what they specifically must do. This requirement can help ensure that the EIA decision is correctly, effectively, and safely implemented.

7.3.2.3 Communications

The EMS must include internal and external communication procedures. ISO 14001 only requires that procedures be established and allows an organization to decide for itself the degree of openness and disclosure of information to the public for such procedures.

7.3.2.4 Operational control

Operational controls must be established to ensure that critical functions related to the policy, significant aspects, objectives and targets, and legal and other requirements are properly identified.

7.4 ISO 14001 certification vs. self-declaration

To obtain ISO 14001 certification, an organization must be externally and independently audited by an accredited auditing body. Certification auditors must be accredited by the International Registrar of Certification Auditors. Once an organization has been independently audited and certified as conforming to ISO 14001 requirements, it may publicly declare that it is "ISO 14001 certified" or "ISO 14001 registered."

Although certification is usually a benefit for large companies or organizations operating in international markets, smaller organizations may view the cost and time required for achieving certification as prohibitive. In such cases, the organization may choose to perform an internal audit and *self-declare* that their EMS meets ISO 14001 requirements.

7.4.1 Benefits of ISO 14001 certification

Table 7.5 summarizes some of the principal benefits of an ISO 14001 certification. A manager who is "too busy managing the business" to consider the benefits that an EMS may bring to an organization could actually be costing the business plenty. Consider the lost opportunities for achieving benefits like those listed in Table 7.4, particularly in terms of reducing the number of accidents, the potential for large penalties, or even the potential for criminal liability such as jail time for a severe violation.

Table 7.5 Principal Benefits of ISO 14001 Certification

Enhanced public image—Organizations benefit from better communication concerning environmental issues inside and outside the organization. An EMS can provide an organization with an avenue for publicly advertising to business partners, regulatory agencies, and consumers that their organization is environmentally responsible.

Increased communications and DCS—The DCS captures important institutional knowledge. An EMS helps ensure that such critical operational information is documented, communicated, and retained. The cyclical nature of the EMS further ensures that such information is updated.

Reduced cost and increased profitability—An EMS can result in streamlined operations leading to monetary savings and increased profits through potential improvements in operational efficiency such as reduced materials usage, energy conservation, and reduced production of hazardous materials that are costly to manage and dispose of.

Enhanced safety and reduced number of accidents—There are safety benefits. A review of operational procedures, including a review of emergency preparedness and response measures, allows an organization to identify and implement safety improvements. This can reduce operating costs through potentially lower insurance rates.

Regulatory compliance—Reduces risk of environmental liability. An EMS helps demonstrate that the organization's systems are complying with environmental laws and regulations. Organizations that implement an EMS often report improved relations with government regulatory agencies, and report that regulators are quicker to provide technical support. Many organizations also report that the regulators are more supportive. This can result in fewer surveillance visits by regulators.

7.5 Complementary benefits of integrating sustainability with a consolidated NEPA/ISO 14001 EMS

As described in this section, an ISO 14001 EMS also provides an ideal mechanism for integrating a sustainable development program into an organization's functions. Moreover, an EIA process such as NEPA provides an ideal tool for identifying and assessing environmental impacts that can then be managed using an EMS. This section builds on the EMS concepts presented in Section 7.4 by describing the complementary benefits that exist among an EMS, an EIA, and the global environmental policy goal of sustainable development. This section provides the basis for the integrated EMS/EIA/sustainable development process described in Section 7.6. The intent of this section is not to repeat the EMS concepts presented in Section 7.4, but to emphasize the similarities, differences, and general complementary nature that exist between an EMS and EIA process.

An EIA, an EMS, and the goal of sustainable development provide three separate and independent approaches for protecting the environment. An EIA process, such as NEPA, provides a scientifically based process for rigorously and objectively evaluating alternatives to a proposal or plan. In contrast, an EMS provides an ideal system for implementing and monitoring an agency's EIA plan and final decision. This section explains why an EMS and EIA share many common features and shows that the weaknesses of one process frequently tend to be counterbalanced by the strengths of the other. Properly combined, an integrated EIA/EMS system provides an efficient mechanism for evaluating and implementing agency actions. An approach for integrating these two processes has been published in the companion text *NEPA and Environmental Planning.*[3]

Meanwhile, the goal of sustainable development is gaining international attention, yet it lacks a general-purpose system for identifying, evaluating, and implementing a sustainable development plan. This section further expands upon the earlier concept, by describing a fourth-generation system that uses an integrated EIA/EMS system to develop and implement a sustainable plan or program.* The advantage of this consolidated process is that it draws from the synergistic strengths of an integrated EIA/EMS system to identify, plan, evaluate, and implement sustainable measures for proposed plans, projects, or programs.

7.5.1 Historical development of the integrated EIA/EMS

At the request of Dr. James Roberts (former president of the National Association of Environmental Professionals [NAEP]), the author was asked to prepare a report in 1997 investigating the commonalities, strengths, and weaknesses that existed between an integrated NEPA and an ISO 14001–consistent EMS. This effort was in support of the U.S. Council on Environmental Quality's (CEQ's) Improving NEPA Effectiveness Initiative. A final report discussing the synergistic relationship that existed between an integrated NEPA/EMS process was issued to the NAEP in 1997.† The report proposed a system for integrating NEPA with an ISO 14001 EMS, including a detailed analysis of the complementary relationship between NEPA and an EMS. This approach was reviewed by over a dozen NEPA

---

* This section is based on an article published by the author: "Integrating sustainable development with a consolidated NEPA/ISO 14001 EMS, redux," *Journal of Environmental Practice* 2010, 12(1): 18–34.
† The original report, titled "A Conceptual Strategy for Integrating NEPA with an Environmental Management System," was prepared by the author in 1997 and issued by the president of the National Association of Environmental Professionals to the Council on Environmental Quality (CEQ) in 1999/2000.

and ISO 14001 specialists from around the United States and was presented at national environmental conferences. Beginning in the year 2000, the author began teaching workshops at NAEP conferences that outlined this process. Comments received from these workshops were incorporated to improve the integrated system.

The integrated NEPA/ISO 14001 approach was published in the first of several papers beginning in 1998.* With the assistance of Ron Deverman (current president of the NAEP), this report was reviewed and approved by the NAEP board of directors in the year 2000. The NAEP president issued the final report to the CEQ with a recommendation that it be promoted to all U.S. federal agencies. The reader is referred to the author's text, *NEPA and Environmental Planning*, for a detailed explanation of how NEPA and an ISO 14001 EMS can be effectively integrated into a single complementary system.[4]

In 2002, this approach was generalized to describe a process for integrating any EIA process with an ISO 14001–consistent EMS.[†] Later still, this approach was expanded to incorporate adaptive management (AM).[‡] Eventually, the CEQ issued guidance for integrating NEPA with an ISO 14001 EMS. Some of the key advantages of an integrated NEPA/EMS system are depicted in Table 7.6.

Table 7.6 Key Advantages of an Integrated EIA/EMS

Development of an internal policy and plan, which meets the expectations of both an EIA and an EMS
Enhanced planning, consideration, and analysis of environmental aspects that can adversely impact the environment
Inclusion of robust EIA procedures with the process of identifying the environmental aspects under an EMS
Integration of EIA documents and schedules into EMS objectives and targets
Incorporation of the EIA documents and administrative record into an EMS records management system
Incorporation of EIA mitigation commitments with other related regulatory requirements and EMS objectives and targets
Monitoring of the selected EIA alternative (e.g., course of action) and implementation of applicable mitigation and/or monitoring commitments

* The original report to the NAEP was published as a lead article: "A strategy for integrating NEPA with an EMS and ISO-14000," *Environmental Quality Management Journal* (New York: John Wiley & Sons Inc. Spring 1998).

† The paper was published as the lead journal article: C. H. Eccleston and R. Smythe. Summer 2002. "Integrating environmental impact assessment with environmental management systems," *Environmental Quality Management Journal*, John Wiley & Sons Inc. 11(4).

‡ See the author's companion text: C. H. Eccleston, *NEPA and Environmental Planning: Tools, Techniques, and Approaches for Practitioners* (Boca Raton, FL: CRC Press, 2008). 47–52.

Table 7.7 Unique and Essential Functions Contributed by EIA, EMS,
and Sustainability to an Integrated Environmental Planning
and Implementation Process

Component	Function
EIA	Planning and assessment process: An EIA process such as NEPA provides a robust, comprehensive, and general-purpose environmental planning process that can be used to evaluate the impacts of and alternatives to proposed actions.
Sustainability	Environmental goals: Provide an overarching and unifying environmental goal applicable to most, if not all, programs and projects.
ISO 14001 (consistent) EMS	Management systems: Provide an internationally accepted system for managing environmental policies, procedures, and requirements.

7.5.2   *Integrating an EIA process, an EMS, and sustainable development*

This section further expands the aforementioned concepts by generalizing the process to address and incorporate sustainable development into a synergistic EIA/EMS process. Before describing how an EIA process such as NEPA can be integrated with an EMS and the concept of sustainability, it is instructive to explore the fundamental function of these three distinct environmental elements or processes (i.e., EIA, EMS, and sustainable development). As depicted in Table 7.7, each of the three environmental components contributes a unique and essential function to the integrated system.

To simplify the approach, this section does not describe how this process can be integrated with AM. For an in-depth explanation of how to integrate AM with a consolidated EIA/ISO 14001 system, the reader is referred to the text *NEPA and Environmental Planning*.[5]

7.5.3   *How an EIA process and an EMS complement each other*

Chapter 5 describes the principles for an EIA process such as NEPA. Table 7.8 compares and contrasts some of the principal strengths and weaknesses of EIA, EMS, and sustainability. As depicted in the table, EIA, EMS, and sustainable development each possesses inherent strengths and weaknesses; moreover, a weakness in one component of an integrated

Table 7.8 How an Integrated System Consisting of EIA, Sustainability, and an ISO 14001–Consistent EMS Complement Each Other

Characteristics	Sustainability	EIA	EMS
Goal	Achieve and maintain a sustainable system or component of a system.	The goal of an EIA is to provide environment protection by ensuring that environmental impacts are considered during the early decision-making process.	The goal of an ISO 14001–consistent EMS is to provide a system for managing actions (environmental aspects) that affect the environment. Its continual-improvement system can further help in reducing environmental aspects.
Environmental policy	The EIA/EMS environmental policy can be developed to incorporate elements of sustainability.	NEPA's policy goals (highlighted in Section 101 of NEPA) provide a high-level commitment to protect the environment. For instance, NEPA regulations state that NEPA analyses should be prepared for new federal policies that may significantly affect environment quality. Consistent with such guidance, an EIA can be prepared to develop an environmental and sustainable development policy.	In conjunction with the EIA process, the EMS must state its commitment to environmental protection and compliance.
Substantive mandate	A sustainability plan provides a goal or direction for achieving substantive and sustainable environmental performance.	Under an EIA process such as NEPA, impacts, alternatives, and mitigation measures must be rigorously investigated to identify actions and alternatives that can protect the environment. However, most EIA processes lack a legally binding substantive mandate to choose an alternative that protects the environment.	Under an EMS, substantive actions are "expected" to be taken, which lead to continual improvement in environmental performance (and thus environmental protection). Targets and objectives also provide tangible criteria for measuring the success in improving environmental performance.

Planning function	The EIA planning process provides a mechanism for defining and assessing the effectiveness of potential sustainable development plans.	An EIA provides a rigorous and comprehensive environmental planning process (sustainable development plan), but lacks an environmental system for ensuring that planning decisions are properly executed.	Requires a planning function, and provides a system for ensuring that the plan is appropriately implemented. However, ISO 14001 does not prescribe a detailed process (like that in an EIA) for performing the planning function.
Impact assessment requirements	An EIA planning process can be used to assess the effectiveness of a sustainability plan, and can be used to discriminate between alternative sustainability plans.	Most EIA processes specify detailed direction for performing an analysis of direct, indirect, and cumulative impacts.	An EMS must identify environmental aspects or actions that can impact the environment. However, little specificity is provided regarding the requirements for performing this investigation. Moreover, the assessment of environmental aspects is generally much less rigorous than most EIA processes such as NEPA's requirement to assess environmental effects.
Objectives and targets	An integrated EIA/EMS system can be used to establish objectives and targets for a sustainability plan.	The EIA analysis can be used to identify, access, and choose objectives and targets.	Under an EMS, an organization is expected to adopt environmental objectives and targets to address significant aspects.
Significance	An EIA can be used to identify, evaluate, and focus on significant sustainability issues.	Most EIA processes have detailed direction for determining the significance of an impact. For instance, in addition to context, 10 specific factors are detailed in the NEPA, implementing regulations for assessing significance.	Unlike most EIA processes, ISO 14001 lacks detailed direction for interpreting or determining the meaning of significance.

(Continued)

Table 7.8 How an Integrated System Consisting of EIA, Sustainability, and an ISO 14001–Consistent EMS Complement Each Other (*Continued*)

Characteristics	Sustainability	EIA	EMS
External input	An EIA provides a mechanism for the public to provide input in developing a sustainability plan.	Most EIA processes have well-defined public participation procedures; they specify a detailed public participation process and a formal public "scoping process" for identifying actions, impacts, and alternatives (sustainability plan) and for eliminating nonsignificant issues from further review.	An EMS simply requires that procedure (not necessarily public) be used to record and respond to external parties; however, ISO 14001 does not prescribe detailed requirements for accomplishing this task.
Other environmental requirements	An integrated EIA/EMS system provides a means for identifying and incorporating environmental requirements into a sustainable development plan.	Most EIA processes have extensive direction for performing the analysis. For instance, CEQ guidance and executive orders direct federal agencies to integrate P2 measures, environmental justice, biodiversity, and other considerations with NEPA.	A top-level environmental policy is required, including a commitment to P2, which is very broadly defined.
Mitigation	Under most EIA processes, mitigation measures can support sustainable development measures whereas an EMS provides a mechanism for implementing such measures.	Most EIA processes require that mitigation measures be identified and analyzed, but many do not require that such measures be implemented.	An EMS provides a system that can be used to ensure mitigation measures are properly executed.

Emergency preparedness	By itself, there is no built-in regulatory mechanism to ensure that a sustainable development plan can address potential emergency situations; however, an integrated EIA/EMS system provides a mechanism for doing so and mitigating the risk of potential incidents in a way that is consistent with a sustainability plan.	An EIA process can provide a rigorous planning mechanism for identifying potential incidents, and assessing the impacts, alternatives, and measures for mitigating potential threats.	EMS procedures must provide measures for preventing and responding to emergencies.
Nonconformity, and preventive and corrective action	By itself, there is no built-in regulatory mechanism to ensure that a sustainability plan is correctly implemented; however, an EMS provides such a mechanism. An EMS can also include an adaptive process for improving the implementation of a sustainability plan. For instance, NEPA's concept of AM provides an efficient corrective-action mechanism for dealing with uncertainty or changing circumstances.	Under most EIA processes, organizations are responsible for ensuring that decisions and commitments are carried out. However, many EIA processes such as NEPA lack a rigorous system or procedure for ensuring such compliance once the EIA process has ended. However, a NEPA AM system can provide an effective management process for implementing corrective actions as a result of new information or changing circumstances.	An EMS must include procedures for identifying and correcting nonconformance. ISO 14001 states specific procedures that can be used to (1) identify circumstances in which EIA commitments or mitigation measures are being incorrectly implemented, (2) correct nonconformities, (3) mitigate the impacts of nonconformities, and (4) develop plans for avoiding nonconformities.

(Continued)

Table 7.8 How an Integrated System Consisting of EIA, Sustainability, and an ISO 14001–Consistent EMS Complement Each Other (*Continued*)

Characteristics	Sustainability	EIA	EMS
Records and documentation	An EMS provides a mechanism for managing a sustainable development plan and other important records.	Many nations such as the United States require NEPA documents to be maintained as part of the administrative record. However, most EIA processes do not specify how such a system should be maintained. An EMS can be used for maintaining EIA records.	An EMS specifies detailed procedures for controlling and maintaining records needed to demonstrate conformance with the EMS standard.
Monitoring	An EMS provides a mechanism for monitoring the progress of a sustainability plan.	Many EIA processes such as NEPA encourage (and sometimes require) postmonitoring measures. However, little direction is provided in terms of how monitoring should be performed.	Monitoring is mandated as part of the EMS continual-improvement cycle. Specific direction is provided on how this element is to be performed.
Continual improvement	An EMS provides a mechanism for continually improving the implementation of the sustainability plan.	Most EIA processes provide no direction for performing a continuous improvement process. However, under NEPA, the CEQ has promoted a "cyclical" process known as adaptive management.	A continual-improvement process is a basic concept inherent in an EMS.

Audits	An EMS provides a system for auditing the success of the sustainable development program and compliance with a sustainable development program. It also ensures that the project is implemented according to the selected course of action (EIA process), EIA impacts remain within designated parameters, and mitigation measures are correctly implemented so as to promote sustainable commitments.	Most EIA processes lack a well-defined auditing process. However, EIA conformance and commitments may be efficiently reviewed and audited where such commitments are linked to EMS objectives and targets. An EMS audit provides a means for ensuring that the EIA process and commitments are correctly implemented.	ISO 14001 defines specific internal auditing requirements for periodically assessing conformity with the EMS; the audit results must be presented to management for review.
Management review	ISO 14001 requires top management to review the organization's EMS progress; this could also include progress in implementing the sus-tainable development plan.	Most EIA processes require the responsible decision maker to review the EIA document and choose a course of action. Beyond this direction, there is often no requirement that management should periodically review the implementation of the selected course of action.	Under ISO 14001, top management is required to periodically review the organization's progress in meeting EMS requirements.

system often tends to be offset by the strengths of one of the others. Some of the succinct characteristics outlined in Table 7.8 will now be described.

7.5.3.1   Developing policies and plans

Many EIA processes, including NEPA, recognize four broad categories of activities (i.e., policies, programs, projects, and plans) as being potentially subject to a detailed impact and alternatives analysis. For instance, under NEPA, establishment of federal policies and plans are "actions" potentially subject to a full NEPA assessment. Thus, policies and plans established as part of an EMS may potentially be subject to EIA requirements, particularly in cases where a policy or plan entails potentially significant environmental impacts or issues.

Although an environmental planning function is a mandatory element within an EMS, the ISO 14001 standard provides only limited specifications for performing the planning function. For example, specific procedures and requirements regarding scoping, investigating environmental aspects, defining temporal and spatial bounds, interpreting significance, and other requirements, are at present only vaguely inferred or defined.

In contrast, most EIA processes provide highly prescriptive direction and requirements for ensuring that an accurate and scientifically defensible planning and analysis process is followed, which provides decision makers with sufficient information to reach an informed decision. This can also include investigating, analyzing, and comparing alternative sustainability plans. Moreover, these requirements are in many cases reinforced by decades of experience gained by engaging in diverse missions and environmental issues. Properly integrated, a combined EIA/EMS system can provide a synergistic process for planning sustainable actions and implementing decisions in a manner that protects and enhances environmental quality and sustainability, while reducing cost, generation of pollutants, and consumption of strategic resources.

7.5.3.2   Substantive vs. procedural requirements

As described, most EIA processes such as NEPA are not obligated to select an environmentally preferable alternative or to demonstrate that their decision conforms to the environmental goals (i.e., *substantive* mandates) such as those established in Section 101 of NEPA. Thus, the EIA's contribution is derived not from a substantive mandate to choose an environmentally beneficial or sustainable alternative but from its *procedural* provisions, which require agencies to rigorously evaluate and seriously consider the impacts of potential actions in their final decision, just as they would balance other more traditional factors such as cost and schedules.

In contrast, an ISO 14001–consistent EMS involves a general "expectation" that some type of substantive action(s) will be taken to improve environmental quality. Not only are environmentally beneficial actions *presumed* to be taken, but they are also implemented as part of a cycle of continual improvement in environmental management practices. To this end, an EMS could provide a mechanism for enacting some of the substantive environmental mandate that most EIA processes lack.

Similarly, most EIA processes require analysis of mitigation measures but place no substantive burden on decision makers to choose or enact such measures. ISO 14001, in contrast, requires organizations to establish objectives (goals) for improving environmental performance. Similarly, environmental targets are established for measuring and achieving those objectives.

Achieving these objectives could involve implementing actions similar to NEPA's mitigation measures. Again, most EIA processes prescribe rigorous requirements for planning and investigating mitigation measures, whereas an EMS provides a mechanism for implementing such measures. An integrated EIA/EMS system could be used to continually improve upon the implementation of an adopted sustainability plan.

7.5.3.3 Analysis requirements

Most EIA processes such as NEPA have regulations that provide practitioners with highly prescriptive requirements for ensuring that an accurate and defensible analysis is performed, and they provide a decision maker with sufficient information to support informed decision making. Most EIA processes are more demanding than ISO 14001, requiring not only identification of environmental aspects but a comprehensive analysis of the actual direct, indirect, and cumulative impacts on environmental resources.

As described, NEPA (and some other EIA processes) practice is reinforced by more than four decades of experience accumulated by a diverse range of federal agencies, each with its own mission and often unique environmental issues. EIAs provide a more comprehensive and rigorous planning process than ISO 14001 for ensuring that environmental impacts are identified, evaluated, and considered before a decision is made to pursue an action. The EIA analysis can be used to evaluate and compare the advantages and disadvantages of, as well as compare and discriminate between, various sustainable development plans.

7.5.3.4 Assessing significance

Both EIAs and ISO 14001 specify requirements for assessing significance. The concept of significance permeates most EIA processes, such as NEPA's regulatory provisions. For instance, the NEPA regulations include a detailed definition of significance, in addition to context, and provide 10 specific factors that decision makers are required to consider while making such determinations (40 CFR §1508.27).

In contrast, under ISO 14001, significance is defined vaguely and the standard contains little direction for use in reaching a significance determination. Again, most EIA processes bring many years of experience to bear on the problem of how best to determine significance. This EIA analysis can be used to evaluate and discriminate in terms of significance and the advantages and disadvantages of various sustainability plans.

7.5.3.5   Public involvement

Public input, review, and participation are elements essential to nearly every EIA process. For instance, NEPA's public scoping process is designed to solicit comments from the public, potentially affected parties, government agencies with special expertise, and subject matter experts. The public is consulted with respect to the scope of the NEPA analysis; the federal agency must also allow the public to review and submit comments on the NEPA analysis.

In contrast, ISO 14001 provides no requirement for public scoping and participation, but only a requirement to develop a plan (not necessarily public) for external communications and inquiries. Lack of such requirements can be viewed as a weakness in many parts of the ISO 14001 standard. Thus, the EIA experience combined with transparency, public participation, and scoping helps to balance the weaknesses of an EMS. An integrated system can facilitate the public's ability to shape and participate in the development and implementation of a sustainability program.

7.5.3.6   Incorporating pollution prevention measures

CEQ has issued guidance indicating that, where appropriate, P2 measures are to be coordinated with and included in the scope of a NEPA analysis.[6] Other EIA processes also have issued similar directives. ISO 14001 speaks to the merits of P2, but primarily from the standpoint of establishing a top-level policy committed to P2. ISO 14001 provides a top-down policy for ensuring that P2 is actually incorporated at the operational level.

In comparison, most EIA processes provide an ideal framework for planning and evaluating the effectiveness of a comprehensive P2 strategy or plan. An integrated system may help facilitate the development of a sustainability plan that reduces pollutants, while encouraging recycling and other beneficial environmental practices. The author's companion book, *NEPA and Environmental Planning*, provides direction for combining P2 with an integrated NEPA/EMS.*

7.5.3.7   Incorporating other environmental requirements

To the extent feasible, federal agencies are instructed to integrate NEPA with other environmental reviews (e.g., regulatory requirements, permits, agreements, project planning, and policies) so that procedures run concurrently

---

* See the author's companion text: C. H. Eccleston, *NEPA and Environmental Planning: Tools, Techniques, and Approaches for Practitioners* (Boca Raton, FL: CRC Press, 2008). 47–52.

rather than consecutively; this requirement reduces duplication of effort, delays in compliance, and minimizes the overall cost of environmental protection.[7] Specifically, NEPA requires federal agencies to

- *Identify* other environmental review and consultation requirements... prepare other required analyses and studies *concurrently* with, and *integrated* with, the environmental impact statement ... (40 CFR 1501.7[a][6] emphasis added)
- *Integrate* the requirements of NEPA with other *planning* and environmental review procedures ... (§1500.2[c], emphasis added)
- [combine] any environmental document in compliance with NEPA... with any other agency document ... (§1506.4, emphasis added)

Consistent with NEPA's regulatory direction, ISO 14001 expects organizations to identify applicable legal and regulatory requirements. The intent is to ensure that organization activities meet applicable legal and regulatory requirements. To this end, an EMS provides a system that can identify regulatory and other requirements and ensures that these requirements are incorporated into an integrated EIA/EMS process. As detailed in the three aforementioned regulatory provisions, a combined EIA/EMS system can be used to develop and implement a sustainability program.

7.6 Developing an integrated EIA/EMS/ sustainable development process

As shown in Figure 7.2, the EIA planning process is generally triggered through the identification of a need for a proposed action (i.e., a new policy, program, plan, or project). The EMS is capable of managing a full range of construction and operational activities (i.e., proposals, operations, and services). To develop an efficient and effectively integrated system, EIA, EMS, and sustainability practitioners must collaborate closely with one another. By establishing applicable environmental objectives and targets, an EMS can help ensure environmental protection and implementation of sustainability commitments and mitigation measures by means of a monitoring program.

Conceptually, the consolidated system (see Figure 7.2) is composed of three integrated stages or phases:

Phase 1: EIA/EMS and sustainable development policy phase
Phase 2: EIA phase
Phase 3: EMS phase

Under Phase 1, the EIA/EMS processes are used to develop a high-level environmental and sustainability policy (Figure 7.2, blocks 1 and 7).

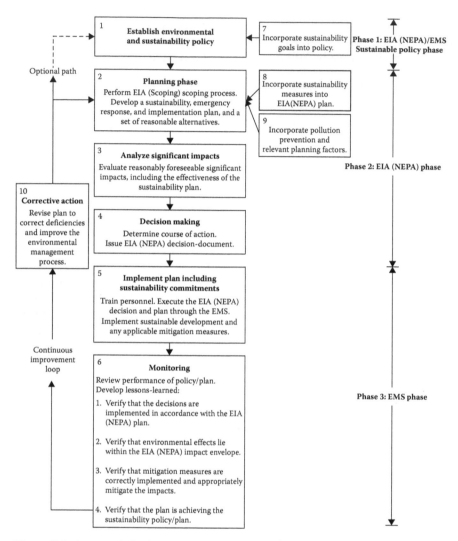

Figure 7.2 Approach for incorporating sustainability development commitments into an integrated EIA/NEPA or EMS process.

In some instances, this could be accomplished by preparing an EA that investigates and compares the impacts and effectiveness of an environmental and sustainable development policy. Under Phase 2, the EIA process is used to identify, plan, evaluate, and compare alternatives. This phase is also used to select a course of action, including a sustainable development plan (Figure 7.2; blocks 2, 3, 4, 8, and 9).

Under Phase 3 (Figure 7.2, blocks 5, 6, and 10, as well as the continuous improvement loop), the EMS is used to implement and effectively manage the course of action selected during Phase 2. EMS elements would be used to monitor, audit, and continuously improve the course of action, including a sustainability plan, that was selected in Phase 2. The principal components of this integrated system are described next.

7.6.1 Policy

The integrated system described in Figure 7.2 (Block 1) is initiated with the establishment of a high-level environmental policy, including a commitment to environmental quality and sustainable practices. The EIA process is used to scope and define the environmental and sustainability policy. The policy will also include a commitment to P2. EMS procedures will be followed in making this policy publicly available.

The sustainable development policy will vary with the context of the proposal under consideration. For example, potential sustainable development policies for a proposal, which affects a biological system or a renewable resource, might involve one or more of the following concepts:

- Managing water resources so that they do not exceed the safe (replenishable) yield
- Sustainably managing the harvest of a forest
- Maintaining and enhancing biological diversity
- Preserving and rehabilitating soil resources

The reader should note that developing a sustainable policy for a facility, city, or nonrenewable resource may require special consideration and forethought with respect to developing policies and plans.

Note that the concept of sustainable practices varies with the context of resources. Sustainable practices may need to be applied differently to problems involving nonrenewable resources as compared to the way they are applied to renewable resources. For instance, consider this sustainable development example: "The program will strive to increase the longevity of nonrenewable sources of silver by ensuring that minerals are mined and used in acceptable ways, both economically and environmentally."

Many sustainability specialists would question if the aforementioned statement strictly meets the underlying goal of sustainability. It could be argued that mining any ore body is an unsustainable practice. This is because mining cannot continue indefinitely and is therefore incompatible with a strict definition of sustainability. An alternative approach might involve developing a sustainable policy or plan that emphasizes the substitution with renewable resources or recycling of nonrenewable resources. A second approach might involve developing a policy or plan in which mining is performed in a manner designed to ensure that environment quality is sustainable over time.

7.6.2   Planning

The EIA planning phase (Figure 7.2, block 2) essentially begins with a formal public scoping process. The EIA scoping process is used to obtain public input and separate potentially significant issues from those that are nonsignificant. The scoping process identifies the scope of actions, alternatives, potentially significant environmental aspects (which will then be "mapped" into environmental impacts for detailed analysis), and mitigation measures. If appropriate, this analysis can also include development of an emergency response plan (ISO 14001 requirement), thus satisfying an important EMS requirement.

With respect to sustainability, this phase is also used to develop specific measures that can be taken to support the environmental and sustainability policy (block 1). The final result is a detailed EIA plan (description of the proposed action and alternatives), which also includes specific measures that would be taken to achieve the stated sustainability policy. Consistent with an EIA process, different scenarios and alternatives are investigated. For example, under NEPA, this detailed plan might involve preparing a description of the proposed action and alternatives. As applicable, the plan should also include measures for controlling or reducing pollution.

Figure 7.3 shows some of the factors (e.g., public input, technologies, and cost versus benefits) that are considered while developing an integrated environmental or sustainability plan. Once a proposed sustainability plan is prepared, it should be reviewed to ensure that it is consistent with other policies including laws and regulations. If it is not consistent, the plan should be revised so that it becomes consistent (see the loop in Figure 7.3).

7.6.2.1   Determining the appropriate sustainability
scale and context

Virtually all human activities influence sustainability to some degree. With respect to the planning function, sustainable goals can be studied over an array of different time periods, contexts, and scales (levels or frames of reference) of environmental, economic, and social organization.[8] This focus can range from the global context down to the sustainability of a national program, city, community, facility, or an individual project. The context can also range from settings as varied as entire ecosystems, forests, oceans, or agricultural areas. During the EIA scoping process, it is essential that the planning team choose the appropriate context and scale for the sustainability study.

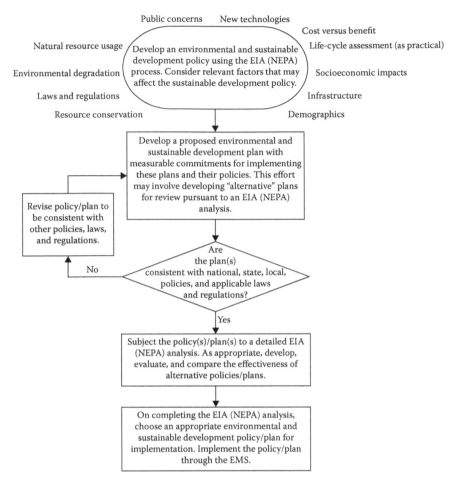

Figure 7.3 Integrating a sustainability development plan with the EIA/NEPA or EMS process.

7.6.2.1.1 Measuring sustainability As part of the aforementioned effort, the EIA analysis should define criteria for instituting sustainability practices. Environmental objectives are documented in the sustainability plan. For instance, depending on the scope and nature of the action (policy, program, plan, or project), potential environmental objectives might include the preservation or maintenance of one or more of the following resource elements:

- Biota (e.g., minimizing impacts, achieving a "carrying capacity," and promoting sustainable yields)
- Land usage (e.g., achieving sustainable development or minimizing urban sprawl)

- Energy (e.g., conservation of nonrenewable and renewable energy sources)
- Natural resources (e.g., conservation of nonrenewable and renewable natural sources)
- Water (e.g., conservation and recycling programs)
- Materials (e.g., conservation and recycling of nonrenewable and renewable materials)
- Food (e.g., sustainable agricultural practices)
- Waste (e.g., recycling, P2, and waste minimization)

A large number of sustainability indicators, metrics, indexes, and benchmarks have been established.[9] These include environmental, economic, and social measures (separately as well as integrated together) over many contexts, as well as spatial and temporal scales. Measurable environmental targets would then be defined to monitor progress in achieving these objectives.

7.6.3   *Analysis, significance, and decision making*

The proposed plan is then subject to an EIA analysis that evaluates the impacts, effectiveness, and various alternatives, including sustainable development options. For instance, under NEPA, an EA or environmental impact statement would be prepared to assess the potential impacts and compare the alternatives to the proposal (Figure 7.2, block 3).

The analysis also investigates the beneficial and adverse impacts associated with the sustainability plan, as well as the probable effectiveness of these planned activities. Such impacts should be investigated in terms of the three pillars of sustainability: environmental, economic, and social factors. Most EIA processes, such as NEPA, define significance. For instance, in addition to the environmental context, NEPA provides 10 significance factors (40 CFR §1508.27[b]) for assessing significance and reaching a final decision. The agency's responsible decision maker then reviews the proposal (including the sustainability plan) and forecasted impacts and reaches a final decision regarding the course of action, including a sustainable development plan (Figure 7.2, block 4).

7.6.4   *Implementation*

Once a final decision has been reached using the EIA process, the course of action, including mitigation and sustainable development measures, can be implemented through the EMS (Figure 7.2, block 5). ISO 14001 ensures that key personnel (with environmental responsibilities) receive and possess job-appropriate experience or training.

7.6.5   Monitoring, enforcement, and corrective-action phase

Depending on the size, complexity, and scope of the proposal, a centralized oversight office might be assigned the responsibility of implementing the proposal and monitoring compliance (Figure 7.2, block 6). As appropriate, an environmental compliance officer (or the equivalent) should be assigned the responsibility of preparing and transmitting input and status reports to the oversight office.

Monitoring data are evaluated to verify compliance with established policies, plans, and the agency's decision. As appropriate, the organizational policy or plan is revised to correct deficiencies (Figure 7.2, block 10; see the continuous improvement loop branching to the box labeled "Corrective Action"). As dictated by the results of the EMS monitoring and auditing steps, changes may need to be made to the methods used for managing and implementing the EIA decision and sustainable commitments.

Ultimately, the ISO 14001 expectation is that environmental aspects (impacts) will dissipate with time, such that subsequent revised plans might address issues different from those addressed by the existing plan. Such a process ensures a continuous improvement cycle, which is the hallmark of an EMS, and also promotes the EIA policy goals and paradigms, such as an integrated NEPA/AM process.

7.7   Summary

In summary, an integrated EIA/EMS system provides an ideal system for scoping, evaluating, and developing a sustainable plan or program. Once a sustainable plan or program has been developed as part of the EIA process, the EMS element provides a particularly effective mechanism for implementing the agency's plan or program, ensuring that it meets the sustainable development criteria evaluated in the EIA plan. A properly integrated system can provide agencies with a synergistic process for protecting and preserving environmental quality.

Discussions, problems, and exercises

1. What does the term "ISO" mean?
2. What is the difference between ISO 14000 and 14001?
3. Do you believe that the ISO 90001 quality assurance system provides a good model for ISO 140001? Defend your answer.
4. Describe the basic elements of an ISO 14001 EMS.
5. Develop an environmental policy for your university, a local government agency, or a small business. Based on the information provided

in this chapter plus outside reading, develop an EMS for implementing this policy.
6. Develop a plan for integrating an EIA and EMS process with a sustainable development policy for your university, a local government agency, or other entity.

Notes

1. BSI Management Systems, http://www.bsi-emea.com/Environment/ CaseStudies/SamsungUSA.pdf (accessed September 16, 2010).
2. R. B. Pojasek, "Introducing ISO 14001 III," *Environmental Quality Management* 17, no. 1 (Autumn 2007): 75–82.
3. C. H. Eccleston, *NEPA and Environmental Planning: Tools, Techniques, and Approaches for Practitioners* (Boca Raton, FL: CRC Press, 2008), 38–47.
4. Ibid.
5. Ibid., 47–52.
6. CEQ, *Guidance on Pollution Prevention and the National Environmental Policy Act*, 58 FR 6478, (January 29, 1993).
7. CEQ, *Regulations for Implementing the Procedural Provisions of the National Environmental Policy Act*, 40 *Code of Federal Regulations*, pts. 1500–1508, (1978).
8. Conceptual Framework Working Group of the Millennium Ecosystem Assessment, "Ecosystems and Human Well-Being: A Framework for Assessment." (London: Island Press, 2003): Chapter 5 "Dealing with Scale,"107–124.
9. T. Hak and others, *Sustainability Indicators* (London: Island Press, 2007).

section three

---

*Environmental ethics
and economics*

chapter eight

Environmental ethics

Relativity applies to physics, not ethics.

Albert Einstein

Case study

The execution of Ken Saro-Wiwa and eight other Ogoni activists on November 10, 1995, raised a storm of outrage across the world.[1] Their deaths highlighted the suffering of the Ogoni people, one of the many marginalized ethnic groups living in the oil-rich delta of the Niger River. Ken Saro-Wiwa had fought for an end to the environmental damage that was turning his homeland into what he described as a "wasteland," endangering the people's health and livelihoods. Fifteen years later, oil spills still blacken the land and pollute the waterways. Hundreds of gas flares burn day and night, filling the sky with soot and fumes. The diverse communities living among the oil flares continue to live in extreme poverty—70% of the local populace subsists on less than $1 a day.

The military government that executed Ken Saro-Wiwa was replaced in 1999 by a civilian government. Yet government security forces are still killing people and razing communities with impunity. Human rights defenders and journalists, including foreign television crews, have been harassed, detained, and sometimes beaten for investigating oil spills or security violations. The federal government has rejected calls for independent and impartial inquiries into abuses by these forces, which operate under its direct control. As Eghare Ojhogar, chief of the Ugborodo community in the Niger Delta, states:

> It is like paradise and hell. They have everything.
> We have nothing ... If we protest, they send soldiers.
> They sign agreements with us and then ignore us.
> We have graduates going hungry, without jobs. And
> they bring people from Lagos to work here.

Oil revenues account for over 98% of Nigeria's foreign exchange earnings. However, little of this wealth is distributed within the Niger Delta, or to the Nigerian people as a whole. Economic and social rights, such as the right to health and the right to an adequate standard of living, remain unfulfilled for many Nigerians.

Thousands died in mass killings when conflict erupted over the control of oil in the late 1990s and again in 2003 and 2004. Growing numbers of human rights activists within the Niger Delta hold the oil companies responsible for human rights violations by Nigerian security forces—both those connected to oil operations and those committed to protecting oil interests.

In many regions of the world, indigenous peoples face deeply entrenched discriminatory laws and policies. Because they are denied adequate protection of their right to live on and use the lands vital to their cultural identity and daily survival, indigenous communities are often driven to extreme poverty and ill health. Amnesty International is working alongside indigenous peoples' movements from around the globe to help advance effective international human rights standards to address their specific needs and circumstances. This includes calling for the immediate adoption of the United Nations (UN) Declaration on the Rights of Indigenous Peoples.

> **Problem:** After reading the chapter, return to this problem. Consider this example in light of what you have read. This problem may be pursued either as an individual assignment or as a class project. Also search on the Internet for "Nigeria oil conflict," and read at least two relevant citations in addition to the one from Amnesty International.
>
> What are the ethical principles at issue? What can modern society learn from the lesson of Nigeria's oil development? What are the root causes of the violence? Which parties are ultimately responsible? To what extent do you believe the lessons of Nigeria are applicable today? What are the ethical policy options for the Nigerian government, the oil companies, and the international financial organizations that support economic development in Nigeria and elsewhere in Africa and the world? In light of these questions and the essay by Segun Ogungbemi discussed in Section 8.3.1, outline the policy approaches that could have avoided the violence and environmental destruction.

Learning objectives

Understand how ethical issues directly and indirectly affect the human and environmental impacts of an economic development program. After you have read the chapter, you should understand the following:

- The nature of ethical issues
- The lapses in ethical performance that may occur among the various stakeholders
- How business, social, and environmental ethical issues are interrelated

8.1   Can, may, and should we?

Virtually every new advancement, every major technological break-through, has resulted in beneficial and adverse effects. Take fire and nuclear energy for instance. Fire has been called the single greatest dis-covery in human history. Using it, we stay warm at night, cook our food, forge our steel, and power electrical plants that electrify the world. But fire also has its darker side—it has been used to burn entire cities to the ground. Take nuclear energy: it can power a city, an entire nation for that matter; it can also be used to annihilate life from the face of the Earth. And so it will be with genetic engineering, nanotechnology, artificial intelli-gence, and every other major technological revolution that lies just around the corner.

Some of the technological breakthroughs described in this book, which we will witness in our lifetimes, may prove to be so powerful and seductive that we will be hard-pressed to resist the temptation to exploit them. But, as we know all too well, a seductive mistress might just turn out to be a Trojan horse in disguise. And we all know what happened to the city of Troy after the citizens opened the gate and led that magnificent wooden horse in; they were warned, but no one paid any attention.

With this thought in mind, consider future economic growth, devel-opment, and technologies from the following three distinct perspectives:

1. *Scientist's perspective: Can it be done?* By nature, scientists are intel-lectually curious. They tend to focus not on moral or legal issues but simply on whether something is possible and how it can be achieved.
2. *Political and legal perspective: May it be done?* The political and legal perspective focuses not on moral questions or scientific curiosity but on policies and law. *May* and *how* should society move forward in a particular direction or with a new technology?
3. *Moral and ethical perspective: Should it be done?* While the scientific and political or legal arenas focus on specific issues germane to their pro-fession, neither tends to ask the question—*should* we pursue a par-ticular technology or course of action? Is it wise or *right* to pursue a given direction or technology simply because it is possible or does not violate the law?

8.1.1   Some philosophical considerations

Ethics has been simply defined as the discipline dealing with what is good and bad, and moral duty and obligation.[2] "Environmental ethics" has been defined as the discipline that studies the moral relationship of human beings to the living environment and also the value and moral status of the environment and its nonhuman contents.[3] The most prominent philosophies of environmental ethics are generally grounded on one of the following two fundamental and opposing perceptions of ultimate value:

- For many environmental community advocates, the global ecosystem and all its living plant and animal creatures, including humans, are of ultimate value.
- For most of society, including policymakers, the welfare of human beings is of ultimate value.

The former emphasizes the intrinsic value of all living things. Some advocates deny the utility of economic measures as they promote a wide variety of ideological and spiritual beliefs. However, as public policy advocates, they necessarily have a voice in policymaking and implementation.

The latter emphasizes the economic costs and benefits associated with meeting human needs and desires, and also seeks to quantify intrinsic values along with economic values. Nevertheless, policymaking often encompasses intrinsic values of humans and their environment apart from their economic measure. Garrett Hardin, who described the tragedy of the commons (see Chapter 2, Section 2.2), wrote the following:

> Ethical literature is almost totally individualistic: it is addressed to private conduct rather than to public policy. The standard ethical dialogue is between people who stand face to face with each other, seeking a reasonable basis for reciprocal altruism ... Contemporary philosophy still evades the hard problem of caring for posterity's interests.[4]

The tragedy of the commons points to a lack of adequate ethical dialogue on ecological environments at all scales of the public commons. Environmental ethics, as a field of study and public policy, is designed to fill that gap. Perhaps the most influential definition of environmental ethics is contained in the preamble to the National Environmental Policy Act:

> To declare a national policy which will encourage productive and enjoyable harmony between man and his environment; to promote efforts which will prevent or eliminate damage to the environment and biosphere and stimulate the health and welfare of man; to enrich the understanding of the ecological systems and natural resources important to the Nation.

The key phrase here is "harmony between man and his environment." This encompasses both economic and intrinsic values that affect human quality of life. Only in modern times have some policymakers come to understand that since the Industrial Revolution, the scale and intensity of economic development have adversely impacted the "the health and welfare of man" across most of our planet with devastating effects that are both economic and intrinsic.

8.1.2 Origin of ethics

The people of the Stone Age evolved ethical attitudes by applying their instincts for empathy and altruism to kin and allies in support of personal and group survival. At the same time, hunger, predators, and enemies produced emotional responses that led to the killing of other people and animals. The former motivated humanistic ethics, and the latter authoritarian ethics. Mark Hauser, a psychologist, argues that human ethics is grounded on an instinct for creating a "moral grammar"[5]:

> We evolved a moral instinct, a capacity that naturally grows within each child, designed to generate rapid judgments about what is morally right or wrong based on an unconscious grammar of action ... I argue that our moral facility is equipped with a universal moral grammar, a toolkit for building specific moral systems. Our moral instincts are immune to the explicitly articulated commandments handed down by religions and governments ... Just because we can consciously reason from explicit principles—handed down from parents, teachers, lawyers, or religious leaders—to judgments of right and wrong, doesn't mean that these principles are the source of our moral decisions. On the contrary, I argue that moral decisions are mediated by an unconscious process, a hidden moral grammar, that evaluates the causes and consequences of our own and other's actions. This account shifts the burden of evidence from a philosophy of morality to a science of morality.

Moral grammars employ religion and ideology to express the full range of possible ideas about the social order. At the heart of moral grammars are habits of social thought and personal responses programmed into our brains long before language and religion ever evolved. Like language and dreams, religious thought is a cognitive tool for expressing drives, emotions, and feelings in a variety of ways, including rules governing social relationships.

The earliest modern humans attributed events in nature to the actions of gods and spirits that had human-like motives that had to be taken seriously. The earliest ethical thoughts were expressed in terms of behaviors that the spirits desired. Thus, the earliest social rules governing human communities focused on people-to-people and people-to-spirit relationships. As human social order evolved, such rules progressively evolved into public policies for villages, city-states, and the great temple-states of the ancient world beginning with Egypt and Sumeria. Shamans and priests were the experts on the will of the spirits, and hence of societal ethical norms.

Plato and Aristotle were the first to formulate ethics directly in terms of the logic of human needs, although they did not deny the role of gods. For Plato, human well-being is the highest aim of moral thought and conduct; ethical virtues are requisite skills and character traits of human beings.[6]

Aristotle, on the other hand, viewed ethical theory as a field distinct from the theoretical sciences. According to him, its methodology must match its subject matter—good action—and must respect the fact that in this field many generalizations hold only for the most part. We study ethics in order to improve our lives, and therefore, its principal concern is the nature of human well-being. Aristotle follows Socrates and Plato in considering virtues to be central to a well-lived life. Like Plato, he regards ethical virtues (justice, courage, temperance, etc.) as complex, rational, and emotionally motivated social skills.[7] The views of Aristotle on ethics represent the prevailing opinions of educated and experienced men of his day. They are not like Plato's views, impregnated with mystical religion.[8]

Although the Old and the New Testaments lack the formal analysis of the ethics of these ancient Greek philosophers, its authors were deeply concerned with moral behavior, as well as with the human social order and the public policies of ancient Israel and Judah. Since the Council of Nicaea established Christianity as the official religion of the Roman Empire in AD 325, the Jewish notion that God defines ethical conduct in His scriptures has dominated the Western world. It was not until the seventeenth century that Baruch Spinoza, living in Holland, which tolerated ideas that the church had branded heresy, established a secular basis for ethical assessment. In many societies, secular ethical thought has ultimately led to democratic societies that resolve competing religious ideas with a common secular ethical foundation of state governance.

8.1.3   Challenge of environmental ethics

Let us assume that extinguishing natural forest fires, culling feral animals, or destroying some individual members of overpopulated indigenous species is necessary for the protection of the integrity of a certain ecosystem.[9]

- Will these actions be morally permissible or even required?
- Is it morally acceptable for farmers in nonindustrial countries to practice slash-and-burn techniques to clear areas for agriculture?

Consider a mining company that has performed open-pit mining in some previously unspoiled area.

- Does the company have a moral obligation to restore the landform and surface ecology?
- And what is the value of a humanly restored environment compared to the original natural environment?

It is often said to be morally wrong for human beings to pollute and destroy parts of the natural environment and to consume a huge proportion of the planet's natural resources.

- If that is wrong, is it simply because a sustainable environment is essential to (present and future) human well-being?
- Or, is such behavior also wrong because the natural environment and/or its various contents have certain values in their own right so that these values ought to be respected and protected in any case?

These are among the questions investigated by environmental ethics. Some of them are specific questions faced by individuals in particular circumstances, while others are more global questions faced by groups and communities. Still others are more abstract questions concerning the value and moral standing of the natural environment and its nonhuman components.

8.2 Judeo-Christian ethics and ecology

Pojman summarizes Lynn White's views on the Christian origins of our current ecological crisis[10]:

> White sets forth the thesis that the roots of our ecological crisis lie in the Judeo-Christian idea that humanity is to dominate nature. By seeing nature as alien, as a mere resource to be exploited, we have wreaked havoc on Earth and are reaping the consequences. Either a new religion or a revision of our old one is called for to get us out of this mess. He suggests that St. Francis of Assisi (1182–1226) is a proper example of a suitable attitude towards nature.

Pojman shares White's negative views regarding the Christian contribution to environmental ethics[11]:

> The limitation of the religious injunction is that it rests on authority, and we are not always sure of or in agreement with the credentials of the authority, nor on how the authority would rule in ambiguous or new cases. Since religion is not founded on reason but on revelation, you cannot use reason to convince someone who does not share your religious views that your view is the right one.

However, Pojman also cites the views of Patrick Dobel, a Christian ethicist[12]:

> Dobel disagrees with White's thesis that the Christian attitude toward nature is one of arrogance and dominance. He argues that the Judeo-Christian attitude is an ethic of stewardship and that humility toward God regarding nature, not arrogance, is enjoined by our religious heritage.

Dobel himself writes as follows[13]:

> The unique contribution that Christian ecology can make to the Earth is the assertion that we can insist on a reasonable harmony with our world without abandoning our commitment to social justice for all members of our unique and self-consciously alienated species. We can love and respect our environment without obliterating all ethical and technological distinctions, and without denying the demand that we cautiously but steadily use the Earth for the benefit of all humanity.

Dobel also maintains that the task of Christian ecology is to discover the enduring sacredness of nature's diverse being[14]:

- All cultures, regardless of religion, have abused or destroyed large areas of the world because of economic or population pressures or from simple ignorance.
- The ethical consequences of nature worship, neopantheism, and the militant assertion of the equality of all creaturehood pose grave problems for establishing any prior claims of worth or inherent dignity for human beings.

- A sort of mindless ecological imperative based on such notions is ultimately reactionary and antihuman as well as anti-Christian.

Pojman and White are among those who blame Western religion for the ills of our civilization, while Dobel and others credit religion for what civilization has achieved. In truth, environmental ethos has also varied in different societies and cultures around the world. For example, individuals from many other religions such as Islam have also imposed significant destruction on the environment. In the past, American Indian tribes have been cited as examples of cultures that live in harmony with nature. But on closer inspection, it is clear that many tribes did not have any special relationship with or regard for the environment, and some even participated in unsustainable practices such as heading hundreds or thousands of bison over cliffs, of which only a small number were actually eaten. Some archaeologists even blame the extinction of many north Mercian species such as mammoths on the unsustainable practices of some tribes.

The truth of the matter is that people use religion to express and act on their own understanding of the natural and social worlds. Before the advent of science historical Judaism, Christianity, and Islam were ignorant of the Earth's ultimate limits to support human life for future generations. This has clearly changed, as evidenced by the many specifically Christian organizations dedicated to the ethics of ecosystem sustainability. In fact, a Web search on "Christian environmentalism" yields several organizations like the Evangelical Environmental Network, Center for Theology and the Natural Sciences, Earth Ministry, and Christian Ecology Link.[15]

Edward Wilson, a biologist, believes that Christian environmental activism can make an important contribution to worldwide environmental ethics. The flyleaf of his book, *The Creation—An Appeal to Save Life on Earth*, written in the form of an impassioned letter to a Southern Baptist preacher, describes its purpose:

> The Creation demonstrates that science and religion need not be warring antagonists. Explaining that there are compelling environmental and spiritual reasons to be alarmed about pollution, global warming, and Earth's rapidly declining biological diversity, Wilson suggests that if these two powerful social and political forces can be combined in a mutually respectful alliance, with basic metaphysical differences put aside in the service of real-world goals, some of the greatest problems of the 21st century might be solved quickly.

8.3   Sustainability movement

Garrett Hardin expresses alarm at our collective ethical neglect of the Earth's capacity to sustain a quality life for its increasing population[16]:

> Hardin argues that the proper metaphor that characterizes our global ecological situation is not "spaceship Earth" but "lifeboat." The spaceship metaphor is misleading since Earth has no captain to steer it through its present and future problems. Rather, each rich nation is like a lifeboat in an ocean in which the poor of the world are swimming and in danger of drowning. Hardin argues that affluent societies, like lifeboats, ought to ensure their own survival by preserving a safety factor of resources. For a society to give away its resources to needy nations or to admit needy immigrants is like taking on additional passengers who would threaten to cause the lifeboat to capsize. Under these conditions, it is our moral duty to refrain from aiding the poor.

Mylan Engel, Jr., refutes Hardin's moral policy, but not his situational pessimism[17]:

> The argument shows that the things you currently believe already commit you to the obligation of helping to reduce malnutrition and famine-related diseases by sending a nominal percentage of your income to famine-relief organizations and by not squandering food that could be fed to them. Being consistent with your own beliefs implies that to do anything less is to be profoundly immoral.

Hardin and Engel share common ethical concerns, but they advocate radically different policies for addressing those concerns. However, their policy suggestions are neither helpful, nor are they adequately informed by issues such as:

- Recent trends show that the rate of population increase is in decline and in some countries it has actually reversed.
- The world's most populous nations, China and India, along with many others in Asia and the Americas, have increased economic productivity, enabling them to compete in world trade and technology markets, while drastically reducing their vulnerability to famine in

the short term—even while global environmental quality is declining at an accelerated pace.

- Global-scale international attempts to cooperate in policies seek to prevent and reverse further damage to the environmental commons.
- There is a growing scale of research, development, and applications that reduce energy consumption and its attendant pollution of all media, such as research on higher fuel efficiencies along with solar, wind, and geothermal power.

8.3.1   Africa's environmental ethics crisis

Although there are some positive trends, there is clearly little room for complacency in the case of Africa, as Segun Ogungbemi explains. Ogungbemi says that three principal factors contribute to the African ecological crisis: "ignorance and poverty; misuse of science and technology; and political conflict including international economic pressure."[18] He describes how these factors adversely impact the land, water, and air environments. He places much of the blame on African leaders who are afflicted with what he calls the "national security syndrome." Ogungbemi's view of this ecological and ethical crisis is detailed next.

8.3.1.1   Land

The ecological and economic importance of forests to indigenous populations is not usually considered when they are cleared for mechanized agriculture and related industrial uses. For example, mechanized farming can cause a breakdown in environmental stability through severe erosion, increased turbidity that silts up streams, and flood disasters. The consequences include degradation of forests first to grasslands and then to desert. But Ogungbemi argues that mechanized farming can be designed to complement subsistence farming without such consequences. Such methods should avoid the overuse of lands for short-term high crop yields that depend on high-cost manufactured fertilizers and other chemicals that result in downstream pollution. This can be accomplished by directing agricultural produce to local markets and minimizing industrial-scale agriculture in projects designed to improve subsistence economies.

Ogungbemi describes how contact with industrialized nations has created unrealistic expectations for many Third World people:

> Furthermore, modern Africa does not want to eat and drink only what is produced locally; it has been infected with a desire for the Western lifestyle, and to cope with this modern lifestyle it has to import goods from the industrial nations … However, the more agricultural products modern Africa has to produce,

the more fertilizers and other chemicals are used. The end result is that when it rains these chemicals are washed away into our rivers, lakes and oceans. The health hazards that these chemicals cause affect not only humans but also other species in our waters.

With the extensive exportation of agricultural goods to Western Europe and the Americas, modern Africa has boosted the economy of the developed nations at the expense of its own ... It is because of this sort of cycle that increasing numbers of grass-roots activists in Southern countries are regarding "development" as dangerous to and exploitative of poor people in poor countries.[19]

8.3.1.2 Water

Some waste management companies have offered to accept toxic wastes from some countries for shipment to and disposal in Africa to avoid expensive treatment and disposal in their countries of origin. For instance, Ogungbemi reports that certain U.S. companies:

were negotiating with some countries in Africa to allow the disposal of such wastes in their territories, in exchange for handsome foreign exchange for the national development needs of those African countries. Clearly, it must take a very gruesome sense of humor to accept that as an approach to development. But two African countries were reported to be actually involved in the negotiations.

While the industrialized nations have polluted our waters, it must be noted that where African business people have developed industries, similar environmental problems have been created. Water pollution is not limited to industrial areas, but is also found in our rural centers. What is happening to Lake Victoria is morally reprehensible.[20]

8.3.1.3 National security syndrome

Ogungbemi also describes how African government officials often divert excess funds to military uses at the expense of economic development and environmental protection:

Modern African leaders have engaged themselves in what might be called a national security syndrome. The end result is the constant buildup of modern military equipment. As a matter of fact, as Tolba observes, "Of 37 countries for which data are available, only 10 spend more on agriculture than the military. This does not achieve meaningful security." Since African leaders do not find any external aggressors who could be put in check militarily, they therefore turn against their own people. In other words, African countries are not presently threatened by any foreign aggressors so the leaders who want to consolidate their powers have been using military arms for their own protection. There are others who have engaged in oppressing their peoples simply because they want to remain in power. ... Apart from the pollution of the biosphere, the loss of human, animal and fish species are clear indications of leadership-madness of Africa.[21]

8.4   Ethical dimensions of coping

8.4.1   Innovative schemes with uncertain effects

This leads us to the question of how best to cope with ethical problems and dilemmas. According to a *New York Times* article by Cornelia Dean[22], innovative companies have proposed some innovative schemes that could create environmental impacts on a global scale (also see Chapter 13 on Global Climate Change). Dean describes concerns raised about innovative proposals to reverse global warming by exotic means such as "fertilizing" the ocean with iron, injecting certain chemicals into the atmosphere, launching reflecting mirrors into space to deflect sunlight from the earth and other speculative measures.

Critics are concerned that such schemes could overwhelm our ability to adequately overcome their adverse planetary environmental effects. Critics are also concerned that such effors to tamper with global atmospheric and oceanic systems could be expensive and ineffective at best, and backfire at worst with disastrous effects on a global scale.

There is currently no effective global policy instrument to address and control such looming threats to the world commons.

Although the United Nations Environment Programme has made immense progress in many areas, it has not yet matured sufficiently to intervene to prevent threats to the global commons from the mixed blessings of aggressive scientific and technological innovation.

8.4.2   Vision of sustainability

Sustainability is one of the core concepts of environmental ethics and hence a recurring theme in this book. Chapter 4 assesses sustainability in detail. The section on sustainability in Chapter 2 includes discussion regarding Agenda 21 from the United Nations Conference on Environment and Development held in 1992. Agenda 21 consists of 27 specific elements of an ethical policy to enable the world community to work effectively for global sustainability. Chapter 3 has additional examples of global environmental cooperation with clear ethical intent. Chapter 9 examines the economic dimensions of sustainability.

Lester Brown is a leader among the world's most influential environmental thinkers. He heads the Earth Policy Institute. Brown also created the Worldwatch Institute in 1974, and in 1991, he published his vision of a sustainable world as a target to be met by 2030. Table 8.1 includes selected elements of this vision.[23] Brown's 1991 vision remains relevant as an 18-year-old benchmark. Some of the burning issues, priorities, and technology opportunities have changed. However, for the most part, the same challenges remain even if history has altered many details.

Table 8.1 Vision for a Sustainable World by 2030

The basics

In 2030, the world economy will not be powered by coal, oil, and natural gas. The choice then becomes whether to make solar or nuclear power the centerpiece of energy systems. The human path to sustainability by 2030 requires a dramatic drop in birthrates.

Dawn of a solar age

The solar age is where the coal age was when the steam engine was invented in the eighteenth century.
Solar panels will heat most residential water around the world. Photovoltaic technology will constitute a major source of decentralized electric power.

Efficient in all senses

Automobiles will achieve 100 miles per gallon. Incandescent bulbs will be replaced by new lighting technologies with greatly reduced demand—along with the technology of many appliances. Homes and other buildings will be weathertight and highly insulated to minimize energy use. A much more diverse set of transportation options will exist.

Nothing to waste

Waste reduction and recycling industries will have largely replaced current practices. A hierarchy of options will enable (1) avoidance of nonessential materials and products, (2) direct reuse of products such as beverage containers, (3) far more extensive recycling of all wastes, and (4) incineration of combustible residues after recycling to generate power.

Table 8.1 Vision for a Sustainable World by 2030 (*Continued*)

How to feed 8 billion people

Land-use patterns will abide by the basic principles of biological stability: nutrient retention, carbon balance, soil protection, water conservation, and preservation of species diversity. Harvests will rarely exceed sustainable yields. Rural landscapes will exhibit greater diversity: different patterns and strains of crops grown in different ways (according to local ecological conditions) to maximize sustainable output.

Adaptation of crops and methods to changes in the global climate. Increased forests and decreased cattle-grazing lands.

Healthy respect for forests

Forests and woodlands will be valued more highly and for many more reasons. Clearing of tropical forests will have ceased. Tropical forests will be exploited for food and fiber in ways that enable their sustainability. Stewardship requires that people have plots of land large enough to sustain them without abusing the land's productivity and the right to pass it on to their children.

Economic progress in a new light

Achieving the many elements of the vision will create many new economic opportunities in new technology industries, and new patterns of rural and urban living. A sustainable economy will be one without planned technological obsolescence. As a yardstick of progress, the gross national product will be seen as a bankruptcy indicator. National military budgets in a sustainable world will be small fractions of their current levels.

New set of values

The public will adopt simpler and less materially consumptive lifestyles.

The gap between the haves and the have-nots will gradually close, thereby reducing social tensions.

Ideological differences will diminish as nations adopt sustainability as a common cause.

Building and maintaining a sustainable global economy will create full employment.

Source: Based on Lester Brown, Christopher Flavin and Sandra Postel, *A Vision of a Sustainable World* (1991); Sec. 82 in Louis P. Pojman, "Environmental Ethics: Readings in Theory and Application," 4th Ed. (Belmont, CA: Wadsworth-Thomson, 2005).

8.4.3   Ethical leadership

In his Nobel acceptance speech, former U.S. vice president Al Gore said[24]:

> In the last few months, it has been harder and harder to misinterpret the signs that our world is spinning out of kilter.
>
> - Major cities in North and South America, Asia, and Australia are nearly out of water due to massive droughts and melting glaciers.

- Desperate farmers are losing their livelihoods.
- Peoples in the frozen Arctic and on low-lying Pacific islands are planning evacuations of places they have long called home.
- Unprecedented wildfires have forced half a million people from their homes in one country and caused a national emergency that almost brought down the government in another.
- Climate refugees have migrated into areas already inhabited by people with different cultures, religions, and traditions, increasing the potential for conflict.
- Stronger storms in the Pacific and Atlantic have threatened whole cities.
- Millions have been displaced by massive flooding in South Asia, Mexico, and 18 countries in Africa. As temperature extremes have increased, tens of thousands have lost their lives.
- We are recklessly burning and clearing our forests and driving more and more species into extinction. The very web of life on which we depend is being ripped and frayed.

We never intended to cause all this destruction, just as Alfred Nobel never intended that dynamite be used for waging war. The penalties for ignoring this challenge are immense and growing, and at some near point would be unsustainable and unrecoverable. For now we still have the power to choose our fate, and the remaining question is only this: Have we the will to act vigorously and in time, or will we remain imprisoned by a dangerous illusion? There is an African proverb that says, "If you want to go quickly, go alone. If you want to go far, go together." We need to go far, quickly. At the same time, we must ensure that in mobilizing globally, we do not invite the establishment of ideological conformity and a new lock-step "ism."

That means adopting principles, values, laws, and treaties that release creativity and initiative at every level of society in multifold responses originating concurrently and spontaneously. By facing and removing the danger of the climate crisis, we

have the opportunity to gain the moral authority
and vision to vastly increase our own capacity to
solve other crises that have been too long ignored.

In his work and in his speech, Gore clearly advocates that policy thinking
and debate about environmental and other public policy be informed by
an emotional concern for "moral authority and vision."

8.4.4   Ethics and decision making

Since the late 1960s, there have been scores of environmental initiatives,
and new ones are being launched as we speak. However, real environmental solutions are frequently watered down and assigned appealing
vocabulary such as "sustainability" and "eco-friendly." But can such ideals, many of which verge on the intangible, really replace hard choices that
come with selecting one environmental policy over another? Consider the
following *Wall Street Journal* report:

> [Activists] fret that a flood of well-meaning but
> inadequate gestures gives people a false sense of
> progress, lulling them into complacency just when
> the world needs more environmental action and
> less talk—not the other way around.[25]

Moreover, environmental meetings with top executives (if they take
place at all) are often limited to discussing the status of cost versus
benefits, reports on current projects, and legal noncompliance issues.
Communications and meetings with executive management need to be
more robust. Corporations and government agencies need to ensure that
they have appropriate decision-making mechanisms in place that allow
them to consider all relevant information.

It is easy to conceal issues in a deluge of conflicting research and
expert opinions. One can often "shop" for the experts until they get their
desired answer. This can all be done under the context of accuracy and
thoroughness. Professor Brown notes that an ethical analysis should be
"complete and fair and leave no gaps. You do not have to know everything, but you are obligated to know when you do not know."[26] He further
observes that "Ethics is about making informed decisions through quality
reasoning, not about being paralyzed by uncertainty."

No decision-making process, however fair, can ensure that each
and every stakeholder receives everything they want. Compromises are
an inevitable part of the decision-making process; there will be losers.
Perhaps the most important point, however, is that winners and losers
must believe that the decision-making process is fair.

Now more than ever, corporations and government agencies find themselves confronting environmental issues that raise ethical dilemmas. Often, the "correct" decision is less than crystal clear. When confronted with such dilemmas, some managers may simply brush off such discussions as "irrelevant" to business decisions. The best path may simply be for top management and environmental professionals to have a frank discussion about the core values and principles that must be adhered to.

8.4.5 Some thoughts for environmental professionals

Those who are part of public and private enterprises that promote and execute any number of "good ideas" to benefit society often do not fully understand the risks that they are taking with the environmental commons and with the lives of people who depend on it. Since the Industrial Revolution, forests have been cleared, earth has been mined, valleys filled, rivers dammed, air basins polluted by toxic fumes, waterways poisoned by toxic mining and industrial residues, and wetlands drained for housing and other profitable enterprises.

Those of us who invest in and perform work for different organizations do not ordinarily think of the ethical choices made by the organizations we serve. Many who try find themselves marginalized at best, and sometimes fired. The fact of the matter is that for the most part, leaders vested with decision-making powers at the highest levels may be ignorant of environmental ethical issues, believe there are no alternative ways to meet the organizational intent, or simply do not care. Those who actually care may find no institutional support.

The good news is that our nation and the world's governments have made significant progress in responding to ethical concerns for a healthy environment. They have instituted policies that raise awareness of responsible institutions and corporations and implemented laws that address some of the worst abuses, and in so doing, they have actually stimulated business opportunities to develop and provide solutions.

Although political corruption is a significant factor that aids environmental degradation, especially in developing countries, (as mentioned in the Case Study section) there are ample opportunities for ethical environmental professionals to make a difference. To begin with, we can choose to apply our expertise in settings that enable outcomes that improve people's lives and their environment. We can use our expertise to influence higher management in ways that serve the stated purposes of the institution or corporation. To do this, we need to refrain from assuming that all bad policies and decisions stem from unethical intent. Even when we have evidence of such intent, few are prepared to become "whistle-blowers," although some will choose that path.

Environmental ethics cannot be boiled down to a simple 10 commandments. Real-world ethics are often fuzzy and mutually contradictory. The most important lessons are experiential. Unless we become hermits, we live in a world of families, personal friends, communities, workplaces, corporations, and governments. There are both good and bad ethical behavior models in all of them. Our task is to become fully aware as well as fully engaged.

The single most important tool of environmental ethics is awareness of how every action in the environment has a reaction that is constructive, destructive, or neutral to an ecosystem at some scale. Ecology is the science of how species interact with each other in myriad ways. This knowledge is essential in formulating policies that ensure the health of ecosystems at all scales. Ecologist Manuel C. Molles writes as follows[27]:

> Our species is rapidly changing earth's environment, yet we do not fully understand the consequences... human activity has increased the quantity of nitrogen cycling..., changed land cover... and increased the atmospheric concentration of CO_2. Changes such as these threaten diversity of life ... and may endanger our life support system.

The primary goal of environmental ethics is to help shape the policies that affect the Earth's ecosystems. This means whatever else a policy does, it must ground itself on an understanding of how human activities impact the Earth's ecosystems for our benefit or detriment.

8.5 Policy options for reducing the environmental footprint

With respect to ethical performance in economic development projects, consider the following ethical policies:

- Ensure that the indigenous population fully supports the project. Work with government and community leaders to obtain written instruments backed by corporate enterprises, host governments, and appropriate international agencies that will ensure benefits and protections to the indigenous population.
- Ensure that host governments and corporate developers fully respect the rights of the indigenous people affected by the project, see that the environment such people depend on for a livelihood is protected from harm, and see that they significantly benefit from the investment. Provide independent social, economic, and environmental

monitoring to identify and resolve problems before irreparable harm is inflicted that motivates indigenous people to violent action.
* Provide independent social, economic, and environmental monitoring to identify and resolve problems before they get out of control. This is a major gap in environmental policy implementation worldwide because of cost issues. Benefit-cost analysis can be employed to decide on what to monitor for specific conditions and how often.

Problems

1. Briefly describe the principles of environmental ethics.
2. Discuss the origins of ethics in human societies.
3. Discuss the positive and negative aspects of how religions approach ethics.
4. What is the role of ethics in sustainable development?
5. Assess the ethical elements in Agenda 21.
6. Assess the ethical elements in Lester Brown's vision for a sustainable world.
7. You are in charge of an emergency relief program charged with feeding starving people in the province of Needymore in Southern Africa. Your program provides $100 million a year in aid and is credited with saving hundreds of thousands of lives. However, an influential environmental resource expert argues that in providing such aid, your program simply allows more people to survive, further increasing the unsustainable demand on an already overtaxed carrying capacity; feeding these people in the short run merely increases the overpopulation problem, which will further increase the death and suffering in the long run. Let us assume this is a factual and reasonable conclusion. What should you do, let the people starve in the short term but reduce their long-term misery, or act on your ethical beliefs and feed them knowing that this may simply result in even larger misery in the future? Are there any alternatives to this dilemma?

Notes

1. Abridged from Amnesty International (AI), Nigeria: Oil, Poverty and Violence. August 2006 AI Index: AFR 44/017/2006, Public http://www.amnesty.org/en/library/asset/AFR44/017/2006/en/dom-AFR440172006en.pdf. Also see AI's update report: Oil Industry has brought Poverty and Pollution to Niger Delda, http://www.amnesty.org/en/news-and-updates/news/oil-industry-has-brought-poverty-and-pollution-to-niger-delta-20090630. Finally, see "Nigeria Shell Oil's Response to Amnesty International's Report: *Petroleum, Pollution and Poverty in the Niger Delta* is disappointing: http://www.amnestyusa.org/document.php?id=ENGAFR440252009&lang=e

2. Merriam-Webster.com, "Ethics Definition," http://www.merriam-webster.com/dictionary/ethics (accessed July 9, 2010).

3. Stanford Encyclopedia of Philosophy, "Environmental Ethics," http://plato.stanford.edu/entries/ethics-environmental/ (accessed July 9, 2010).

4. G. Hardin, *The Limits of Altruism* (Bloomington, IN: Indiana University Press, 1977).

5. M. D. Hauser, "Moral Minds: How Nature Designed Our Universal Sense of Right and Wrong." (New York: Harper Collins, 2006): xvii–xviii.

6. Stanford Encyclopedia of Philosophy, "Plato's Ethics: An Overview," http://plato.stanford.edu/entries/plato-ethics/ (accessed July 8, 2010).

7. Stanford Encyclopedia of Philosophy, "Aristotle's Ethics," http://plato.stanford.edu/entries/aristotle-ethics/ (accessed July 8, 2010).

8. B. Russell, *A History of Western Philosophy* (New York: Simon & Schuster, 1972): 172–173.

9. See note 3 above.

10. L. White, "The Historical Roots of Our Ecological Crisis," in *Environmental Ethics—Readings in Theory and Application*, 4th ed, ed. L. P. Pojman, (Belmont, CA: Thompson-Wadsworth, 2005), chap. 2, 19.

11. Ibid., 7.

12. L. P. Pojman, ed., *Environmental Ethics—Readings in Theory and Application*, 4th ed. (Belmont, CA: Thompson-Wadsworth, 2005), 31.

13. P. Dobel, "The Judeo-Christian Stewardship Attitude," Christian Century, October 1977. Quoted in *Environmental Ethics—Readings in Theory and Application*, 4th ed, ed. L. P. Pojman (Belmont, CA: Thompson-Wadsworth, 2005), 31.

14. Ibid., 33.

15. The following Web site lists 16 such organizations. http://dir.yahoo.com/Society_and_Culture/Religion_and_Spirituality/Faiths_and_Practices/Christianity/Organizations/Environmental/ (accessed July 9, 2010).

16. L. P. Pojman, ed., "Summary of Garrett Hardin's 'Living on a Lifeboat,'" Bioscience 24 (1974) in *Environmental Ethics—Readings in Theory and Application*, 4th ed. (Belmont, CA: Thompson-Wadsworth, 2005): 412–430.

17. Pojman Sec. 52 Abstract from 2003 essay by Mylon Engel, Jr., *Hunger, Duty and Ecology—On What We Owe Starving Humans*. Commissioned for "Environmental Ethics" in 2003.

18. S. Ogungbemi, "An African Perspective on the Environmental Crisis," in *Environmental Ethics—Readings in Theory and Application*, 4th ed. (Thompson-Wadsworth, 2005), 305. Commissioned for the first edition of the book.

19. Ibid., 306–7.

20. Ibid., 307–8.

21. Ibid., 308.

22. Cornelia Dean, "Handle With Care." New York Times August 12, 2008. http://www.nytimes.com/2008/08/12/science/12ethics.html

23. Based on Lester Brown, Christopher Flavin and Sandra Postel, "A Vision of a Sustainable World," *The World Watch Reader on Global Environmental Issues*, Editor Lester R. Brown (New York: W.W. Norton & Company, 1991): 299–315; Pojman Sec. 82.

24. Official Web site of the Nobel Prize. "Nobel Lecture—Al Gore," http://nobelprize.org/nobel_prizes/peace/laureates/2007/gore-lecture_en.html (accessed July 9, 2010).

25. J. Ball, "Burning Question: Is Earth Day Bad for the Planet? Some Activists Fear Bold action Is Lost amid the Annual Hype," *Wall Street Journal*, April 22, 2008, A19.

26. R. MacLean, "Environmental Leadership: The Road to (Environmental) Hell Is Paved with Good Intentions," *Environmental Quality Management* (Winter 2009): 93–9.

27. M. C. Molles Jr., *Ecology: Concepts and Applications*, 5th ed. (New York: McGraw Hill, 2010).

chapter nine

Environmental economics

> Anyone who believes exponential growth can go on forever in a finite world is either a madman or an economist.
>
> **Kenneth Boulding**

Case study

Many critics claim that environmental policies, laws, and regulations add billions to the cost of doing business and make countries less competitive in the world marketplace. But is this necessarily true?

Consider the example of the U.S. Department of Energy's (DOE's) Hanford Site in Richland, Washington. For nearly 50 years, it produced plutonium for nuclear weapons. This site has been characterized as the "most contaminated" area in the western hemisphere. Radioactive groundwater is migrating toward the Columbia River. Radioactively contaminated facilities and soil sites are located across much of the 600-square-mile site. Enormous underground tanks have leaked high-level radioactive waste into the soil column. Some local citizens claim that their health has been jeopardized as a result of past activities.

As of this writing, approximately $2 billion is being allocated annually to the management and cleanup of the site; it has been estimated that the total price tag could exceed $100 billion over perhaps 40 years. The scope of this cleanup has been described as a "monumental environmental problem."

Rather than spend huge funds and other resources cleaning up the aftermath of ill-conceived mistakes, modern environmental policy acts such as the U.S. National Environmental Policy Act (NEPA) require that federal officials examine the potential consequences of their proposed action (and alternatives) before taking a final decision to proceed. Had NEPA and other modern environmental statutes been on the books and enforced back when these activities were taking place, they could have saved taxpayers untold billions of dollars in later cleanup costs.

Thus, much of the enormous Hanford cleanup bill and those at hundreds of other sites around the world could have been largely avoided if environmental policy and planning acts such as NEPA had simply existed at that time, and had been used to prudently plan projects and waste management strategies.

> **Problem:** After reading this chapter, return to this problem. Consider this example in light of what you have read. This problem may be pursued either as an individual assignment or as a class project. Many critics complain about the cost of complying with modern and strict environmental laws and regulation. But, often, critics only point out the direct costs of complying with these regulations. Frequently, little consideration is devoted to the amount that such laws actually save taxpayers in terms of deferred environmental remediation and loss of natural resources. The Hanford Site was constructed and operated largely in an era governed by little or no environmental regulations. Taxpayers are now paying the tab for the cleanup costs. Research the environmental damage and history of the Hanford Site on the Internet. Based on Hanford's experience, how would you reply to criticism? How would you consider the economic concept of *externalities* in your argument? How do you think contingent valuation could be used to accurately estimate the total damage and cleanup costs?

Learning objectives

After you have read the chapter, you should understand the following:

- How economic principles affect environmental policy
- The principal schools of thought on how economics influences environmental policy
- The difference between environmental and resource economics (ERE) and ecological economics (EE)
- Why the concept of externalities is one of the underlying principles in the discipline of environmental economics, and how it plays a key role in determining whether the benefits of a decision are worth the costs
- The law of supply and demand and how it affects environmental policy

9.1 Introduction

Economic considerations and factors are integral to the goal of formulating any environmental policy. Surprisingly, many environmental advocates, as well as professional practitioners, and even some policymakers have little understanding of the fundamental environmental–economic principles that can affect an environmental policy. This chapter is designed to provide the reader with a foundation in economic principles that can shape the development, implementation, and performance of environmental policies.

This chapter assumes that the reader has no formal studies or grounding in economics. To this end, it begins with first principles and progressively introduces the reader to more advanced topics.

9.1.1   Environmental economics and the tragedy of the commons

As described in more detail in Chapter 2, the "tragedy of the commons" is a metaphor for the dilemma that haunts those who seek to develop environmental policies that ensure the Earth's natural resources will meet the needs of a fulfilled life for all humans. Environmental economics seeks solutions to the tragedy of the commons. Strategies include balancing global supply and demand by implementing policies that address fair allocation by market systems, while assuring the environmental and ecological viability of the natural systems on which these markets depend. Unfortunately, in its current maturity level, the principles, strategies, and theories of environmental economics have not yet earned the consensus among the world's political communities that the laws of Newton, Darwin, and Einstein have earned in the scientific community. Nevertheless, there has been major progress and momentum, as the world's policymakers have come to understand the magnitude of threat. We will introduce some of the major tools that environmental economics makes available to policymakers and policy implementers in government, education, advocacy, and the private sector.

9.1.2   Applying environmental economics

Environmental economics in all its diversity and internal disagreements is essential in shaping essential policies in the interests of our global civilization. Economics can also defeat policy goals that seek to improve the environment. The objective of this chapter is to convey an understanding of the principal relationships between environmental and economic policies and actions by considering the following questions:

- *What are the economic consequences of failing to invest in environmental protection and improvement?* The course of taking "no action" is a de facto policy decision to accept long-term and often devastating physical and socioeconomic costs such as vital resource depletion; and pollution and degradation of human and natural environments with attendant costs in terms of mortality, health, food, mineral supply, and quality of life in general. The economics of lost environmental opportunities require that we go beyond pricing principles in a short-horizon marketplace.

- *How do we determine when the investments required by environmental policy are effective and economically well-grounded?* Standard economic measures in the modern economy are tied to market pricing, which varies with supply and demand, and with expected return on an investment. Historical experience has amply demonstrated that dependence on market pricing alone can result in large-scale (including global) degradation of the land, water, air, and biotic environments. Nevertheless, the traditional principles of market economics continue to play a significant policy role.
- *How do we determine the effects of a given policy on different social classes and communities?* Economic and environmental policies can have differential effects on human communities. Historically, industrialization has degraded the quality of life of people who cannot afford to live outside environmentally degraded areas. When common water, land, and air environments are degraded by uncontrolled activities such as mining and industry, these people can no longer support themselves by farming, fishing, hunting, and gathering.
- *How does environmental economics estimate the costs of environmental degradation from policies, programs, and projects that are outside the market framework of the investor?* Cost–benefit analysis (CBA) is perhaps the most widely used decision tool in private and public investments. Its central problem is how to take into account the economic and social costs of quality of life degraded by the adverse environmental consequences of these investments.

These questions express the basic problems we address in this chapter. Their solutions embody equity and ethical dilemmas. Hence, environmental professionals are called upon to measure and predict how different courses of public policy can result in different consequences. As a practical matter, public policies require that the concepts, theories, and tools of economics be applied to decision making at many levels, such as

- Local areas such as towns, villages, and counties concerned with the quality of human habitat, and environmental resource usage that may be essential to people's livelihoods
- Local, regional, national, and global land, as well as atmospheric and water systems, essential to a large variety of human needs and a healthy ecosystem
- Ecosystems involving all media, which encompass diverse communities of living plants and animals, mutually dependent on each other via food chains and other biological and physical environmental mechanisms

The dilemma that we will encounter repeatedly is that different stake-holders are economically or materially impacted in different ways. For example, the Western rancher or farmer whose water allocation in a regional management scheme is reduced in order to enable the expansion of cities and suburbs may oppose a policy, whereas urban and suburban real estate developers, builders, merchants, and bankers will support the policy. This is an example of competing economic interests between two parties seeking environmental benefits from the same resource.

The job of environmental policy professionals is to assess and explain all of the costs and benefits that accrue to a given policy, how they are allocated among competing stakeholders, and whether or not the policy's effects can be priced in the market. Ultimately, policy decisions are made by elected and appointed officials, and sometimes by courts or directly by voters.

Among the most important functions of an environmental professional is the issue of "sustainability." Sustainable policies and resource systems will continue to produce benefits (no matter how distributed) for the indefinite future. Unsustainable policies may benefit short-term investors, but harm beneficiaries who depend on long-term economic and environmental survival. By developing fact-based and credible scenarios, the economist, working with the environmental professional, can greatly enhance the quality of such decision outcomes.

The sections that follow describe a wide variety of ways in which economics and environmental professionals think about problems and advise policymakers. Experts in these methods do not always agree on what approaches will actually work. However, environmental and economics professionals generally share the following four policy principles:

1. The ecosystems and environments upon which the quality of human life depends must somehow be sustained for essential human uses.
2. Price and market mechanisms are sometimes (but not always) sufficient to ensure protection of environments and ecosystems.
3. When price and market mechanisms are not sufficient, the next logical step is to consider policies that create artificial prices and investment incentives that will ensure a sustainable outcome.
4. When these principles do not ensure sustainability of essential environmental and ecological resources, a variety of direct control policies such as prohibition of certain activities and limits on allowable degradation may need to be imposed.

9.1.3   Schools of thought

There are two broad schools of thought regarding the ways in which economic considerations can be applied to environment problems: ERE and EE.

In terms of their theories, views, and approaches, these two groups are often diametrically opposed. Each is represented by a professional organization, discussed next.

The Association of Environmental and Resource Economists (AERE) was founded in the United States in the year 1979 as a means for exchanging ideas, stimulating research, and promoting graduate training in environmental and resource economics.[1] Its European counterpart, the European Association of Environmental and Resource Economists, explains in its official journal,[2]

> The primary concern of Environmental & Resource Economics is the application of economic theory and methods to environmental issues and problems that require detailed analysis in order to improve management strategies.
>
> The contemporary environmental debate is in a constant state of flux, with new or relatively unexplored topics continually emerging ... Areas of particular interest include evaluation and development of instruments of environmental policy; cost–benefit and cost effectiveness analysis; sectoral environmental policy impact analysis; modeling and simulation; institutional arrangements; resource pricing and the valuation of environmental goods; and indicators of environmental quality.

The International Society for Ecological Economics (ISEE) was founded in the year 1989. Its Web site explains the following[3]:

> Ecological economics exists because a hundred years of disciplinary specialization in scientific inquiry has left us unable to understand or to manage the interactions between the human and environmental components of our world. While none would dispute the insights that disciplinary specialization has brought, many now recognize that it has also turned out to be our Achilles heel. In an interconnected evolving world, reductionist science has pushed out the envelope of knowledge in many different directions, but it has left us bereft of ideas as to how to formulate and solve problems that stem from the interactions between humans and the natural world. How is human behavior connected to changes in hydrological, nutrient or carbon cycles? What are the feedbacks between the social and natural systems,

and how do these influence the services we get from
ecosystems? Ecological economics as a field attempts
to answer questions such as these.

Examples of the questions pursued by EE economists include the
following:

- How is human behavior connected to changes in hydrological, nutri-
  ent, or carbon cycles?
- What are the feedbacks between the social and natural systems?
- How do these factors influence the value we obtain from ecosystems?

Kenneth E. Boulding, Herman Daly, Nicholas Georgescu-Roegen, and
others founded the discipline of EE in the late 1980s in response to the
aforementioned concerns.[4] Boulding is best known in EE for an article
that contrasts the "cowboy economy" with the "spaceship economy."[5]
The cowboy economy is a metaphor for a local or national open economy
having little regard for a holistic view of environmental degradation;
environmental problems are dealt with only on a limited and local scale.
Conversely, the spaceship economy is a metaphor for the holistic view of
the environment characterized by limits on its ability to meet food, energy,
and natural resources demands in a sustainable manner. The survival
strategy in this case would involve wise management of these resources
and maximizing conservation and recycling, while minimizing resource
consumption to ensure sustainable levels.

Generally, ERE economists are trained to apply traditional economics tools in addressing environmental resource constraints. ERE places a
stronger emphasis on more standard market mechanisms as both ethical
and effective ways for ensuring sustainable economies while protecting
the environmental resources that they depend upon.

In contrast, EE economists are generally trained in ecology and other
environmental sciences and seek both market and nonmarket tools to protect ecological systems. They focus on achieving long-term ecological sustainability by applying whatever will work. EE economists place a stronger
emphasis on understanding the limits of nature and on ways to constrain
unnecessary exploitation of living ecological systems for economic purposes. Issues such as uncertainty of long-term outcomes, intergenerational
equity, and sustainable development are the guiding principles of EE.[6]
Methodologies include measuring a society's "metabolism" (i.e., flows of
energy and materials that enter and exit the economic system).[7]

ERE and EE economists frequently come into conflict over their opposing views of how noneconomic factors should be integrated into environmental policy considerations. ERE economists are sometimes considered
more "pragmatic" in that they depend on the economic supply, demand,

and pricing system to guide environmental policymaking. EE economists tend to assign intrinsic value to ecosystems and human quality of life and rely less on market forces in shaping environmental policies.

Nevertheless, they still overlap in scope and are both important contributors to the formulation and implementation of environmental policy. A basic difference between EE and ERE involves their principal goals. ERE focuses on "internalizing" *external costs* that can be measured in the economy. Market-based solutions fit well with this scheme because they deal with real monetary costs that are paid by parties other than the investor who reaps the benefits. Conversely, EE focuses on principles such as sustainability and emphasizes the precautionary principle in dealing with uncertainty and potentially large impacts on the ecosystems. Not all decisions should be made based on market measures alone. ERE analysis typically deals with uncertainty by framing probabilistic scenarios and applying methods that seek to maximize expected benefits. EE analysis takes a decidedly different approach, namely, *adaptive management*,[8] based on the realization that managing complex and uncontrollable systems frequently requires an interaction between performing an activity, monitoring it, and if necessary (based on new monitoring findings) changing the manner in which the activity is implemented. This approach has been applied to international environmental impact assessments, and is consistent with the principles articulated in NEPA.

9.2   *Elementary economic marketplace and investment theory*

To appreciate how investments in the marketplace can affect environmental quality, one must first gain an understanding of some elementary economic principles that are of critical importance in understanding later concepts.

9.2.1   *Law of supply and demand*

The concept of *supply* and *demand* in relation to *price* is to economics what Newton's second law of motion is to Newtonian physics. It provides the unifying and governing operating principle underlying economics. The interaction of buyers and sellers in the marketplace controls the price of products and services. As shown in Figure 9.1, the relationship between price and quantity is universally referred to as the "law of demand"; the relationship between price and supply is called the "law of supply."

The law of demand states that as the price of a product decreases, consumers' demand for that good will increase, and vice versa. The law of supply states that as the price of a product increases, suppliers will produce more of that good, and vice versa. In a dynamic market, producers will drive supply upward as price goes up, while buyers will drive demand downward. An equilibrium point occurs when supply equals demand, achieving a balance—that is, when the quantity of goods supplied is equal to the demand—represented by the intersection of the two curves in Figure 9.1. This intersection represents, in theory, the most efficient allocation of goods and services. Any change that causes the supplier or consumer to alter supply, demand, or price will move the equilibrium point to a new position.

9.2.2   Law of diminishing returns

The *law of diminishing returns* states that at some point, each successive investment in pursuit of a return will yield less than the previous level of investment. This concept is illustrated in Figure 9.2. If demand is held constant, each incremental investment, say in a given abatement technology, tends to yield less and less in terms of an abatement return (e.g., lowered emissions).

On the upward side, each step contributes an incremental increase; but each successive increase is less than that in the previous step. This continues until the top is reached where the cumulative gain is at a maximum. From this point onward, an incremental increase in investment begins to decrease

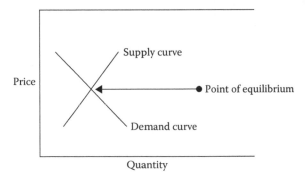

Figure 9.1 Typical price versus supply and demand curves.

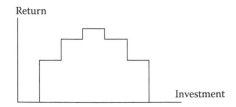

Figure 9.2 Graph shows the typical relationship between investment and diminishing returns.

the return, such that each step subtracts from the peak rate of gain until the cumulative gain is zero. Any further investment produces only losses. This is of course an idealized situation. An enterprise that survives and prospers typically moves up with occasional movement to the downward side. An enterprise that falls to the bottom right will almost surely fail.

9.2.3   Enterprise perspective: Maximize profit

Figure 9.3 shows a simplified model of a typical enterprise, which helps us to discuss how public policies, including environmental policies, affect the enterprise's business performance. An enterprise seeks to maximize its profits by boosting sales and other potential revenue sources. Correspondingly, it seeks to minimize its *capital costs* and *production costs* to what is essential to produce competitive products and services and attract customers. Enterprises, especially large ones, have developed a reputation for opposing public policies that would subtract from the "bottom line." They also profit by

- Lobbying for extraction rights to natural resources such as petroleum, coal, natural gas, and minerals while seeking to minimize or eliminate environmental controls that would protect the environment and people from irreversible harm
- Seeking to dispose of production wastes with little to no effective treatment
- Resisting governmental efforts that would impose costs (internalize) on those who depend on the abused environment for their living and for maintaining their health

Section 9.3 introduces the concept of external costs (costs that are external to the profiting enterprise) in environmental policy.

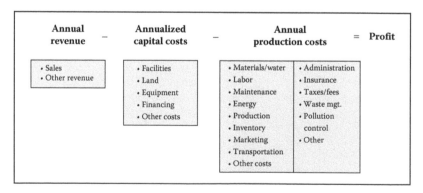

Figure 9.3 A simplified model of a typical enterprise.

In the United States and Europe, there is a long history of economic development policies that have given little or no attention to the adverse effects of unconstrained natural resource extraction and production. Current policy debates generally accept environmental regulation as a given, and focus on the degree and nature of the controls and the best means to achieve them without cooking the "golden goose" of production essential to an adequate standard of living.

To avoid paying the external costs of resource extraction and production, some pollution-intensive industries have sought to operate in Third World countries where environmental regulations are insufficient or nonexistent in terms of protecting indigenous people and the environments they depend upon for survival. However, not all private enterprises engage in such crass abuses (see Chapter 8).

9.2.4   Economy perspective: Maximize public welfare

While enterprises manage business in pursuit of profit, democratic governments (in theory) manage economies in pursuit of public welfare. Of course, what constitutes "public welfare" is a philosophical issue that is highly political and often ideological. For example, some conservatives advocate a purely free-market system, in which public welfare competes with unregulated enterprise; this system constitutes the most reliable guarantor of public welfare. Some liberals advocate government ownership of all environmental resources as a means to internalize external costs by setting the prices and requiring controls on resource extraction. They believe that only government programs and regulation can reliably guarantee public welfare by shifting external costs to the profiting enterprise through a variety of mechanisms.

In reality, there is debate and compromise in a political arena. Conservative policies give more priority to free-market values with lower costs and taxation rates, which favor the enterprise. Liberal policies give more priority to government programs, taxation, and regulation. While the former seeks to minimize taxes and other constraints on the enterprise, the latter seeks to rein in the negative effects of unregulated enterprise to fill gaps in public welfare. This tension frames the policy climate in which environmental professionals must work.

Many Westerners agree that public welfare includes goals such as education, full employment, public health and safety, and a well-developed infrastructure, while ensuring that the environment and its living systems can sustainably meet the community's material needs. However, the policies and strategies implemented for serving the public are often contested. It is the job of the environmental professional, to the best extent possible, to apply an objective scientific approach in predicting and monitoring the effects and estimating the costs of policies that form the ongoing political debate regarding what works best for society as a whole.

9.3 Externalities

If there is a single unifying concept underlying the discipline of environmental economics, it is the concept of externalities. *Internal costs* are those that tend to be easily accounted for and are thus included in an economic analysis such as a Cost–Benefit Analysis (CBA). But not all costs are easily accounted for. This is particularly true when it comes to the assessment of environmental damage and associated costs. One of principal complaints against traditional economic analysis is that they tend to ignore what are referred to as *external costs*.

An externality occurs when one party makes an economic choice that affects other people, but whose effects are not accounted for in the market price; the producers and consumers either do not bear all the costs or do not reap all of the benefits of the economic activity. The term "external" means that the cost or benefit is external to the economic analysis, that is, it is not accounted for in the economic analysis. For instance, it can be particularly difficult to account for environmental damage to environmental assets such as ecosystems or scenic views as such attributes are not traded in the market place. Therefore, such costs are often ignored in standard economic assessments. Yet, external costs to certain individuals or society at large can be very significant. Mainstream economists have been widely criticized for failing to account for externalities in their economic models.

In terms of a negative externality, such as pollution, the product may be overproduced, as when the producer does not take into account external costs imposed on others. The results may include long-term loss of natural capital by polluting and destroying environmental resources, and impairment of public health. Consequently, pollution exceeding a "socially efficient" level can occur. Consider a lead processing factory. The company may impose external costs on the public in the form of air- and water-quality degradation and their adverse health and ecological effects. Additional public costs may include a water treatment plant to remove toxins from contaminated waterways or public expenditures to monitor and maintain a landfill that contains lead waste byproducts. The cost of future remediation (an externality) is often entirely ignored in a standard economic analysis.

An egregious example is the cost incurred to the citizens of New York from the consequences of discharging polychlorinated biphenyls into the Hudson River. What are the costs in terms of medical costs on the surrounding community? What about the health effects in terms of lost or reduced worker productivity?

Still another situation involving externalities occurs when certain nations seek to compete in world markets. Unless regulated by national law and international treaty, competition may cause nations to "race to the bottom" in terms of health and environmental regulation. In many respects, Russia and China have followed such a path as a quick means of increasing their economic wealth or share of world markets.

For instance, under the former Soviet Union, state planners for the Communist Party placed strict quotas on sectors such as annual petroleum production. Petroleum geologists can compute the optimum rate at which petroleum from a reservoir should be pumped. If this optimum value is exceeded, more petroleum can be produced in a given year, but the cost can be that the higher annual pumping rate destroys the hydraulic characteristics of the reservoir, reducing the total amount of petroleum that can be recovered over its life cycle. Because Soviet managers were rewarded for meeting or exceeding annual quotas, with no concern given to long-term productivity, state planners cared little about the health of these petroleum reservoirs. As a consequence, vast Soviet petroleum reservoirs were irrevocably damaged, reducing the total ultimate recovery of this highly prized commodity.

9.3.1 Examples of monetary externalities

Examples of monetary external costs include the following:

- Loss of earnings when land and water resources used by farmers and ranchers are impaired by pollution from mineral extraction, mining, industry, roads, and urban development
- Costs associated with people moving from their lands and homes in search of new economic activity
- Loss of real estate value when environmental degradation occurs in urban neighborhoods
- Loss of future economic benefits to those who suffer the aforementioned losses

9.3.2 Examples of nonmonetary externalities

In this section, cases in which monetary valuation is relatively more subjective are described:

- Impairment and loss of habitat, plant species, and animal species, including endangered ones, that has no market basis for valuation but claims intrinsic value
- Aesthetic impairment of landscapes and neighborhoods, not only visually but also by noise and odors
- Overall loss of cultural attributes when development displaces indigenous people

9.3.3 Internalizing the external effects

Policy, program, and project analysts concerned with environmental protection seek ways to internalize the external costs of their proposals. Approaches include

- Shifting from the enterprise perspective, in which such effects are largely external, to the economy perspective of society at large, in which they are inherently internal
- Seeking to justify decisions, to the extent possible, using CBA monetary measures that are credible to decision makers
- Making the case for intangible and intrinsic values that cannot be reliably measured but can be factored into the decision process
- Using the aforementioned measures to develop approaches for shifting some of the economic costs to corporations, such as
- Laws and regulations creating environmental performance standards
- Taxing corporations to pay for public services that protect or remediate the environmental impacts
- Defining cumulative limits on the annual amount of emissions, effluents, and solid wastes that the environmental commons can accept, requiring that corporate polluters purchase the shares in a competitive market (see Chapter 13)

All these effects are examples of external costs that directly or indirectly harm uncompensated parties who are not paid for by the investor or by project proponents. Much of the work of environmental policy employs various tools of environmental–economic systems analysis to quantify external costs where feasible, as well as to qualitatively describe them with a sensitivity to the ethics involved (also see Chapter 8) Such results are often used by decision makers in formal CBA to help identify feasible policies that internalize some portion of the external costs to the internal enterprise, investors, or project proponents, while compensating affected parties (also see Chapter 6).

9.4 Cost–benefit analysis

External effects such as those outlined in Section 9.3 are fundamental to the discipline of environmental economics and an essential element of CBA. CBA is a decision tool that generally uses monetary units to compare investments and other costs with the corresponding benefits over a specified time. We introduce the basics of CBA by assuming that the external costs have been included in the cost streams used in the examples. Estimation of external costs for CBA is often controversial and difficult to defend. See Section 9.6 for some attempts to measure external costs. For now, let us begin with two general cases of CBA:

- Uniform series of costs and benefits
- Nonuniform series of costs and benefits

9.4.1   Uniform series of costs and benefits

The following example represents a solar power plant completed in year 0 with uniform values of annual costs and benefits earned in each of the following years. (Note: the following examples are provided for illustrative purposes only; those interested mainly in the economic ramifications of the results of CBA for environmental policy may wish to skip them.)

Year	0	1	2	3	4	5
Investments	$1,000,000					
Annual cost	0	$200,000	$200,000	$200,000	$200,000	$200,000
Annual benefit	0	$500,000	$500,000	$500,000	$500,000	$500,000

If money were free (zero interest), we could very simply calculate measures of costs versus benefits as follows:

Project costs (initial investment plus 5 years of annual costs) = $2,000,000
5 years of project benefits at 0% discount = $2,500,000 ($500,000/year ×
5 years)
Benefit/cost ratio = 1.25 ($2,500,000/$2,000,000)
Benefit − costs = $500,000 ($2,500,000 − $2,000,000)
Payback period = years to cumulatively earn or payback the initial
investment

From the table it is clear that the payback period is equal to the difference between the annual cost and benefit ($500,000 − $200,000) or $300,000 per year for a total of $900,000 over the first 3 years. An additional one-third of a year would add $100,000 for a total of $1,000,000, which is required to pay back the initial investment. Thus, the payback period for $1,000,000 is 3.33 years.

Typically, project investors must either borrow money at a given interest rate or choose an alternative investment such as a bond that would guarantee annual interest benefits. The interest rate of borrowing, or for earning income from a bond, is called the "discount rate." The discount rate is applied to calculate the present value (PV) of the five successive years of interest payments.

For example, assume the interest rate is 10%. In this case, to earn a $500,000 benefit in year 1 at the 10% discount rate, we compute,

$$X = \text{Year 0 investment to earn stated benefit in year 1}$$
$$X + 10\% \, X = 1.1 \, X$$
$$1.1 \, X = \$500,000$$
$$X = \$454,545$$

When the streams of costs and benefits are nonuniform, the computation becomes more complex. We illustrate this for our example using a PV calculator available on the Web,[9] which computes the PV of a 5-year stream of uniform net benefits/year on $300,000 at a 10% discount rate.

$$\text{Cost: } (\$200,000)(5 \text{ years}) = \$1,000,000$$
$$\text{PV of future net benefits: 5 years @\$300,000/year} = \$1,250,959$$
$$\text{Benefit/cost ratio: } \$1,250,959/\$1,000,000 = 1.25$$
$$\text{Net PV discounted to year 0: } 1,250,959 - 1,000,000 = \$250,959$$

In principle, this project would be justified, provided that a better return could not be obtained from an alternative investment.

9.4.2   Nonuniform series of costs and benefits

The following example represents a solar power plant with three stages leading up to full production. Hence, the investment is made over a 3-year period, and annual costs and benefits are scaled up as each stage of the solar plant is completed.

Year	0	1	2	3	4	5
Investments	$1,000,000	$500,000	$250,000	0	0	0
Annual cost	0	$50,000	$100,000	$400,000	$300,000	$200,000
Annual benefit	0	$100,000	$300,000	$500,000	$500,000	$500,000

If the discount rate were zero as in the previous case, simple addition would again suffice to compute PV and benefit-cost ratio. Any other interest rate would require the application of a somewhat more complex formula to compute each entry in years 1–5. (See the cited Web site [9] for the general nonuniform series formula and other useful information.) Real-life applications of nonuniform series are common, and project lives are much longer than 5 years.

9.4.3 Opportunity costs

Economists think of the CBA elements in terms of "opportunity costs and benefits" that are lost or gained by a policy, program, project, or regulation. The term "opportunity costs" refer to the economic benefits lost when a public policy investment is profitable but causes economic losses as a consequence. Such costs may be internal or external. An example of internal costs involves the Organization of Petroleum Exporting Countries (OPEC), which seeks to avoid overproduction that would reduce their profit margins regardless of its impact on the welfare of other nations. An example of external costs involves petroleum wells in the Niger Delta that have ruined the indigenous farm and fishery economy of local areas. Opportunity costs can propagate through economies in the form of unemployment, impaired health, increased poverty, or crime and social unrest.

9.4.4 Shadow pricing

Shadow pricing means assigning a price to a product that includes a fair share of the external costs of producing it. Such results can also be obtained by increasing or decreasing taxes, fees, and fines, or by offering various incentives. For example, a policy of agricultural subsidies for tobacco will increase lung cancer, while a policy of a tobacco consumption tax will decrease lung cancer while paying for medical research.

If the policy reduces current environmental effects, it becomes an economic benefit. If it increases current environmental effects, it becomes an economic cost. There are important consequences to human quality of life that have no direct market effects, such as the aesthetic aspects of natural and built environments, ambient noise levels in neighborhoods, and access to social and cultural amenities such as open space and water, parks, playgrounds, churches, and theatres. Moreover, where market effects can be inferred, there are significant controversies regarding their accuracy and distributional effects. For example, the costs of increased human mortality and health impairment, and the impacts on ecosystems that are not commercially exploited, may be understated. Moreover, the burden of the impact often falls on the vulnerable, lower-income indigenous people who have little to no market economy and live off their local environment.

9.4.5 History of controversy

Fiscal conservatives have promoted the wide use of CBA in governmental decision making. However, environmental activists have argued that CBA does not adequately reflect social and economic values that cannot be meaningfully reduced to monetary values, as well as other indirect costs. This viewpoint is reflected in the Council on Environmental Quality (CEQ) regulations for NEPA[10]:

> When a cost–benefit analysis is prepared, discuss the relationship between that analysis and the analysis of un-quantified environmental impacts, values, and amenities … the weighing of the merits and drawbacks of the various alternatives need not be displayed in a monetary cost–benefit analysis, and should not be when there are important qualitative considerations (Sec 1502.23).

In 1981, President Ronald Reagan issued Executive Order 12291 requiring that major regulations, including those with an economic impact exceeding $100 million, be subject to a "regulatory impact analysis," which includes CBA, to be submitted to the Office of Management and Budget (OMB) for review and approval. In 1993, President William Clinton issued Executive Order 12866 to improve the OMB vetting process and to give more consideration to the nonquantitative consequences of major rules. As a result, CBA has been well established in U.S. policy-making since 1981.

9.4.6 Assessment of the pros and cons of cost–benefit analysis in environmental policy

Resources for the Future published an assessment of the role of CBA in federal policymaking.[*] Among its findings is that the actual implementation of CBA has been highly controversial among policymakers, environmentalists, and economists. Table 9.1 depicts the advantages of CBA, while some of its problems and disadvantages are noted in Table 9.2.

9.5 Economic principles important to environmental policy

The following subsection describes basic economic principles and concepts that are fundamental to the development and success of environmental policy. These considerations are important for environmental professionals to understand because they significantly influence, if not greatly complicate, the assessment, development, and efficacy of modern environmental policies.

---

* Winston Harrington, Lisa Heinzerling, and Richard D. Morgenstern, Reforming Regulatory Impact Analysis, RFF Report: March 2009. http://www.rff.org/publications/pages/publicationdetails.aspx?publicationid=20769

Table 9.1 Advantages of Cost–Benefit Analysis

Argues that private economic activity, as well as regulation, can generate value, and hence public policymaking becomes a more balanced process.

Develops metrics for both beneficial and adverse consequences of alternative regulatory approaches, allowing those alternatives to be compared to one another.

Provides transparency and accountability.

Provides a framework for consistent data collection and for the identification of gaps and uncertainty in knowledge.

Employs a monetary metric, the ability to aggregate dissimilar effects (such as those on health, visibility, and crops) into one measure of net benefits.

Forces regulatory designers to think about the physical effects of regulations proposed for public health, environmental quality, and ecosystem health.

Uses benefit and cost measures grounded in economic theory.

Elucidates the various regulatory choices to a broader audience.

Table 9.2 Disadvantages of Cost–Benefit Analysis

CBA is a flawed technique that, among other things, excessively emphasizes the quantification and monetization of risks, trivializes the future through discounting, and fails to meaningfully assess the value of avoiding consequences.

Many critics consider CBA to excessively focus on economic analysis and economic efficiency. For example, CBA inappropriately values impacts on "priceless" species, habitats, and other important difficult-to-quantify resources; discounts future regulatory consequences, including human mortality; treats lives unequally; and trivializes the future.

CBA typically measures health-risk reductions associated with statistical lives. Statistical life is based on surveys of individual's willingness to pay to avoid various levels of risk. This can be meaningless when people conflate their ability to pay with willingness. Thus, life or health in this manner invites inequality.

It is often impossible to quantify many of the important benefits of environmental policy. Even CBA proponents concede that there are no good estimates of the monetary value of many environmental benefits such as the avoidance of many kinds of adverse health conditions and the prevention of harm to species and ecosystems.

9.5.1 Economic growth

Economic growth is widely heralded as the cure-all for the world's developmental, economic, and environmental ills. Its advocates argue that lowering interest rates and investment to stimulate economic growth will lower unemployment, increase economic production, and spur consumer spending, thereby growing the economy. The resulting wealth will eradicate poverty and create large economic surpluses to pay for the reduction of environmental degradation by investing in new technology. This scenario would rely on demographic changes to reduce birthrates, such as those witnessed by industrial nations in the twentieth century. Such market-based solutions have had some measure of success.

9.5.1.1   Thermodynamic systems analysis

The first law of thermodynamics states that energy cannot be created or destroyed, but only changed from one form to another. The second law states that any closed system (i.e., a system that is denied energy exchanges with other or outside systems) always moves from a state of higher order toward disorder (from low to high entropy). For example, smoke from a cigarette (high order) diffuses and mixes with air throughout a room (low order). It never drifts in the reverse direction from being a diffuse gas in a room to a concentrated puff emitted into a cigarette.

By analogy, some economists argue that long-term sustainability of economic production can never be achieved because the Earth itself is a closed system with finite mineral and energy resources (i.e., it is governed by the second law). Modern industrial economies convert low-entropy fossil fuels and concentrated minerals in the environment to largely high-entropy wastes. As minerals are exploited, the concentration of wastes continuously increases (second law) as useless detritus until the minerals are exhausted. Thus, this thermodynamic analogy clearly applies to nonrenewable minerals and fossil fuels. As we will see, another, albeit smaller, school of economists is not very convinced by the closed thermodynamic system arguments.

9.5.1.2   Mainstream economic disagreement

Many standard economists believe that continued growth could indeed be sustained, more or less, indefinitely under certain conditions. For example, if environmental resources were virtually unlimited, economic growth could continue indefinitely. In fact, Earth is actually an open thermodynamic system that receives a vast and constant supply of energy from the sun, which fuels the circulation of air and water and hence powers the entire ecosystem that the human economy depends upon. In this scenario, the only limit to growth is the technological ability to capture more of the sun's energy. However, EE economists counter that there are many other growth constraints, such as the availability of only limited land for agricultural production and even space for human habitation as population increases. Hence, most EE economists argue that at some point, economic growth must cease. Steady-state equilibrium must be reached in which birthrates equal death rates, and resource consumption is reduced to sustainable levels consistent with Earth and solar resources.

Herman Daly argues that as economic expansion encroaches upon resource limits, depletion of natural capital (timber, minerals, fish, and petroleum) will occur. Such resources have inherently greater economic worth than the short-life-span man-made capital (plants, factories, cars, and appliances) that such depletion enables.[11] Such an unsustainable condition would cause uneconomic and unsustainable growth.

Most governments and institutions today have an economic stake in a positive rate of growth. Thus, it is difficult to determine to what extent economic growth can be controlled. Population increase is the major source of social pressure that drives economic growth. China, through the use of draconian methods, has certainly been successful in reducing its rate of population increase. But then, China is an authoritarian government with a history of dealing harshly with dissenters.

The ability to control population depends in part on a host of social and economic factors. Economies in which aging parents depend on their children to support them in old age, such as in the Third World, tend to have relatively high exponential population growth rates. Economies such as those in Europe and the United States that have achieved more or less full employment, general welfare, and price stability, which enable a higher quality of life while virtually eliminating the widespread economic dependency of elderly parents on their children. As a result, population growth in most developed countries has actually been reduced and, in some cases, reversed.

9.5.2   Market failures and environmental protection

One of the key, albeit debated, concepts of environmental economics is that of *market failure*. Market failure refers to the observation that markets sometimes fail to allocate resources efficiently. Hanley et al. explains the concept as follows[12]:

> A market failure occurs when the market does not allocate scarce resources to generate the greatest social welfare. A wedge exists between what a private person does given market prices and what society might want him or her to do to protect the environment.

In the nineteenth century, economist Knut Wicksell described how a marketplace may not provide an adequate supply of public goods, because people may conceal their preferences for such goods but still enjoy the benefits without paying for them.

9.5.3   Coase theorem for external costs

The *Coase theorem* is a somewhat rare species—an economic theory that is both completely counterintuitive and yet generally correct in practice. The basic concept of this theorem is that no matter who is assigned property rights, as long as transaction costs are not too high, an efficient outcome will be reached. Coase's idea was motivated by writing about the problem of how sparks from railroad trains could set entire wheat fields on fire. These fires were costly; hence, the best course of action was to prevent them from occurring in the first place.

The fires can be prevented by (1) adding some attachment to the train that catches the sparks or (2) ensuring that farmers do not plant wheat close to the railroad tracks. The naive view is that if the courts conclude that railroads are responsible for the farmers' losses, the railroad will simply invest in the spark-catching attachment since it is a cheaper alternative than paying farmers for their losses. Conversely, by this logic, if the courts rule that the farmers are responsible for the losses, then they will not plant wheat next to the tracks.

Surprisingly, the Coase theorem states that this logic is wrong. According to the theorem, regardless of who is liable, both parties should negotiate to reach an efficient solution. Assume it is cheaper to stop the fires with a spark-catching attachment; then, even if the railroads do not have to pay for fire damage, the farmer should offer to pay for the railroad attachment, plus some profit margin. The railroad is now better off (it has more money), and the farmer too is better off (the attachment was cheaper than not planting next to the tracks). Everyone is better off, assuming the transaction costs are not too high. The key insight was not that property rights could be traded in a market but that an efficient outcome could be had regardless of the party to whom the property rights were assigned.

This theorem has obvious implications with respect to environmental pollution and degradation problems. One of the problems with the Coase theorem has to do with the assignment of property rights and who "owns" them. A firm may believe it has the right to emit air pollutants. A victim may feel he or she has the right to clean air. Such issues have spurred endless debate with respect to the correctness of this theorem when applied to the environmental commons.

Consider a smelting company that emits lead and other toxic air pollutants across a large region. The Coase theorem would not apply to this situation in which many people could end up with medical ailments. It would be difficult if not impossible for any one person to negotiate with the company, and if they could, there would be large transaction costs involved. The most common alternative is simply to regulate the firm.

In other environmental instances, however, the Coase theorem may be relevant. For example, consider a logging company that has rights to a forested area and is planning to log in a way that would negatively affect a nearby resort. In principle, the resort owner and the logger could negotiate. The resort might pay the logging company to move its operations to an area that would not affect its view. But what if no one clearly owns or controls a forested area (or an air shed)?

Thus, if a community living near a smelter has a right to clean air, or the factory has the right to produce toxic emissions, then either (1) the factory could pay those affected by the pollution or (2) the people could pay the factory not to pollute. Alternatively, under a third option, the citizens could take action just as they would if other property rights were violated. An early example of this approach involves the nineteenth-century U.S. River Keepers Law. This act provides citizens downstream with the right to take action to end pollution upstream if the government itself does not stop the pollution.

9.5.4   Economic inequality—the Kuznets hypothesis

Economists have long been concerned with the problem of inequality of assets and income in modern economies. The *Kuznets hypothesis* states that inequalities such as pollution tend to increase as per capita income grows to a certain level. But as per capita income continues to grow, the inequality peaks and then begins to decrease.[13] Figure 9.4 shows a standard Kuznets diagram as an inverted U curve.

Since 1991, the Kuznets curve has been widely applied to environment resource studies, such as water and air pollution, which also display an inverted U-shaped curve.[14] Environmental degradation appears to decrease once a critical per capita income level is reached. The environmental Kuznets curve (EKC) is one of the most investigated topics in environmental economics.

Previously, it was widely assumed that richer economies degraded their natural resource endowments at a faster pace than poorer ones.

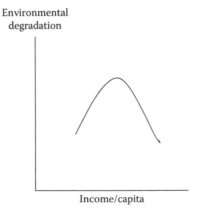

Figure 9.4 U-shaped environmental Kuznets curve.

Environmental quality, it was commonly argued, could only be achieved by escaping the clutches of industrialization and higher incomes. But from its infancy in 1991, EKC revealed a surprising outcome. Some significant environmental problems such as sulfur dioxide levels and air-particulate levels actually improved as incomes and levels of consumption went up.

In 1995, Krueger concluded, "We find no evidence that economic growth does unavoidable harm to the natural habitat."[15] The following pollutants have been examined for evidence of EKC patterns: automotive lead emissions, deforestation, greenhouse gas emissions, toxic waste, and indoor air pollutants. Some generalizations have begun to emerge. For instance, the turning points for individual pollutants differ across nations. Nations first to deal with a pollutant have higher income levels than succeeding ones. It seems that the later nations reap some benefits from more mature and less costly technologies, economic aid, and the experience of more developed nations.

Acceptance of the EKC has important policy ramifications. It suggests that some degree of environmental degradation is inevitable during a nation's economic development, particularly during the early phase of industrialization. Moreover, it implies that as a certain level of per capita income is reached, economic growth helps to remediate the damage from earlier years. It also suggests that international policies that stimulate growth (trade liberalization, free trade, and globalization) in the absence of environmental quality controls should likewise be good for the environment, at least over the long term.

Despite these different schools of thought, most economists appear to take a neutral stance concerning EKC. Many believe the inverted-U curve only captures the *net effect* of income and the corresponding impact, in which income growth is used as an omnibus variable representing a variety of underlying factors whose separate effects are actually obscured.[16] Such a view suggests that there is no automatic linkage between income growth and a reduction in environmental deterioration. Many economists strive to demystify the relationship by providing explanations for it.

Proponents of EKC argue that the curve occurs because at low incomes people tend to value development over environmental quality. However, as greater wealth is achieved, people become more willing to devote more money and resources to improve environmental quality.[17] Greater concern is directed toward environmental protection through strict, modern regulations and enhanced social consciousness. As income increases, individuals are more willing to spend a larger share of their incomes on environmental improvements, reducing the pollution per unit output. Environmental degradation begins to peak and follows a parabolic downward curve.

Developing economies such as China and India have been cited as potential test cases for the EKC.[18] It is also important to note that economic growth without a corresponding change in cultural and institutional values is unlikely to result in environmental improvement. A study of Latin American countries found significant evidence of an EKC relationship for deforestation.

9.5.4.1   U- or N-shaped curve?

A more recent, albeit controversial, refinement of the EKC suggests that the inverse-U shape does not accurately reflect all situations. For example, with longer time scales, the actual relationship may be N-shaped, as shown in Figure 9.5. The N shape tells a different story. It indicates that pollution initially increases as a country develops its industry, then decreases as a certain threshold of gross domestic product (GDP) is reached. From that point on, degradation again increases as national income continues to rise. The importance of this finding is that there is the possibility pollution may actually increase later on.[19]

Many theorists are concerned that economic growth in the long term has to result in environmental degradation as represented by the N shape because the costs of control technology eventually become too large to sustain. This leads to the conclusion that economic growth and pollution are separate phenomena and are only sometimes correlated. Thus, the traditional economic-pollution progression from agrarian communities (clean), to industrial economies (pollution intensive), and finally to service economies (cleaner yet) may be false, if pollution once again increases in the service-based economies.

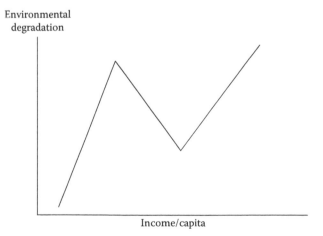

Figure 9.5 N-shaped environmental Kuznets curve.

Arguably, the most troubling problem with the traditional U-shaped model is that a net global reduction in pollution may not be occurring. Developing counties have become "pollution havens." Many wealthy nations have a troubling tendency to simply transfer polluting activities to poorer nations. Critics argue that this transfer is what actually explains the U-shaped function.[20] If true, this implies that as the world's poor nations develop, they in turn will have no place to "export" their pollution. Such a situation for any nation is of course unsustainable and would merely add to the global plight.

9.5.4.2   Environmental Kuznets curve criticisms and controversy

Early studies found convincing evidence supporting the EKC relationship. Beckerman argues that consistent with the EKC, the easiest way to obtain environmental improvement is to continue along the original economic growth path and endure the transient environmental impact.[21] In his controversial book *The Skeptical Environmentalist*, Lomborg describes our future as a "beautiful world."

In recent years, however, the idea has become the subject of intense debate. While the EKC appears to hold for some pollutants (nitrogen and sulfur oxides, lead, chloroflurocarbons, and dichloro-diphenyl-trichloroethane), there is little evidence that it holds for other pollutants or for other types of degradation such as natural resource exploitation and biodiversity conservation. Many ecological "footprints" (land, energy, and resource usage) do not appear to fall with rising income. Although the ratio of energy per real GDP has fallen, total energy consumption is still rising in most developed countries. Similarly, greenhouse gas emissions are much higher in industrialized countries.

This does not necessarily invalidate the Kuznets hypothesis in all cases, as the shape and scale of these curves may vary with different impacts. For instance, it could be that Western nations are still climbing the "upward" leg of energy usage to become richer before further environmental degradation is noticed. Alternatively, increased use of nuclear power and renewable energy might result in a steep drop in fossil energy use and its emissions.

There is yet another dilemma. The impacts of many pollutants are localized to relatively small socioeconomic units like nations. As a country develops, its marginal investments in environmental mitigation measures initially reap relatively large marginal improvements in environmental quality. Over time, as the level of environmental pollution diminishes, the marginal benefit also diminishes.

This tends to discourage future environmental investments. However, national policies that seek to reduce greenhouse gas emissions may provide no local return at all, unless all global localities do likewise. Moreover, if the economic return on investment is not viewed as locally equitable by all parties, such policies may fail. This is precisely the tragedy of the commons discussed earlier. Thus, even in a country like the United States that has a high level of income, carbon emissions are not decreasing in accordance with the EKC.[22]

9.5.5   Environmental policies for enhanced competitiveness—the Porter hypothesis

The Porter hypothesis argues that stricter environmental performance standards can actually spur technological innovations that enhance competitiveness. Properly conceived, environmental and pollution-prevention policies may actually add to long-term productivity, employment, and profits. They do so by taking into account external costs (internalizing) in ways that motivate innovation and investment, which capture the intrinsic value of healthy ecosystems. Porter and van der Linde argue that economists are locked in a "static mindset that environmentalism is inevitably costly." Specifically, they assert the following[23]:

> By stimulating innovation, strict environmental regulations can actually enhance competitiveness ... Fundamentally, pollution is a manifestation of economic waste ... Efforts to reduce pollution and efforts to maximize profits share the same basic principles, including the efficient use of inputs, substitution of less expensive materials, and the minimization of unneeded activities.

In response, Palmer, Oates, and Portney defended mainstream economists and questioned the validity of the Porter hypothesis.[24] They published a mathematical proof that shows that charging polluters a "green tax" (thereby increasing the *Pigovian tax* rate described in Section 9.7[25]) cannot benefit a profit-maximizing enterprise, regardless of opportunities for innovation. They argue that for several decades, EE economists have made their case for incentive-based policy instruments (e.g., effluent charges or tradable emissions), often by emphasizing the incentives for creating innovative abatement technology. Most standard EE texts assert that incentive-based approaches may be more attractive for their efficiency rather than for minimizing the costs of attaining environmental standards. Bauman argues that EE economists have adopted a static mindset, which has been overly focused on end-of-pipe abatement, overlooking opportunities for production process innovations.[26] The validity of the Porter hypothesis continues to be debated.

9.6   Environmental–economic metrics and effects

This section introduces tools for measuring the economic effects of environmental policy (see Chapter 3).

9.6.1   Sustainable economies

Herman Daly defines a *sustainable economy* as one "that can be maintained indefinitely into the future in the face of biophysical limits." This requires one to specifically define what is to be sustained over time. Several economic measures have been proposed.

9.6.1.1   Problem with gross domestic product

GDP is the most common measure of an economy's total output of goods and services, as measured by the monetary values associated with the capital and labor expended. It is a "real-time" measure that does not account for the impacts of future scarcity of natural resources. Those who argue that a sustainable economy can encompass GDP growth assume that technological development will increase the efficiency of resource utilization and avoid scarcity. But such outcomes may be feasible only in the absence of population growth, and even then are by no means assured. Hence, the community of EE economists generally rejects such arguments. A truly sustainable economy must achieve a steady-state balance between the rate of natural resources consumption and their replenishment. Technological innovation that improves efficiency and induces certain lifestyle changes can actually result in GDP reduction over the long term, while quality of life is maintained or improved. Economic development then becomes a continuing process of improving efficiency and resource substitutions.

We have seen that enterprise decisions are based on expectations of profit, while national decisions are based on the larger societal interests. A nation's GDP does not measure the future value of depleted nonrenewable resources or natural capital. Nor does it measure the effects of higher mortality, aesthetics, livability, impaired health, happiness, or loss of ecosystem habitat. In fact, increased mortality and impaired health can add to GDP because they increase the demand for mortuaries, hospitals, and medical care, offset by GDP decreases due to the productivity losses of premature death and illness.

9.6.1.2   Sustainable economic welfare index

Going beyond GDP as a measure of sustainable well-being requires numerous and complicated adjustments (e.g., environmental destruction, uncounted household services, and international debt). By making such adjustments, Daly et al. have developed a *sustainable economic welfare index*.[27] If this index is correct, the negative factors have been increasing faster than the positive ones for the United States since the 1980s. Studies have found similar results for the United Kingdom, Austria, Germany, and Sweden. More to the point, for some (possibly many) countries in recent years, the costs of growth are rising faster than the benefits.

A sustainable economic welfare index would subtract from the GDP the costs to the economy such as air, water, and land pollution; losses of biodiversity; and the consequent reductions in the value of natural capital such as forests, fisheries, and farms. Some of the concepts associated with this index are discussed in the following subsections.

9.6.1.2.1   Throughput   This is the rate at which an economy transfers raw materials from the environment to the marketplace. An economy with a fast throughput is generally considered more efficient. For example, giant trawler fleets achieve very rapid throughput in bringing fish to the market. When fish are plentiful, the economy of scale benefits the consumer with low prices. But when overfishing depletes the fish population, collapses in the fishing economy cause prices to soar. Thus, there is a maximum value at which GDP can be sustained, and the same is true of virtually all economic throughputs that depend on limited natural resources.

9.6.1.2.2   Natural capital   This is the intrinsic value of the environment and its living systems to sustain the needs of human communities. Pollution and resource depletion reduce this capital. Current measures of GDP do not account for the value of natural resource depletion, which sends the wrong economic signals to producers and consumers.

9.6.1.2.3   Weak and strong sustainability   In traditional economic theory, resources are merely a form of capital and include everything from raw materials to factories and products. Such capital can be natural or manufactured. Economists who consider man-made capital substitutable for natural capital (i.e., tradable in a free market) typically argue that the sum of man-made capital and natural capital be maintained at some constant level. This approach is called *weak sustainability*.

Most EE economists maintain that natural capital and man-made capital generally establish the market economy and that they must be freed from increased marketplace pressures. Thus, EE "complements" as opposed to "substitutes." If this is true, natural capital should be maintained constant on its own, because it has become the limiting factor. This approach is referred to as *strong sustainability*.

Consider again the fisheries example. The world's fish catch was once limited by the man-made capital of fishing boats. Today, the world's fish catch is limited by the sea's natural stock of fish. Weak sustainability leads to the ridiculous conclusion that building and deploying more fishing boats could augment dwindling fish supplies. In contrast, strong sustainability recognizes that catches must be limited to maintain a sufficient supply of fish; building and deploying more fishing boats is useless when the catch is limited by an insufficient supply of fish.

Some environmental policies aimed at sustaining natural capital employ *cap-and-trade systems* where the total amount of throughput is limited. Under such a system, the right to deplete resources (e.g., fisheries or forestry) or to pollute air or water resources is no longer free. Resources are assigned a value and become a "scarce" asset that can be bought and sold on the free market. In principle, cap-and-trade systems provide an efficient mechanism for maintaining sustainable resource usage.

9.6.2   *Measuring contingent valuation*

Many environmental resources provide people with utility (a measure of the relative satisfaction obtained from various goods and services), although certain amenities, such as beautiful scenic views, do not have a specific market price because they are not directly sold. Thus, it is difficult to place a dollar figure on them using standard economic valuation methods. Nevertheless, homes having scenic views normally command a higher price than equivalent homes with lesser views.

Research indicates that most people are willing to pay for environmental benefits and the nonuse of resources that otherwise would be lost to future development. However, unless a dollar value is estimated for nonused resources, it is likely that they will be treated implicitly as having no value. This leads to a quandary: How much are nonmarket environmental resources really worth and how can their real value be estimated or afforded protection? Often, the only option for estimating their value is to ask questions through a survey.

The theory of *contingent valuation* was proposed by Ciriacy-Wantrup in the late 1940s to provide a method for assessing the valuation of such resources. His technique provided a method for determining the market valuation (assessing value) of a nonmarket product, service, or good. Davis first applied his technique in the 1960s to estimate the value of a particular wilderness area used by tourists and hunters. His survey results compared favorably with an estimation of value based on travel costs. The method became popular in the 1980s as some U.S. agencies began to sue private companies for damages to natural resources. Recovered damages included nonuse or *existence values*. An existence value is an amenity that cannot be assessed using standard market pricing mechanisms.

The Exxon Valdez petroleum spill in Prince William Sound was the first major test case in which this technique was used to quantitatively assess damages. Today, the technique is widely used particularly with respect to assessing damage caused by pollution and environmental degradation.

9.6.2.1   Surveys

Contingent valuation uses surveys to measure the valuation of such resources. It is often referred to as a *stated preference* model, in contrast to a price-based *revealed preference* model. The method of revealed preference was pioneered by economist Paul Samuelson, as a method for determining the basis of consumer behavior. It assumes that preferences of consumers can be revealed by their purchasing habits. The theory is founded on the premise that consumers make consumption decisions based on their intent to maximize their utility.

The questionnaire used in one study put interviewees in the position of decision makers. Based on the cost and benefit information supplied to them, interviewees were asked questions about their preferences regarding incinerators involving different levels of pollution control technology. The researchers found that most people were willing to pay more for stricter control technologies if they believed that the health benefits to be gained outweighed the costs.[28]

Both the stated preference and revealed preference methods are utility based. Typically, the survey asks how much money people would be willing to pay (or willing to accept) in order to maintain the existence of (or be compensated for the loss of) an environmental resource such as a scenic view, a natural undeveloped habitat, or biodiversity. Early surveys were often open-ended questions such as "How much money would you pay to preserve this hunting area?" or "How much compensation would you want for the destruction of this hunting area?"

Some prominent economists and psychologists question whether the use of stated preferences can accurately gauge the true values people place on nonuse goods. They argue that such surveys suffer from shortcomings such as response bias, group behavior, protest answers, or ignoring income constraints.[29]

A panel of economists assembled by the U.S. National Oceanic and Atmospheric Administration (NOAA) recommended that contingent valuation surveys be carefully designed to address the inherent difficulty in eliciting accurate economic values through survey questions. The underlying recommendation was that the survey investigator has a significant burden of proof to satisfy before the results can be accepted as accurate. Such surveys can be expensive and difficult to design. Thus, the panel recommended a set of reference surveys against which future surveys could be calibrated. The panel also recommended that contingent valuation surveys measure the willingness of a person to pay for protecting a resource rather than the willingness to accept compensation for its loss.

As a result of the panel's recommendations, current empirical tests indicate that modern survey techniques can correct for such bias.[30] The technique has been increasingly used by U.S. agencies in performing CBA of projects that impact the environment. For example, the method has been used to assess the valuation of biodiversity restoration in the Mono Lake, restoration of salmon spawning grounds, and development of recreational opportunities on the river downstream from Glen Canyon Dam.

9.6.3   *Measuring economic surplus*

A methodology known as *economic surplus* provides a theoretical basis for assessing economic benefits. *Consumer surplus* is a monetary measure of the net benefit that a consumer gains from a transaction. For example, consider the purchase of a sports utility vehicle (SUV). The economic surplus is simply the difference between the market price of the vehicle (the amount the vehicle costs) and what the consumer is able and willing to pay. Assume, for instance, that a consumer is willing and able to pay a maximum of $45,000 for a Pluto SUV. Now assume that the actual negotiated price is $37,000. The consumer surplus is thus $8,000, which is the difference between the consumer's actual negotiated price and the actual amount he or she is prepared to pay. If the SUV sells for $40,000 but the consumer is only willing to pay $35,000, the economic surplus is –$5,000.

Similarly, commodities or services that are not purchased in markets (e.g., air quality) also have a consumer surplus. Consider a tourist who is willing (and able) to pay $10 for each additional opportunity to visit a scenic site. If a proposed regulation leads to air quality improvement enabling the tourist to visit three additional sites beyond the one normally experienced under the previously unregulated viewing conditions, the consumer surplus (i.e., monetary measure of sightseeing) increases by $30. However, since the sightseer did not pay directly for the air quality improvement, the $30 increase in consumer surplus is difficult to assess. While this approach is difficult and far from perfect, it at least provides a theoretical model for valuation of environmental quality.

9.7   *Non-market-based policies for environmental management*

Two general nonmarket approaches for managing environmental degradation are described here. *Pigovian taxes* involve various schemes for taxing effluents and emissions at the source. *Ecological taxes* involve ways to account for variability in emitter circumstances and to focus more on controlling the effects of emission and effluents on ecological systems, rather than on uniform taxation of polluters.

9.7.1   Taxing environmental and social costs: Pigovian taxes

The English economist Arthur C. Pigou argued for the imposition of taxes (more generally referred to as a "green tax" or "sin tax") on generators of pollution. Since the social cost of pollution exceeds the private cost to the polluter, the government should intervene with a tax, making pollution more costly to the polluter. A *Pigovian tax* (also pronounced "Pigouvian") is a tax levied on companies to correct for the negative externalities of their activities. For example, a Pigovian tax may be levied on companies who pollute the environment to encourage them to reduce pollution and to provide revenue, which may be used to remediate the impacts of their pollution. If pollution is more costly, the polluter will generate less.

In a true market economy, a Pigovian tax is perhaps the most efficient and effective way to correct for negative externalities, leading to an improved quality of life. Under certain conditions, a double dividend can be achieved: The first is the reduction of environmental degradation. The second is that it can promote environmental technologies and more innovative processes, such as recycling, that otherwise would not be adopted.

Perhaps the biggest problem with this type of tax involves the "knowledge gap," encountered in calculating the level of tax necessary to counterbalance a negative externality; determining the amount of tax required to compensate for negative externalities is shrouded with uncertainty. Political lobbying and other factors by polluters can also diminish the tax levied, which reduces its mitigating effect.

A Pigovian tax can create two problems: First, these taxes can encourage black markets and smuggling, particularly if they create significant differences in the price of products. Second, they can significantly lower the income of individuals who tend to spend a greater portion of their income on products with external social costs such as electricity or gas; in this case, the Pigovian tax becomes a regressive tax.

An alternative to Pigovian taxation appears to be gaining popularity—a market for "pollution rights." Pollution rights markets (e.g., emissions trading; see Chapter 13) are not necessarily more efficient than a Pigovian tax, but are often more appealing to policymakers because giving such rights out for free (or less than the market price) allows polluters to lose less profits or actually realize a profit (emissions trading).

An approach that provides corporations with greater flexibility than strict regulations is to assess a straightforward tax on pollutants. By granting polluters the freedom to determine how best to reduce emissions, such charges can motivate the creativity and ingenuity of individual companies to find efficient ways to meet regulations. Unfortunately, some companies find it in their interests to estimate abatement costs on the high side. This enables them to argue that the proposed abatement approach is too expensive to comply with.

9.7.2 Ecological taxation

Ecological taxation, also known as "green tax" or "ecotax," is a policy that uses taxes and economic incentives to promote ecologically sustainable activities. These taxes can complement or avert the need for regulatory approaches. Such policies may attempt to maintain overall tax revenue by proportionately reducing other taxes (e.g., on human labor and renewable resources), which is known as the "green tax shift."

The goal of the green tax shift is often to implement a complete-cost accounting or establish the true cost of a product or service, using fiscal policy to account for market-distorting externalities. Setting the correct taxation level can be quite difficult and may lead to further distortions or unintended consequences. Amory Lovins advocates a "feebate" approach in which additional taxes and fees on less sustainable products (e.g., large gas-guzzling vehicles) are pooled to fund subsidies for more sustainable alternatives such as hydrogen or electric vehicles.

9.8 Market policies for environmental management

This section explores market-oriented economic policies for protecting the environment. We begin with the concept of emissions trading.

9.8.1 Cap and trade: Emissions trading

A procedure known as "cap and trade" promises to compensate for the problems inherent in the tax-based systems described in Sections 9.7.1 and 9.7.2. Under a cap-and-trade system, Company A could buy an emission allotment from Company B if its emission cap rises above a point where further emission reductions become prohibitively expensive. In the real world, companies can simply buy and sell the right to emit until the most efficient price level is found. There is little need for regulators to calculate the best level for each company. Under a trading quota system, a firm adopts pollution abatement measures if doing so costs less than purchasing an emission credit from another party. This leads to a lower cost for the total abatement effort.

The concept was first tested in a series of American computer simulations in the late 1960s. These simulations were performed on several cities and their emission sources to compare the cost and effectiveness of various strategies. Emission trading uses the standard economic theory of supply and demand (see Section 9.2) to minimize the impact of pollution on ecosystems and the economy that relies on them. A cap-and-trade system provides an approach for controlling pollution by offering economic incentives for achieving reductions in the emissions.

Consider carbon dioxide emissions. These emissions impact the entire planet; thus, their effect on the environment is generally similar wherever in the globe they are actually released. Thus, the location or origin of these emissions does not matter from a climatic standpoint. Under a cap-and-trade approach, the central authority (the government or international body) sets a limit, or *cap*, on the amount of an emission (typically, air emissions). Companies and other entities are issued emission permits. These permits provide credits (allowances), which represent the right to emit a specific amount of the pollutant. The total amount of credits cannot exceed the preestablished cap, which limits total emissions. Companies that find it technically or economically necessary to exceed their emission allowance can purchase credits from those who have fallen under their allowance credit. Thus, the pollutant possesses an economic value that is determined by free-market forces. The transfer of these credits is referred to as a *trade*. The purchaser is effectively paying a fee for polluting, while the seller is rewarded for having reduced emissions by more than their allotted cap.

Trading programs have been established for several pollutants. In the United States, there is a national market designed to reduce acid rain. For greenhouse gases, the largest market is the European Union Emission Trading Scheme. Emissions trading can be cheaper, and politically expedient for existing industries because the initial allocation of allowances are often allocated with a grandfathering provision, that is, granting rights in proportion to historical emissions. The practice of grandfathering has also been widely criticized, as polluters are given free allowances instead of being made to pay for them. Critics suggest that all credits should be openly auctioned.

The total cap can also be lowered over time, aiming for an emissions reduction target. Organizations that do not pollute can often participate, so that environmental groups may purchase and retire credits and hence drive the price of the remainder up according to the law of demand.

One approach is referred to as the *baseline-and-credit* program.[31] In this approach, polluters can create credits by voluntarily reducing their emissions below an established baseline. They can then sell their credits to other polluters for whom the costs of purchasing credits are less than the costs of installing expensive abatement technology to reduce their pollution. In theory, firms will choose the least-costly way of complying with pollution regulation, creating incentives that reduce the cost of achieving a pollution reduction goal.

Reductions in emission can only occur if emissions can be accurately measured in the field and reported to a regulatory authority. Many industrial process emissions can be automatically measured using instruments mounted on chimneys, stacks, and discharge pipes. However, there are other types of activities that rely principally on theoretical calculations.

Emission regulation may include mechanisms such as enforcement of fines and sanctions on those who exceed their emission allowances.

9.8.2   *Theory of marginal abatement costs*

The reason cap-and-trade methods work is that some businesses or operations find it much easier and more cost effective to reduce pollution than others.[32] Figure 9.6 illustrates how Company A's initial investment earns a steep marginal emission reduction. The figure shows the marginal abatement cost at which, as the money spent to reduce pollution goes up, the amount of emissions goes down. Initially, it can easily decrease pollution through pollution abatement investment. However, a point is eventually reached at which increasingly larger amounts of capital are required to produce each incremental decrease in emissions (point of diminishing returns). Thus, the company has a strong incentive to invest up to the point of diminishing returns.

In contrast, Company B (Figure 9.7) has a more linear marginal abatement cost function. Compared to Company A, Company B can make much a larger reduction in emissions for the same amount of investment. That is, to reduce the most emissions for the least cost, Company A should make the investment that allows the initial plunge in emissions; however, beyond the point of diminishing returns, further investments are relatively futile. If a regulatory agency desires further reduction in the total amount of emissions, the investment should be made by Company B to maximize economic efficiency. Incentives can be given to Company B to make further reductions. One approach involves Company B making the investment and selling an emission credit to Company A.

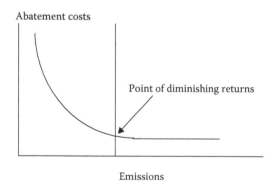

Figure 9.6 Marginal emissions of Company A reduce as abatement investment increases over time; the initially high marginal return on investment diminishes over time.

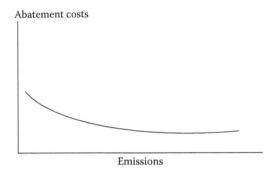

Figure 9.7 Marginal emissions of Company B reduce as abatement investment increases over time; the initially low marginal return on investment remains low over time.

9.8.3   Criticism

The relative merit of *price* versus *quantity* instruments in achieving emission reductions has been the subject of considerable debate. A cap-and-trade system is a quantity-based system because it fixes the overall emission level (quantity) and allows the price to vary accordingly. Many scientists warn of a future threshold in atmospheric concentrations of carbon dioxide beyond which a runaway warming effect causes catastrophic and irreversible damages. Under this scenario, a quantity-based system is a better choice as the total emissions can be capped with a higher level of confidence. Others counter that this may not be true if a risk exists, but cannot be correlated to a scientifically designated level of greenhouse gas concentration.[33]

Critics also maintain that emissions trading does little to solve pollution problems overall. They charge that regulatory agencies have issued too many credits, diluting the effectiveness of regulation and effectively removing the cap. Instead of promoting a net reduction in carbon dioxide, beneficiaries of emissions trading simply do more of the polluting activity because groups that minimize their pollution have sold their conservation efforts to the highest bidder. They argue that total reductions need to come from a sufficient reduction of allowances available within the system. More extreme voices argue that it places excessive emphasis on individual carbon emissions, and distracts attention from the wider systemic, cultural, and political lifestyle changes that need to be made. Some critics view carbon trading as a proliferation of the free market into public spaces and environmental policymaking.[34] They also cite dubious science, accounting failures, and other problems as reasons trading allowances should be avoided.

9.8.4   Safety valve system

A third system, referred to as the *safety valve*, provides a hybrid of price and quantity constraints. This approach is essentially an emission cap and tradable permit system; however, the maximum (or minimum) permit price is capped. Polluters may either obtain permits in the marketplace or purchase them from the administrative authorities at a specified trigger price, which can be adjusted as the market changes. The flexibility to adjust the system as new information becomes available provides a potential mechanism for addressing the fundamental disadvantages of quantity versus price systems. By setting the trigger price sufficiently high, or restricting the number of permits to a low enough level, this system can mimic either a pure quantity or a pure price mechanism.[35]

9.8.5   Welfare effects of capitalism and communism

One of the governing principles of traditional capitalism is that of the *invisible hand*. The invisible hand is a metaphor coined by the father of modern capitalism, Adam Smith. In his book *The Wealth of Nations*, Smith shows that in a free market, a person pursuing his or her own self-interest also tends to promote the good of his or her community through a principle that he labels "the invisible hand." In essence, Smith argues that each individual who attempts to maximize revenue for himself or herself also maximizes the revenue of society as a whole, as the latter is identical to the sum total of individual revenues. Although capitalism has largely been successful at promoting the wealth of societies, it is not without its shortcomings, particularly with respect to environmental protection. As we will soon see, one of these shortcomings involves the concept of environmental market failures.

Under a purely free market, which lacks any government constraints, a polluter who freely commits revenues to prevent environmental degradation loses out to competitors who do not. But socialist systems do not necessarily fare any better. They frequently face political pressures such as mandated quotas to maximize short-run production that may result in equal or even greater environmental damage. Consider Russia and China, poster child examples of the environmental destruction that can occur in an effort to meet state-mandated quotas.

9.8.6   International trade

Traditional economic trade theories are based on differing comparative advantages that increase the welfare of all contributing countries. However, many economists believe that this thinking no longer holds, as capital flow is mobile, that is, nations can effectively "export" their pollution and multinational corporations can simply move operations to other less-regulated countries. For this reason, Daly and Cobb conclude that production, whenever feasible, should take place in the host country.

9.8.7   Environmental regulatory mechanisms

Early environmental regulations limited the levels of effluents and emissions in ways that sometimes violated basic economic logic by failing to account for the differing capabilities and circumstances of polluters. Such "command and control" attempts often backfired when polluters ignored them, went to court, sought political relief, moved elsewhere (including overseas), and sometimes just closed shop. As a result, policymakers currently consider two sets of options in pursuit of environmental–economic goals: nonmarket and market-based policies.

9.9   Interim energy policy?

All too often, the debate over energy tends to be a polarizing one. Free-market advocates argue that we continue increasing unsustainable demands until market forces compel the implementation of draconian measures that, their critics maintain, unfairly penalize a majority of have-nots. Public sector economic management advocates argue for massive investments in theoretically sustainable energies that require massive public sector subsidies that, their critics maintain, cannot be sustained. When it comes to energy policy, it may be that we have been thinking too much in terms of black and white. Consider the following interim energy policy that seeks to shift the debate from ideology-driven motives to a flexible and pragmatic policy option that can respond to the realities of resource limitations and market forces.

9.9.1   Recent economic history of energy

Throughout the 1960s and early 1970s, both economic growth and energy consumption were driven by the mature, developed nations, but this trend no longer holds true. Emerging markets of developing nations are now principally responsible for driving the growth in energy consumption. Economic growth in countries outside the Organization for Economic Cooperation and Development (OECD) increased from about 20% in the early 1990s to 50% today. These same poor but rapidly growing nations are also responsible for a disproportionate share of increase in global energy usage. Much of this disproportionate increase is because developing countries are less efficient in their energy usage. Energy intensity (i.e., the energy needed to produce one unit of GDP) in the developing world is three times greater than that in the developed world. This observation is evidenced by the fact that it takes 3.4 barrels of oil equivalent to produce $1000 worth of GDP in the non-OECD countries versus 1.1 barrels of oil equivalent in the OECD nations.[36]

Developing countries are much less energy efficient than most industrialized nations, partly because of various wasteful practices such as widespread subsidization of energy, which leads to misallocation and inefficient use of energy. Moreover, in emerging economies, growth has caused unprecedented social and structural changes; hundreds of millions of people have left low-energy-intensive activities, such as agriculture, for energy-intensive activities, such as factories and industries—while lagging behind industrialized nations in managing the environmental impacts of such massive shifts.

The growth in power generation within non-OECD countries also tends to exceed the growth of GDP, which is not the case for OECD nations. This increased need for power is a key reason coal has become the fuel of choice in many nations, particularly non-OECD states. In addition to being more widely available than oil and gas, coal is also more abundant and relatively cheap. On average, coal has become the world's fastest-growing major fuel source.

From 1970 through the late 1990s, global air emissions per unit of energy consumed declined. However, reflecting the increasing dependence of non-OECD nations on coal, carbon emissions per unit of energy consumed began to rise in 1999. Since then, carbon emissions have increased by about 2% globally and by almost 3% in the developing world. Much of this increase in carbon emissions can be traced back to the proliferation of coal usage.

There has also been a noticeable shift in the demand for oil from the developed nations to the developing world. Since 2007, oil consumption in the OECD countries has peaked and actually begun to shrink by some 8%. So, the non-OECD consumption is now responsible for the growth in global oil demand.

9.9.2   Oil's roller-coaster ride

In 2006, crude oil prices started falling sharply (declining from $80 per barrel). To halt this price decline, OPEC cut production by nearly 1 million barrels per day. Oil rallied—from $50 a barrel at the beginning of 2007 to $147 by mid-2008.[37] You may wonder how a production cut of a mere 1 million barrels per day out of a market of 81 million barrels per day could cause such a sharp increase.

The global oil market needs a degree of redundancy to operate efficiently. As OPEC's production cuts created tighter inventories, the price decline reversed and began moving upward. Crude prices sailed past $120 per barrel in the early summer of 2008. Sensing that such high prices could lead to global economic turmoil, Saudi Arabia announced a unilateral increase in production. Unfortunately, just as the increased production found its way into the crude oil inventory, the global credit crisis struck in 2008, causing global oil demand to plummet. Driven by rising inventories and falling consumption, prices slid from $147 per barrel in the summer to $34 by the end of 2008.

Racing to catch up with falling demand, OPEC announced they would significantly cut production. But OPEC's move lagged behind events and could not prevent prices from taking a nosedive. Moreover, OPEC's policies caused global environmental impacts. For example, during 2008, the price of coal, like that of its cousin oil, rose through the summer and then collapsed. The price of coal fell longer and further than that of oil because there was no coal cartel to stabilize it, thereby increasing the larger environmental impacts associated with the coal industry.

9.9.2.1   Coal vs. gas

In the 1990s, combined-cycle gas turbines increased the capacity of gas-fired power generation. However, this trend did not last. As gas consumption expanded, supplies tightened and prices rose, eroding the competitiveness that gas had over coal. However, the principal reason coal emerged as the fuel of choice over the last decade, particularly in developing nations, is that it is relatively abundant. While oil production tends to be concentrated in small geographical regions, the same is not true for coal. Moreover, the market for coal is much wider than that for oil.

In contrast, gas is highly concentrated; 60% of the global natural gas supply is consumed by regions that collectively control 14% of the reserves. Yet, geologically, there are vast "unconventional" pockets of gas around the world, which had been impossible to tap until recently. If unconventional natural gas can be developed, it could supplant coal as the local fuel of choice. Increased gas production is partly the result of high-tech advancements in U.S. gas production. Tight supplies and rising prices, in tandem with technological advances (horizontal drilling, which eases access to layers of oil or gas, and hydraulic fracturing), have brought large U.S. gas deposits online. For example, production from unconventional U.S. gas deposits has nearly doubled over the past decade.

9.9.2.2   Implications for global energy production

As gas production has increased, prices have fallen. Gas prices are once more competitive with the prices of coal. Gas-fired power generation in the United States now rivals coal-fired power generation in terms of capacity. Other nations are now beginning to follow in the footsteps of the United States, tapping unconventional gas resources around the globe. Among the major fossil-fuel sources, coal is still the most competitive and coveted. However, natural gas is increasingly becoming more competitive.

If unconventional natural gas is widely developed, it could become the local fuel of choice and eventually supplant coal.

World energy usage is dominated by fossil fuels. For instance, oil currently accounts for 35% of global primary energy usage, coal for 29%, and gas for 24%. Hydropower and nuclear power together account for 12% and renewable energy for less than 1%.[38] Most reasonable forecasts do not foresee a significant shift away from fossil fuels before at least 2030. Thus, fossil fuels are likely to remain the principal fuel sources for some time to come. In terms of fossil-fuel dominance, coal is second to none. Its use is clearly increasing. From an environmental standpoint, this trend does not bode well. Yet, we also know that unconventional sources of natural gas can provide an enormous resource that has the potential to compete with coal. Natural gas burns much cleaner than coal. One kilowatt-hour of electricity produced from natural gas emits a little more than half the amount of carbon as producing the same amount of electricity from coal. The ability of natural gas to serve as a substitute for coal is important for its future.

9.9.3   Pragmatic interim option

Greenhouse emission policies are often portrayed as fostering an expeditious and complete transition from a carbon-based to a carbon-free world. From a technological standpoint, however, the concept of non-carbon-producing energy is still a distant goal. This argues for a fundamental rethinking of international environmental and energy policies. What if natural gas is used to bridge the transitional gap, until new nonpolluting technologies are brought online? Thus, an interim environmental policy that encourages the use of natural gas over coal could be promoted as a means of ensuring energy security while stemming greenhouse emissions until a transition can be made to non-carbon-producing energy sources.

Discussions, problems, and exercises

1. You have been asked to outline a cost–benefit study for an environmental impact analysis of a proposed uranium mine on the Navajo reservation. The Navajos have employed a firm of skilled environmental attorneys and an environmental ecologist. Several environmental advocacy groups have also offered their services. The Navajos and their supporters are concerned about how the project can be configured to improve the economic life of their community while minimizing the risk of harm to their environment and health. Consult the NEPA regulations (available on the Web) for guidance on how to scope your approach.
2. Based on Section 9.9, develop and outline a systematic line of inquiry that would help national and global policymakers assess the issues

associated with an interim natural gas policy, and the specific market and nonmarket policies that could be considered.

Notes

1. Association of Environmental and Resource Economists, http://www.aere. org/index.php (accessed July 9, 2010).
2. http://www.springer.com/economics/environmental/journal/10640.
3. The International Society for Ecological Economics, http://www.ecoeco. org/ (accessed July 9, 2010).
4. Wikipedia, "Ecological Economics," http://en.wikipedia.org/wiki/Ecological_ economics (accessed July 9, 2010).
5. K. E. Boulding, "The Economics of the Coming Spaceship Earth," in *Environmental Quality in a Growing Economy*, ed. H. Jarret (Baltimore, MD: Johns Hopkins University Press, 1966), 19.
6. M. Faber, "How to be an Ecological Economist," in *Ecological Economics* 66, no. 1 (2008): 1–7.
7. C. J. Cleveland, "Biophysical Economics," *Encyclopedia of Earth*, (last updated September 14, 2006).
8. C. H. Eccleston, *NEPA and Environmental Planning: Tools, Techniques, and Approaches for Practitioners* (Boca Raton, FL: CRC Press, 2008).
9. Calculator Soup, "Present Value Calculator," http://www.calculatorsoup. com/calculators/financial/present-value.php (accessed July 9, 2010).
10. 40 Code of Federal Regulations (CFR) Part 1502.23.
11. H. Daly, "Economics in a Full World," *Scientific American*, September 2005, vol. 293, 3.
12. N. Hanley, J. Shogren and B. White, *Environmental Economics in Theory and Practice* (London: Palgrave, 2007).
13. G. Grossman and A. Krueger, "Economic Growth and the Environment," *Quarterly Journal of Economics* 110 (1995): 353–77; World Bank, *World Development Report 1992* (New York: Oxford University Press, 1992).
14. D. I. Stern, "The Rise and Fall of the Environmental Kuznets Curve," *World Development* 32, no. 8 (2004): 1419–39.
15. Krueger, Alan B. 1995. *Economic Growth and the Environment.* Quarterly Journal of Economics 110(2): 370.
16. T. Panayotou, *Economic Development and the Environment*, (working paper, Harvard University and Cyprus International Institute of Management, 2003).
17. B. Field and M. Field, *Environmental Economics: An Introduction*, 4th ed., 2006. (New York: McGraw-Hill, 1994).
18. S. Hayward, "The China Syndrome and the Environmental Kuznets Curve," (American Enterprise Institute, November–December 2005). Available at http://www.perc.org/pdf/china_enviro_kuznets.pdf (accessed July 9, 2010).
19. A. Levinson, *The Ups and Downs of the Environmental Kuznets Curve* (Orlando, FL: UCF/Center conference on Environment, November 30–December 2, 2000), Available at http://www9.georgetown.edu/faculty/aml6/pdfs&zips/ups%20and%20downs.PDF (accessed July 8, 2010).
20. V. Suri and D. Chapman, "Economic Growth, Trade and Energy: Implications for the Environmental Kuznets Curve," in *Ecological Economics* 25, no. 2 (1998): 195–208.

21. W. Beckerman, "Economic Growth and the Environment: Whose Growth? Whose Environment?," *World Development* 20, no. 4 (1992): 481–496.
22. B. Yandle, M. Vijayaraghavan and M. Bhattarai, *The Environmental Kuznets Curve: A Primer* (The Property and Environment Research Center, 2000). Available at http://www.perc.org/articles/article688.php (accessed January 31, 2009).
23. M. E. Porter and C. van der Linde, "Toward a New Conception of the Environment-Competitiveness Relationship," *Journal of Economic Perspectives* 9, no. 4 (1995): 97–118.
24. K. Palmer, W. E. Oates and P. R. Portney, "Tightening Environmental Standards: The Benefit-Cost or the No-Cost Paradigm?" *Journal of Economic Perspectives* 9, no. 4 (1995): 119–32.
25. A Pigovian tax is one levied on companies to correct for the negative externalities generated by their activities. See Section 9.4 "Non-Market-Based Policies for Environmental Management."
26. Y. Bauman, "Paradigms and the Porter Hypothesis," January 28, 2004, http://www.smallparty.org/yoram/research/porter.pdf (accessed December 28, 2008).
27. Daly, H. and Cobb, J. "For the Common Good." (Boston, MA: Beacon Press, 1989).
28. A. Rabl and others, "Impact Assessment and Authorization Procedure for Installations with Major Environmental Risks," Final Report, 1999.
29. P. A. Diamond and J. A. Hausman, "Contingent Valuation: Is Some Number Better than No Number?" *The Journal of Economic Perspectives* 8, no. 4 (Autumn, 1994): 45–64.
30. R. Simons and K. Winson-Geideman, "Determining Market Perceptions on Contaminated Residential Property Buyers using Contingent Valuation Surveys," *Journal of Real Estate Research* 27-2, (2005): 193–220.
31. T. Tietenberg and N. Johnstone, "Ex-post Evaluation of Tradable Permits: Methodological Issues and Literature Review," in *Tradable Permits: Policy Evaluation, Design and Reform* (Paris: OECD, 2004).
32. ProQuest, "Environmental Economics: Basic Concepts and Debates," http://www.csa.com/discoveryguides/envecon/review2.php?SID=h5560nikuk62o2drcl43qa7830#v2#v2 (accessed July 9, 2010).
33. C. Philibert, "Certainty versus Ambition—Economic Efficiency in Mitigating Climate Change, International Energy Agency," (working paper series, IEA, Paris, 2006).
34. Carbon Trade Watch, http://www.carbontradewatch.org (accessed July 9, 2010).
35. D. H. Jacoby and A. D. Ellerman, "The Safety Valve and Climate Policy," *Energy Policy* 32 (2004): 481–491.
36. C. Rühl, "Global Energy after the Crisis," *Journal of Foreign Affairs* 89, no. 2 (2010).
37. Ibid.
38. Ibid.

section four

---

*Critical global
environmental issues*

chapter ten

Coming water wars

> Not all chemicals are bad. Without chemicals such
> as hydrogen and oxygen, for example, there would
> be no way to make water, a vital ingredient in beer.
>
> **Dave Barry**

Case study

As a diplomatic means of courting Egypt, the Soviet Union provided
funding and technical expertise for constructing the Aswan Dam. Work
began in 1958. On its completion, the mighty Aswan High Dam spanned
a length of 3830 m and a height of 111 m.

In terms of environmental damage, damming the Nile River was
a nightmare. Much of the lower Nubian flatland was flooded, and over
90,000 people were displaced. The dam created Lake Nasser, which
flooded valuable cultural and archaeological sites.

For thousands of years, silt had been deposited on the flanks of the
river during yearly floods. The silt, rich in nutrients, made the Nile flood-
plain fertile. This inflow of nutrient-rich soil is now blocked by the dam.
Because it has nowhere else to go, the silt is now deposited on the floor of
Lake Nasser. No longer renewed by the Nile silt, the delta has lost much
of its fertility. This lack of silt has also accelerated desertification because
farmers are abandoning fields that are no longer cultivable.

Another impact on the environment is the disruption of the local ecosystem. The building up of silt on the floor of Lake Nasser is also lowering
the water storage capacity of the reservoir. Mediterranean fishing has also
witnessed a steep decline because nutrients that used to wash down the
Nile to the Mediterranean Sea are now trapped behind the dam.

The construction of the dam has significantly increased the salinity in
the delta soils and groundwater. The salinity increased because the flow
of the Nile is not strong enough to hold back the Mediterranean saltwater,
which is now migrating back up the Nile and seeping into the delta soil
and groundwater.

As the river replenishes what is left of its sediment load, downriver it
also erodes farmland. Erosion of coastline barriers will eventually cause
the loss of brackish lake fishery, which currently provides Egypt with its

largest source of fish. Meanwhile, the Nile Delta is undergoing subsidence, which will eventually lead to the inundation of seawater into the northern portion of the delta that is now used to raise rice.

The red-brick construction industry, which uses delta mud, is also severely affected. There is significant erosion of coastlines (due to lack of sand, which was once deposited by the Nile) all along the eastern Mediterranean. Because the rich soil is no longer being replenished, farmers have had to turn to artificial fertilizers. This switch has resulted in chemical pollution of the Nile water and adjoining soils.

Traditionally, the marine ecosystem depends on the rich flow of phosphates and silicates into the Mediterranean Sea. Following the construction of the dam, Mediterranean fish catches have decreased by almost half. The dam has also been implicated in a rise in cases of schistosomiasis (bilharzia), spread by the thick plant life in Lake Nasser that hosts the snails that carry the disease. The dam has also increased the salinity of the Mediterranean Sea, which has affected its outflow current into the Atlantic Ocean.

> **Class problem:**   After reading this chapter, return to this problem. Consider this example in light of what you have read. If you had been advising Cairo, what policies would you have recommended for mitigating the impact of this dam? Consider this project in terms of the international environmental impact assessment (EIA) processes requirements described in Chapter 5. Do you think a thorough EIA process could have mitigated some of the project impacts? How would you have defined an EIA process so as to focus on impacts and mitigation measures that could reduce these impacts? Should the Aswan Dam have been built? What would have been the socioeconomic impacts if it had not been built (positive and negative)?

Learning objectives

A well-developed environmental policy will shape specific environmental legislation and controls that follow in its wake. After you have read the chapter, you should be able to

- Develop an appreciation for pressures that are being placed on Earth's finite sources of freshwater.
- Understand why demand factors are being placed on freshwater resources.
- Understand the factors that could lead to armed combat over these resources.
- Understand how water scarcity can affect food scarcity.

- Learn how water pollution is contributing to freshwater scarcity.
- Investigate some of the principal technological solutions and their viability in terms of reducing water scarcity.

10.1   Introduction

Water has been described as the elixir of life. A bottled-water company described its importance in the following ways[1]:

> It's why your heart beats.
> It's how your blood flows.
> It helps your lungs breathe, your joints move, and
> your muscles work.
> It's your body's cooling system.

As you will soon see, water's life-sustaining capability is much broader than what this advertisement might lead us to believe. But before we explore the coming water crisis, here is a trivia question: Who claims to have among the most prized drinking water in the world? Why, the residents of Bydgoszez, Poland, of course. One morning in 1973, these residents turned their water taps on and got beer! What happened? As it turned out, a damaged valve in a local brewery had diverted thousands of gallons of beer into the city's water supply. I have been told it took them several days to fix a simple broken valve. Humorous? Yes, but as we will see shortly, things like broken valves can be among the causes for a looming crisis.

The ancient Chinese understood all too well that the purposes for which water could be used were limited only by human imagination and inventiveness. In the sixth century BC, Lao-Tzu wrote the following[2]:

> In the world, there is nothing more submissive and
> weak than water. Yet, for attacking that which is
> hard and strong, nothing can surpass it.

Lao-Tzu certainly realized the importance that water plays in virtually every aspect of our lives—from military strategy to city water supply, navigation, bathing, and sustaining all life. The world's river basins have been called "cradles of human civilization." Some analysts have observed that the first civilization emerged only after people learned how to efficiently use large amounts of water. Somewhere around 5000 BC, the Mesopotamians observed that the periodic flooding of rivers fertilized the soil with nutrient-rich silts, sprouting bumper crops. You can almost picture in your mind's eye—the day some long-forgotten but enterprising Mesopotamian farmer dug the first ditch to divert water from the Euphrates River to his outlying farmland. Irrigation was born! I sometimes wonder what his first words were. Because this event, for its time, may have been as significant as Neil Armstrong's first walk on the moon. Mesopotamian farmers went on to form the world's first irrigation-based civilization. Farming flourished as communities began reaping bumper crops. Mesopotamian civilization began to evolve, as many people were finally freed from the drudgery of working the land—allowing them to become artisans, teachers, and traders.

Millennia came and went. People eventually learned to transport water from increasingly remote sources and harness its power. Engineering efforts, such as dams and aqueducts, became ever more sophisticated. As the Industrial Revolution blossomed, the world's populations mushroomed. Demand for water began to explode. Conflicts and skirmishes ensued. Tens of thousands of monumental water projects have now been completed. Today, water is central to the religious practices and cultural identity of untold millions around the world. But these marvelous accomplishments have left us with a paradox. The global scale of technologies that control water may come to jeopardize the stability of our very society.

Many of us were raised in an era in which we behaved as if our "cup runneth over." But this era of plenty is coming to an end. The looming crisis is so severe that it has led pundits to predict that water may become the socioeconomic equivalent of oil in the twenty-first century. Indeed, water scarcity may be to the coming decades what the oil-price shocks were in the 1970s. Within a few short decades, we may be identifying countries not as "developed" and "developing," but as "upstream" versus "downstream." Let us see what leads to such a startling conclusion, and how it will affect our lives.

10.2   Is the glass half full or half empty?

From 2500 BC, when King Urlama of Lagash diverted water in the Tigris/Euphrates Valley in a border dispute with nearby Umma,[3] to 1924, when California farmers blew up part of an aqueduct that served Los Angeles, societies have fought, negotiated, and rearranged geographies to get this badly needed resource. When populations and economies grow, the demand for water reaches exploding levels. Ironically, this is happening at a time when water resources are being rapidly degraded and polluted. Inequalities in the distribution of water have also been increasing, exacerbated by poor water management. Consequently, the world's food supply, human welfare, ecological health, and economic prosperity are in jeopardy. And as if this is not enough, water scarcity now threatens even national security! Many might ask, "Our planet is covered with water. So what is the big deal?"

10.2.1 Water, water everywhere

We would like to believe there is an infinite supply of fresh water. Unfortunately, this is not the case. With the exception of a little water added now and then from colliding "snowball" comets, most of the water has been on this planet for millions of years—the product of outgassing from volcanic eruptions and collisions with comets. Although water may appear to be abundant, covering over 70% of the Earth's surface, fresh water comprises only a small 2.5% of the Earth's total water content. Even this statistic is misleading, because freshwater is distributed unevenly across our planet. Nearly two-thirds of the world's freshwater is locked away in inaccessible glaciers and the Earth's ice caps. Only a minuscule amount—far less than 1%—is drinkable, easily accessible, and renewed each year through precipitation. That is all we have to supply the needs of 6 billion people, growing at the rate of nearly 80 million per year.[4]

This is not the end of the story. Many rivers flow erratically, running high when water is needed least and low when it is needed most. Almost one-fifth of the water in the Earth's rivers flows through remote areas such as the Amazon basin and the Arctic tundra. In many of the poorer countries, monsoons bring between 70% and 80% of the year's rainfall in just 3 months. Much of this water is lost through runoff and evapotranspiration.

10.2.2 Will the mere mention of water soon elicit fear?

Today, the seriousness of water scarcity issue is increasingly coming to the forefront of international debate among leading scientists and world leaders alike.[5] In a recent survey, water scarcity was ranked second out of 37 pressing environmental issues, by 200 leading scientists from 50 different counties; fresh water pollution was ranked fourth among these concerns.[6]

Soon, the mere mention of the word "water" may cause the hair on the back of your neck to stand up. When Americans (and many other Westerners for that matter) think about water shortages (assuming we think about them at all), we tend to perceive them as "localized" or as Third World problems far removed from our affluent society. Most of us do not regard such problems as particularly worrisome. We are confident that such problems can be handled with the proper mix of investment, technology, and infrastructure. We expect that when water feuds arise, they will be resolved by negotiation or at worst in the courtroom. While this has generally been the case, it may not be so in the future. We are headed for a crisis on both the domestic and international fronts.

10.2.3   Demand is exploding

For thousands of years, water has been relatively plentiful and often virtually free. In the last century, we humans vastly expanded our water usage
to meet the needs of agriculture, industry, and a burgeoning population.
Between 1900 and 2000, demand for fresh water increased sixfold—twice
the rate of population growth. Although fertility rates have dramatically
slowed, causing the average age to dramatically increase, the Earth's overall population is still increasing, albeit at a diminishing rate; currently,
nearly 80 million people are added to the planet each year, increasing the
demand for this scarce resource.[7]

By some estimates, the global population is hurtling toward 10 billion
or more by the middle of this century. Yet the quantity of water available
today is approximately the same as that was available when civilizations
first arose thousands of years ago. This means the amount of water available to *each person* has declined dramatically with time.

In addition to this dilemma, the World Bank states that global water
usage is doubling approximately every 20 years, more than twice the
rate of human population growth.[8] As the global population continues to
increase and cities and industries mushroom, it is unlikely that the global
freshwater supply will keep pace with demand. Soon, it will be time to
pay the piper. And here is why: most cost-effective sources of water have
already been tapped.

One of the strongest conclusions reached in the most recent report by
the Intergovernmental Panel on Climate Change (IPCC) states that global
warming will lead to "changes in all components of the freshwater system." It concludes that "water and its availability and quality will be the
main pressures on, and issues for, societies and the environment under
climate change."[9]

10.2.4   What the experts have to say

The Ceres is a group of sustainable investors that urges firms and investors to undertake water risk assessments as a matter of urgency. According
to a new report from the Ceres group, global water shortages are a greater
and more imminent threat to businesses than oil supplies running out.
The report *Water Scarcity & Climate Change: Growing Risks for Businesses
and Investors*, notes that water shortages are already causing disruption to
large numbers of businesses and warns that the situation will worsen as
the combination of population growth and climate change places further
pressure on water supplies.[10] Nestlé's chairman, Peter Brabeck-Letmathe,
has called water scarcity a bigger challenge than energy security: "I am
convinced that, under present conditions and with the way water is being
managed, we will run out of water long before we run out of fuel."[11]

Christine Whitman, former director of the U.S. Environmental Protection Agency (EPA), called water quantity and quality "the biggest environmental issue that we face in the twenty-first century."[12] According to Sandra Postel, director of the Global Water Policy Project in Amherst, Massachusetts,[13]

> A number of areas could enter a period of chronic shortages … including much of Africa, northern China, pockets of India, Mexico, the Middle East and parts of western North America.

Postel predicts that providing drinking water for the 2.5 billion people expected to be added to the world's population over the next 30 years would consume a volume equivalent to 20 Nile Rivers or 97 Colorado Rivers. She notes that "It is not at all clear where that water could come from on a sustainable basis."[14] Robert Wetzel of the University of Alabama, a leading expert on international water issues, has the following to say[15]:

> Globally, the single greatest environmental crisis that faces humankind in the next century will be the unavailability of drinking water. Food, energy, and everything else depends on it.

To educate the populace about the coming crisis, deceased former U.S. Senator Paul Simon authored a book titled *Tapped Out: The Coming World Crisis in Water and What We Can Do about It*. Simon sounds the alarm when he writes the following:

> Within a few years, a water crisis of catastrophic proportions will explode upon us—unless aroused citizens … demand of their leadership actions reflecting vision, understanding and courage.

Ah, but pundits will retort, "We can simply desalinize water from the ocean's inexhaustible supply." As we will see, it is unlikely that this technology will "save" most countries.

10.2.5 Minimal water requirements

Many people are surprised to learn how much water is required to sustain a person, even at a minimal lifestyle. An absolute subsistence level necessitates approximately 1700 m^3 (61,000 cubic feet) of water annually per person. A country is generally considered to be experiencing *water stress* if it averages less than this amount. A country is considered to be experiencing a *water shortage* if it averages less than 1000 m^3 (36,000 cubic feet).

Most Westerners, particularly Americans, consume an amount of water far greater than this. For example, it takes 291,000 gallons of water to supply a single person with a modest, low-meat diet for a year.[16]

10.3 It is bad, and about to get much worse

Already 450 million people in 29 countries suffer from water-shortage problems.[17] According to United Nations (UN) population projections, and assuming that the renewable water resources will remain unchanged, some 34 countries are projected to face water stress by the year 2025. Such large numbers can only spell problems for all of us. By 2025, the population living in water-scarce countries is projected to rise to between 1 billion and nearly 3 billion. Africa and parts of western Asia appear particularly vulnerable to increasing water scarcity. By 2025, it is estimated that 1.1 billion Africans, or three-quarters of the continent's population, could be living in water-stressed countries.[18] The list could also include northwestern China, western and southern India, large parts of Pakistan and Mexico, and the western coasts of the United States and South America."[19] China, home to 21% of the world's population, has just 7% of the water. Further, 9 in 10 Chinese city groundwater systems are fouled by human waste, industrial toxins, and pesticides.

In Asia and Latin America, some people are so desperate that they must buy their water from street vendors at inflated costs, paying as much as 30% of their income for water. In contrast, people in the United States pay about 1–2% of their household income for water supply and treatment.[20] According to a recent UN report, more than a billion of the world's people lack a safe supply of drinking water; this number is expected to double in the next 30 years.

In the United States, rain has saturated Silicon Valley for the last several years. However, the area's high-tech businesses are worried about what will happen when a drought inevitably strikes. Justin Bradley, director of environmental programs for the Silicon Valley Manufacturing Group (a 170-member trade association), cautions,[21]

> Even though we've had very wet weather, our main
> water suppliers have barely met demand ... given
> our projections for growth, and our history of dry
> years, we're on the verge of difficulties.

In the future, water will become increasingly scarce, particularly throughout the Southwest. Las Vegas, Nevada, is the most arid area in the United States. Approximately 70% of the state's population resides near Las Vegas. The city has been growing at the phenomenal rate of more than 1500 residents a week. State officials know that a severe water shortage is looming sometime in the future. They are desperately taking measures to postpone the day of reckoning as long as possible.

To compound the problem, much of the American water-system infrastructure is breaking down. Fissures are spreading in the 700,000 miles of pipes that deliver water to consumers. This also increases the chances of outbreaks of waterborne diseases. Cost estimates for overhauling the system range from EPA's figure of $150 billion to a stratospheric price tag of $1 trillion as estimated by a water industrial coalition. This could result in costs as high as nearly $7,000 per household in some areas. One thing is for sure: the price of water will increase in the future.[22]

Now, let us look at our friends to the north. Canada and the United States have quarreled over water for decades. Every time a plan emerges to divert water southward to the United States, Canadian citizens rise up and stop the proposal. In the 1960s, a proposal was floated to divert the flow of the MacKenzie River to the United States. Not surprisingly, Canadians greeted this proposal with much less enthusiasm than their American counterparts.

Yet such disputes pale in comparison to those between United States and Mexico, who have been squabbling over water usage for much longer. In the 1890s, they came to blows over rights to the Rio Grande River. Ironically, both countries have failed to protect the river. It has become a flowing "latrine" for raw sewage and industrial waste. The Rio Grande has now been declared America's most endangered river.

10.3.1   Are the world's aquifers running dry?

Much of the Earth's water supply resides in underground geologic formations called "aquifers." Underground aquifers contain 60 times more groundwater than all of the world's surface waters (lakes, rivers, etc.) combined. These underground aquifers are recharged and purified as water percolates from the surface into the underlying soil and rock. This water resource can be depleted if groundwater is pumped out at a rate that exceeds sustainable replenishment. Indeed, a number of the world's most important aquifers are being pumped faster than nature can replenish them. Severe depletion of an aquifer can even permanently destroy its underground storage capacity. Water tables are falling on every continent. The International Water Management Institute warns about groundwater depletion[23]:

> Many of the most populous countries of the world—
> China, India, Pakistan, Mexico, and nearly all of the
> countries of the Middle East and North Africa—have
> literally had a free ride over the past two or three
> decades by depleting their groundwater resources.
> The penalty of mismanagement of this valuable

> resource is now coming due, and it is no exaggera-
> tion to say that the results could be catastrophic for
> these countries, and, given their importance, for the
> world as a whole.

But one does not have to travel to the remote corners of North Africa or the Middle East to see what lies ahead. Let us take the case of United States' Ogallala Aquifer, one of the world's largest sources of groundwater; the Ogallala runs from South Dakota to Texas. It is so immense that it lies beneath parts of eight U.S. states. This aquifer is being depleted so rapidly that some farmers, who once depended on it, are now being forced to rely on rainwater and dryland farming practices, which has significantly lowered their yields. In less than a decade, the amount of farm acreage supported by this aquifer fell by nearly 20%.

What is happening to the Ogallala Aquifer pales in comparison with similar problems in India. India is exhausting its underground storage of water faster than any other large country in the world. Throughout most of India, withdrawals take place at twice the rate of natural recharge, causing water levels to drop by a precipitous 1 to 3 m per year. Some Indian households are reported to be spending a staggering 25% of their income on water. In India, a 1996 national assessment commission found that water tables in critical farming regions are dropping at such an alarming rate that it jeopardizes as much as one-fourth of the country's grain harvest.

10.4   Wars and rumors of war

With respect to water problems encountered in the American Old West, Mark Twain once mused, "Whiskey's for drinking. Water's for fighting!" With this in mind, it should come as no surprise to find that the words "rival" and "river" share the same Latin root; a rival is someone who "shares the same stream." Enough said.

Water scarcity is a complex issue subject to competing interests, which are often entrenched along historic, ethnic, and social divisions. This can fuel preexisting tensions, and create new tensions where previously there were none. Writing in a recent issue of the *Straits Times,* Alan DuPont of Australia's Strategic Defense Studies Center poses a provocative question: "Will future wars be fought over increasingly scarce freshwater resources?"[24] Let us see what the future holds.

10.4.1   Upstream vs. downstream

Some observers have warned that in the near future, water could replace oil as the focal point of disputes. In the twentieth century, national security was primarily characterized by the Cold War, which was largely defined along ideological lines. Many experts believe that in the twenty-first century, national security will be more influenced by failing states, environmental issues, and economic conditions, with water and petroleum scarcities at the forefront of environmental issues.[25] As the world population increases and more people strive to increase material wealth, the accessibility of water is likely to be the subject of great competition and future conflict. The leaders of countries facing water crises will be faced with options and threats that could affect every man, woman, and child around the globe.

A staggering 40% of the world's population relies on shared river basins. In the Middle East and African powder kegs, this dependence jumps to 50%. Approximately 260 of the world's rivers flow through two or more countries. In the majority of such cases, no treaty exists governing how water should be shared between the countries. Countries located upstream usually have a distinct advantage over their downstream neighbors. Tensions are mounting as demand taxes the supply of water in these regions.

Postel writes,

> Tensions over water security have the potential to incite civil unrest, spur migration, impoverish already poor regions, and destabilize governments—as well as to ignite armed conflict.[26] In five water hot spots (Ganges, Aral Sea region, Jordan, Nile, and the Tigris and Euphrates), the population of the nations in each basin is expected to increase by 30% to as much as 70% by 2025. The question that we are left with is whether we can resolve future disputes peacefully through equitable allocation. Unfortunately, history does not have a good track record in this regard.

Former Secretary General of the UN and Egyptian Minister of State for Foreign Affairs, Boutros Boutros-Ghali, in addressing U.S. Congress, said,[27]

> The next war in our region will be over the waters of the Nile, not politics … The national security of Egypt is at the hands of eight other African countries bordering on the river Nile.

Former Israeli Prime Minister Yitzhak Shamir has also indicated that water, or the shortage thereof, may spark the next war between his state and its neighbors.[28] The late King Hussein of Jordan echoed similar sentiments in 1990 when he said that water was the only issue that could prompt a war between Jordan and Israel.

10.4.1.1   Around the world

Consider the plight of Peru. Rising demand for irrigation and drinking water is draining the local aquifer faster than it can recharge, and a scheme to channel more water from the Andean highlands, which receive seasonal rainfall, is still in the planning stage. However, with an expanding agricultural base and growing cities throughout South America, experts fear that climate change will exacerbate water scarcity, pitting city residents against rural residents and people in arid regions against those in areas with abundant rainfall. The conflict will be further intensified as a result of the dispute that exists between Andean mining companies and farm communities over this increasingly scarce resource. Plans for rerouting rivers or drilling raise problems for which neither scientists nor policymakers have ready solutions. Brazil plans to support its increasing energy needs by damming the Amazon River, which will further disrupt the region's hydrology.[29] How much water can be diverted without destroying fragile ecosystems? Who should be given priority: thirsting farms or swelling cities? Cities or the hydroelectric dams that power them? Communities or mining operations?

Compounding the problem, three-quarters of Peru's wastewater is discharged untreated into rivers, lakes, and the Pacific Ocean; dozens of rivers are polluted with cadmium, arsenic, mercury, lead, and other metals from mining operations. Large downstream farming operations are calling for a major infrastructure project to channel water from the highlands and disperse some of it through desert canals to recharge the aquifer. Upstream, small farmers challenge this scheme, arguing that it could dry up Andean bogs, an ecosystem that is not well understood.

10.4.2   First blood

In 1964, an Arab summit in Amman, Jordan, decided to divert the headwaters of the Jordan River, depriving Israel of its main water supply. Despite Israel's warning that it considered such a diversion to be an infringement of its national rights, work soon began on the Syrian and Jordanian side of the border. Air raids and artillery duels soon erupted. Eventually, Israel launched air strikes deep into Syria, destroying the proposed dam site and forcing the Arab states to call off their scheme.

In 1997, Malaysia, which supplies about half of Singapore's water, threatened to cut off that supply because Singapore criticized its governmental policies. In Africa, relations between Namibia and Botswana have been severely strained by Namibia's plan to build a pipeline to divert water from the shared Okavango River to eastern Namibia.[30] South African troops invaded Lesotho, a tiny independent mountain kingdom, in 1999. Ostensibly, the attack was launched to restore order in the face of a coup. However, regional newspapers reported a different story. A principal objective of the attack was to protect a South African–built system that pipes water from Lesotho into South Africa's arid industrial heartland. This has been called the region's "first water war."[31]

10.4.2.1   Middle Eastern water law

To underscore the importance that water holds in the Middle East, the term for Islamic law (or *shari'aa*) stems from a word meaning the "sharing of water." Many Arab countries have rules governing water, which predates Islam itself. For example, people who dig a well have the first right of use; however, they cannot deny the use of the well for drinking to either man or beast. A man lowering a container into a well has a right to only the amount of water that fills it at that precise moment. Laws pertaining to the sharing of rivers and aquifers across borders are less clear. Recently, Islamic extremists have begun to include water issues in their radical literature. Their publications are of course carefully calculated to further fuel conflicts along Islamic versus non-Islamic lines.

10.4.2.2   Middle Eastern powder keg

Although they represent 5% of the world's population, the Middle East and North Africa contain less than 1% of global freshwater resources.[32] Israel, Algeria, Qatar, Saudi Arabia, Somalia, Tunisia, the United Arab Emirates, Yemen, Jordan, and Kuwait are currently on the list of water-scarce countries. Paradoxically, the oil-rich Arab countries of Saudi Arabia, Kuwait, Qatar, Bahrain, and the United Arab Emirates are also five of the nine countries in the world that have the least water resources per capita.[33] The soaring population in the Middle East only aggravates an already ominous problem. By 2025, Egypt, Ethiopia, Iran, Libya, Morocco, Oman, and Syria are also expected to be added to the list.

Conflicts over Middle Eastern water resources are already coming to a head. Most of the water flowing through the Middle East and North Africa is supplied by just three shared river basins: the Nile, Jordan, and the Tigris and Euphrates. Now let us consider the real problem. The population depending on these rivers is expected to double within the next two to three decades.

Nearly all of the underground aquifers in these countries have been exploited; there are very few new reservoirs available to meet future needs. If these aquifers continue to be used at the current rate, they may be exhausted within the next 10–20 years. In Saudi Arabia, for example, 90% of the water resources used for agriculture come from nonrenewable underground aquifers.[34] In Yemen, where the per-capita usage is 15% of the U.S. average, groundwater aquifers are being pumped at a rate 30% greater than they are being recharged.[35]

Countries located upstream along the three principal rivers mentioned earlier tend to believe that they have the right to control the flow of these rivers. Not surprisingly, the countries lying downstream think this is a very poor idea, and they may not be shy in exercising military force if the situation reaches a flash point.

As an example, in 1989, Ethiopia's ambassador was summoned to the foreign office in Cairo. He was asked to explain the presence of Israeli hydrologists and surveyors studying areas near the Blue Nile for the possibility of building dams. The Blue Nile contributes about 85% of the annual water flow reaching Egypt. Egyptian officials made it clear that such action would be dealt with harshly. Egyptian parliament members began making speeches indicating that military action would be taken against Ethiopia if the Blue Nile was diverted.[36] Ethiopia halted carrying out its ambitions.

Former Soviet President Mikhail Gorbachev, who now heads Green Cross International, an environmental think tank, met with Arab and Israeli leaders to discuss the water problem and regional stability. Gorbachev had the following to say[37]:

> I can say that concern was expressed in particularly strong terms by Mr. Barak, who said that unless we solve the problem of freshwater resources in the Middle East, in the next 10 to 15 years we could see a conflict ... that could be worse than all the conflicts we have seen here in this region.

10.4.2.3   Gaza Strip

Upon venturing into the Gaza Strip (a journey that nearly ended my life), I was unprepared for the conditions that I found. The conditions were appalling, among the worst I have seen in my travels. The water-shortage problem only reinforces this observation. A recent World Bank report cites Gaza as[38]

> One of the most extreme examples of the water crisis ... where each Palestinian now has access to less than 15 gallons of water per day, compared to 800 gallons of water per average American. ...

In Gaza, aquifers are pumped at a rate more than double the annual rainwater recharges. Obviously, such a rate is unsustainable. Without substantial changes, the situation can only worsen.[39] Says Aaron Wolf, a Middle East water expert at Oregon State University in the United States,[40]

> The question of Palestinian water rights has to get
> resolved before the peace process gets too far ...
> The situation in Gaza is critical ... There is a level of
> shortage that needs immediate attention, regardless
> of costs.

10.4.2.4 Turning seawater to fresh water: The Achilles' heel

A few wealthy Middle Eastern countries rely heavily on *desalination* of sea-water to meet their needs. So much so, that 60% of the world's desalination capacity is concentrated in the Persian Gulf. Some of these desalination plants are nearly the size of a small town. Simple logic dictates that since they have lots of money, they will simply continue to expand their desalination capacity. But here is the problem. What concerns Saudi officials is that huge desalination plants, which supply a product more precious than oil itself, are also the Achilles' heel of such countries. They are the perfect targets for a potential enemy. A few well-orchestrated attacks on these "soft" facilities might bring millions of people to their knees in a matter of days. During the 1981 Gulf War, the Allies targeted Baghdad's water supply system, while the Iraqis attacked Kuwait's desalination plants. This is the nightmare scenario facing several oil-rich Persian Gulf countries.

10.5 Can we feed ourselves?

Many people believe that the biggest impediment to feeding the world is simply a scarcity of farmable land. Chapter 12 explains why a shortage of water will be one of the principal culprits responsible for future famines. According to the World Commission on Water, which includes eminent scientists, Nobel laureates, and policymakers,[41]

> Water scarcity, not shortage of land, will be the main
> constraint to increased agricultural production in
> developing countries in the coming years.

With this in mind, consider the following observation: about 1000 tons of water are required to produce 1 ton of grain. Yes, this means a ratio of 1 to 1000. It is little surprise that many experts believe that the impending water crisis may cripple the world's ability to feed itself.

Sandra Postel concludes that by the year 2025, the number of people living in water-impoverished countries is projected to climb from 470 million to 3 billion![42] In her view, there will not be "enough exportable grain to meet increased import demands at a price the poorer countries can afford." Can poor nations really depend on the generosity of grain-rich nations to fill the coming food void? What will happen if the poorer nations cannot secure sufficient food supplies through peaceful approaches? What are the risks to global stability when food suppliers are eventually forced to choose between supplying increasingly hungry countries (China, Middle Eastern countries, India, Pakistan, and others) when they all place demands on the world's limited grain supply at the same time? Will hungry nations wage wars to feed their people? Will Americans and others around the world accept rationing of their own food to free up what is needed to feed billions of others worldwide?

10.5.1   Cropland

The bulk of the world's food supply is derived from three major sources: rangeland, fisheries, and cropland. Because of overgrazing, approximately 20% of the Earth's rangeland has lost significant productivity. Likewise, the world's fisheries are being decimated by overfishing.[43] This implies that the primary food source that will be exploited to feed the additional 3 billion or more humans expected by the middle of this century will be cropland. But here is the problem. The choicest cropland is already under cultivation. Most of the remaining land available for cultivation will probably be only marginally productive. Environmental degradation will take its toll, such that any newly added cropland may be roughly offset by urbanization, erosion, and salination. Now the clincher: Where will the additional food necessary to feed several billion people come from?

It has been estimated that groundwater resources are being *overpumped* to the tune of at least 160 billion m^3 per year—an amount equal to the annual flow of two Nile Rivers! Remember that it takes a mind-numbing 1000 m^3 of water to produce 1 ton of grain. Some 160 million tons of grain—nearly 10% of the global food supply—depends on a practice that is depleting groundwater aquifers worldwide.[44] Obviously, this is an unsustainable practice that cannot continue indefinitely.

10.6   Water runs dirty

Pollution has contaminated much of the world's water resources to the point where they are now unsafe to use. Mismanagement and pollution of our oceans, lakes, rivers, and groundwater resources are severely damaging ecosystems, threatening fish species with extinction, and intensifying hostilities between countries vying for clean water. For example, three-fourths of Poland's river water is now too contaminated for even industrial use, let alone drinking.[45]

Agriculture is one of the biggest, if not the single largest, sources of water pollution, even more than industries and municipalities. For example, in virtually every country where agricultural fertilizers and pesticides

are used, they have contaminated surface waters and groundwater aquifers. Over 90% of Europe's rivers have high nitrate concentrations, mostly
from agrochemicals.[46]

It is estimated that an astounding 95% of the world's cities are still
dumping raw sewage into their waters. But the story does not end here. In
the developing world, hazardous chemicals and industrial waste are routinely dumped into what were once pristine water bodies. What was once
clean water has been polluted beyond use. Polluted water often becomes
a medium for spreading dangerous diseases. Precious resources that will
be vital in heading off the coming water crises are being destroyed before
our very eyes.

10.6.1   Wastewater treatment

Around the world, more than 1.5 billion people have no safe water and
sanitation facilities. Eighty percent of the people in India (700 million
people) defecate into buckets or onto open land. In Bangladesh, the picture is even worse—the figure is 90%. Contaminated groundwater is then
drawn from feces-polluted aquifers. Where there is no sanitation, contaminated water is stored in contaminated containers (because they cannot be
washed), open to insect-, parasite-, and animal-borne diseases.

Then there is Latin America; despite official rhetoric, only about 6%
of the wastewater generated in Latin America is properly treated and disposed of. Nearly 3 billion people worldwide lack even minimal sanitation
facilities. The problem is so severe that a child dies from a water-related
disease every 8 seconds. The total investment costs for modern wastewater treatment plants and management practices are nearly astronomical in
poor countries. Most developing countries will find it extremely difficult
to implement such infrastructure anytime within the foreseeable future.[47]

10.7   Dammed if they do!

The total volume of water stored in the world's dammed reservoirs is truly
awesome—nearly five times the volume found in all the world's rivers put
together. Although it may sound unbelievable, dams have redistributed
so much weight around the globe that some geophysicists actually believe
they have slightly altered the tilt of the Earth's axis, its speed of rotation,
and the shape of its gravitational field.[48]

Worldwide, approximately 40,000 large dams and some 800,000
smaller ones have been constructed to store and control freshwater. In
fact, a new dam is being added daily! In North America, Europe, and
North Asia, dams influence more than three quarters of all streamflows.[49]
In the United States, only 2% of the country's rivers and streams remain
free-flowing and undeveloped.[50]

10.7.1   The good, the bad, and the dammed ugly

On the surface, dams appear to be a splendid way of securing tremendous sources of water for the good of humanity. Unfortunately, in addition to the great good they can bring, dams can also precipitate a host of harmful environmental problems. For example, dam reservoirs trap nutrients. Algae thrive on these nutrients and reproduce. The growing algae consume the lake's oxygen, turning the water acidic. As acidic water is released downstream, it scours the riverbed. This, in turn, affects habitats, fish, and other animal life.

Salmon, steelhead, trout, and other migrating fish frequently find their paths obstructed by dams. The path of juvenile fish swimming downstream to mature is often disrupted, as is the path of adults swimming upstream to spawn. Many believe this problem is adversely affecting the survival of the juvenile salmon in the Columbia and Snake Rivers of the American Pacific Northwest.[51]

10.7.1.1   That dammed pollution

As natural water flows through a saline (salty) watershed to a river, it collects salt along its path. Rivers naturally wash small amounts of salt out to sea. In fact, a few billion years ago, the oceans consisted of fresh water. Slowly over aeons, the oceans turned saline as the rivers emptied their salt cargo into them. Because reservoirs behind dams expose so much water to the sun, a huge quantity of water can be lost by evaporation. The salt concentration increases in the water that remains behind. Many irrigated lands around the world are slowly being poisoned with this salt. Each year, farmers are forced to abandon 1 million hectares of land to this pollution. Neither the ecosystem nor urban areas are spared the crippling effects of salt; it kills aquatic organisms in rivers and corrodes pipes.

However, there is a silver lining. Over the past several years, around 500 old, dangerous, or environmentally harmful dams have been removed from U.S. rivers.[52] No, I am not against all dams, or even most for that matter. After all, dams have contributed greatly to our economic well-being, some more than others. However, when the social or environmental costs become too great, we need to rethink the wisdom of building new dams or leaving existing dams in place.

Fortunately, we do not necessarily have to face a future haunted by doom and gloom. Thanks to strict legislations incorporated in the Clean Water Act, most American rivers and streams are cleaner today than they were 30 years ago. This demonstrates what is possible when there is both a will and a way. The question we are left to ponder is whether the world will have both the will and the way to combat the emerging water crisis.

10.8 Will technology come to the rescue?

Many observers argue that technological innovations will come to the rescue before this issue festers into a disaster.

10.8.1 Desalination

Desalination of ocean water is frequently suggested as our lifesaving silver bullet. Salty ocean water is heated. The evaporated water is then trapped and distilled to produce fresh water. The method is simple only in principle. In reality, a large-scale desalination system is very complex and energy thirsty. The cost of a very large plant can range from $3 to $5 billion and that is only for the initial construction. Then there is the problem of its operation, which consumes huge amounts of expensive electric power. The cost is prohibitive for nearly all but the wealthiest of nations.[53] Although the price of desalinated water has dropped, it is still quite expensive, costing between $1 and $2 per cubic meter. Currently, desalination accounts for less than 1% of the human water supply.[54]

A potentially cheaper technology referred to as *membrane desalination* uses a principle known as *reverse osmosis*. A thin, semipermeable membrane is placed between a volume of saltwater and a volume of freshwater. Because the saltwater is under higher pressure, water molecules (but not salt and other impurities) are literally pushed through the membrane to the freshwater side. While the method is more energy efficient than traditional desalination, early membranes had many problems and were susceptible to clogging. New composite membranes are currently being developed that may eliminate some of these problems.

10.8.2 Drip irrigation

The first line of defense is to increase the efficiency of irrigation systems. Today, most farmers irrigate their crops by channeling water through long furrows or by flooding their fields. Approximately 60% of fresh water goes to irrigate crops through flooding.

Plants absorb only a small fraction of this water; the rest is lost. Drip irrigation systems, pioneered by Israeli farmers, rank near the top of measures that can be taken to conserve water. These systems are actually quite simple in principle. They consist of a network of perforated plastic tubing, installed on or below the soil surface. The tubes deliver small amounts of water directly to the roots of plants. Thus, losses through evaporation or runoff are minimized. When drip irrigation is combined with soil moisture measure and other monitoring measures to assess a crop's water needs, the system delivers up to 95% of its water to the plant. It can also substantially boost crop yield and quality because it enables farmers to maintain a nearly ideal moisture environment for plants.[55]

The potential for water conservation is huge, since only 1% of the world's irrigated lands now use drip and other efficiency methods. However, it comes at a price. It is relatively expensive technology compared to more traditional systems and often presents difficult operational problems. Farmers in developing countries generally cannot afford this technology.[56]

10.8.3 Recycling and conservation

It is time to seriously rethink how we use water. Agricultural, industrial, and urban water systems are notoriously inefficient (e.g., prone to leakage), so much so that it is estimated that 60% of the water used for agricultural purposes is lost through inefficiencies (leaky pipes, percolation from canals into the underlying soil, evaporation, etc.). There is room to significantly increase efficiency.

Innovative ideas such as granting farmers special water rights might produce dramatic results. For example, if farmers could make more money selling their water to urban centers than wasting it, they would have an incentive to irrigate more efficiently so that they can sell the surplus water; both cities and farmers could benefit alike.

Why should cities raise all water to drinking-water standards for uses as diverse as flushing toilets or watering lawns? For example, 70% of Israeli municipal wastewater is partially treated and then reused, mainly for agricultural irrigation of nonfood crops.[57]

10.8.4 Exotic technology

Some companies have turned to another possibility—moving fresh water in ships or even giant plastic bags to places around the globe suffering from water scarcity. For example, an English company has towed freshwater from mainland Greece to nearby resort islands in enormous polyurethane bags. This method of transportation is cheaper than using conventionally modified oil tankers. However, such approaches may have serious economic and technical constraints.[58] For example, designing large bags capable of withstanding the strains of a long ocean voyage may prove to be a formidable task.

There have even been proposals to tow fresh water icebergs from the Arctic to seacoast cities where they could be melted. Unfortunately, such schemes are plagued with technical and economic constraints.

10.8.5 Pricing

The final wild card is not so much technological as economic in nature. Because water is still relatively "cheap" in much of the world, there has been little economic incentive to explore for groundwater on a large scale. As prices rise and exploration increases, we may find that national and global reserves are higher than the currently estimated values.

For example, in 1973, when the first major energy crisis hit the world, proven oil reserves amounted to 640 billion barrels. As oil prices increased, oil exploration became economically more attractive. By 1996, the proven reserves had increased to 1030 billion barrels.[59] Groundwater exploration might follow a similar pattern.

Discussions, problems, and exercises

1. Why do we pose the problem in terms of "fresh water scarcity" instead of simply referring to it as a "water scarcity" problem?
2. Describe some of the principal factors that are placing increased demands on water resources.
3. What is the water to wheat grain ratio?
4. How much water is required for a person to maintain an absolute subsistence level of existence over a period of 1 year?

Class problem

Suppose you were a policy director for a county that is in competition with another country over a contested river that borders both countries. Describe the policies that you would propose and agreements that you would recommend for achieving an equitable usage of this resource. What policies would you recommend for eliminating pollution and protecting the quality of the river? What initiatives would you propose for reducing the total demand on this resource?

Notes

1. An excerpt form an Evian bottled-water commercial.
2. St. Xenophon Library, "Tau Te Ching," http://wayist.org/ttc%20compared/chap78.htm (accessed August 13, 2010).
3. P. H. Gleick, "Water Conflict Chronology," Database on Water and Conflict (Water Brief, November 10, 2008), www.worldwater.org/conflictchronology.pdf (accessed July 12, 2010).
4. Wikipedia, "World Population," http://en.wikipedia.org/wiki/World_population (accessed April 3, 2010).
5. A. K. Biswas, "The Water Crisis: Current Perceptions and Future Realities," D+C Development and Cooperation, No. 1 (January/February 2000): 16–8.
6. UNEP, GEO-2000. The Scientific Committee on Problems of the Environment of the International Council for Science (2000).
7. See note 3 above.

8. D. Randle, "Blue Gold: The Global Water Crisis and the Commodification of the World's Water Supply," http://www.ipcnsr-peace.org/water.htm (accessed January 2010).

9. B. C. Bates, Z. W. Kundzewicz, S. Wu and J. P. Palutikof, eds., "Climate Change and Water," Technical Paper VI of the Intergovernmental Panel on Climate Change (Geneva: IPCC Secretariat, 2008).

10. Ceres and Pacific Institute, *Water Scarcity and Climate Change: Growing Risks for Businesses and Investors* (Oakland, CA: Pacific Institute, 2009).

11. "A Water Warning": Peter Brabeck-Letmathe, chairman of Nestlé, argues that "water shortage is an even more urgent problem than climate change," *The Economist*, November 19, 2008, http://www.economist.com/theworldin/ PrinterFriendly.cfm?story_id=12494630 (accessed July 12, 2010).

12. M. Lavelle and J. Kurlantzick, *Water Crisis* (U.S. News and World Report, 2002), 21–30.

13. E. Robbins, "Water, Water Everywhere: Innovation and Cooperation are Helping Quench the World's Growing Thirst," *The Environmental Magazine*, September-October, 1998.

14. S. Postel, *Last Oasis: Facing Water Scarcity.* (New York: W.W. Norton, 1992, rev. 1997).

15. "Global Needs Put Great Lakes at Risk," *Rochester Democrat and Chronicle*, 2000, http://www.rochesternews.com/extra/lakes/0206lakes.html.

16. J. Motavalli, "Down the Drain: The Coming World Water Crisis," In *These Times*, April 14, 2000, http://www.alternet.org/print.html?StoryID=56.

17. The World Commission on Water for the 21st Century, "The World Water Gap: World's Ability to Feed Itself Threatened by Water Shortage," press release, March 20, 1999, http://www.hoffmanpr.com/Press_Releases/ Archived_Press_Releases/VisionUnit/WaterGapPR/WaterGapPR.html.

18. See note 15.

19. Geneva: International Conference on Hydrology, 1999.

20. R. Goldberg, *Reuse Water to Prevent a Water Crisis* (Academy of Natural Sciences, 1994), http://www.acnatsci.org/erd/ea/reuse_water.html.

21. W. Royal, *Industry Week, High and Dry: The World May Be Running out of Fresh Water, and Manufacturers Are Feeling Parched* (2009).

22. See note 11 above.

23. J. Leslie, "Running dry," *Harpers*, July 2000.

24. *Straits Times* (a Singapore newspaper), Alan Dupont of Australia's Strategic Defense Studies Center.

25. Seyyed Shamseddin Sadeqi, "Water Crisis: Future Challenges in the Middle East and the Persian Gulf," *Ettela'at Siasi va Eqtesadi* (Political and Economic Edition of Ettela'at) Bimonthly, April–May 1997, 115–6, http://www.netiran .com/Htdocs/Clippings/FPolitics/970415XXFP01.html.

26. S. Postel, *Pillar of Sand: Can the Irrigation Mireacle last?* (1999).

27. P. Gleick, "Water, War, and Peace in the Middle East," *Environment* 36, no. 3, (1994): 14.

28. See note 24 above.

29. B. Fraser, "Water Wars Come to the Andes," *Scientific American*, May 19, 2009, http://www.scientificamerican.com/article.cfm?id=water-wars-in-the-andes (accessed May 30, 2009).

30. See note 7 above.

31. K. Horta and L. Pottinger, *U.S. Can Help Stop Brewing Water Wars* (Christian Science Monitor, 1999), http://www.clo2.com/reading/waternews/world bank.html (accessed January 2010).

32. World Bank report, "From Scarcity to Security: Averting a Water Crisis in the Middle East and North Africa," 1996, http://www.usembassyisrael.org. il/publish/press/worldbnk/archive/march/wb1_3-20.htm.

33. See note 24 above.

34. See note 24 above.

35. See note 31 above.

36. A. Darwish, "The Next Major Conflict in the Middle East: Water Wars" (lecture, Geneva Conference on Environment and Quality of Life, June 1994), http:// www.mideastnews.com/WaterWars.htm (accessed September 16,2010).

37. D. Alexander, "Mideast Peace Could Start with a Glass of Water," The DAWN Group, 2000, http://www.dawn.com/2000/04/25/int9.htm.

38. See note 31 above.

39. See note 31 above.

40. See note 36 above.

41. See note 16 above.

42. See note 22 above.

43. See note 22 above.

44. S. Postel, "Troubled Waters," *Science World*, March 2000.

45. S. Postel, *Last Oasis: Facing Water Scarcity.* The World Watch Environmental Alert Series. (New York: W.W. Norton & Co., 1992), 21.

46. J. Newcomb, "Hopkins Report: Water Crisis Looms as World Population Grows," August 27, 1998, http://info.k4health.org/pr/press/082698.shtml (accessed September 16, 2010).

47. See note 4 above.

48. See note 22 above.

49. German Advisory Council on Global Change, Annual Report 1997, "Germany Must be Active in Shaping International Water Policy," http://www.wbgu .de/wbgu_jg1997_presse_engl.html (accessed July 12, 2010).

50. See note 7 above.

51. P. H. Gleick, "Making Every Drop Count," *Scientific American*, February 2001, 41–5.

52. See note 50 above.

53. See note 12 above.

54. See note 22 above.

55. S. Postel, "Growing More Food with Less Water," *Scientific American*, February 2001, 46–51.

56. See note 22 above.

57. See note 50 above.

58. See note 50 above.

59. See note 4 above.

chapter eleven

Peak oil, alternatives, and energy policy: The looming world oil crisis*

> If it weren't for electricity, we'd all be watching television by candlelight.
>
> **George Gobel**

Case study

The United States was hit with its first energy crisis in 1973. The price of oil skyrocketed from $3.50 to more than $12 per barrel. Within a period of a few weeks, the price of gasoline at the pumps soared from $0.30 to $0.65 per gallon. A temporary gap of only 5% between supply and demand caused the price of gasoline to more than double. For the first time in history, Americans sat in lines that circled entire blocks, for periods of hours, just to fill the tanks of their cars.

The huge increase in the cost of this energy source pushed the inflation rate up. The cost of everything from gasoline to eggs shot up. Factories began to shut down. At the same time, economic activity decreased. Simultaneous inflation and economic stagnation (stagflation) was a phenomenon that most economists believed impossible.

Eventually, the price of oil stabilized and the economy began to right itself. Then it happened again. This time, the price of oil soared to $35 a barrel. Gasoline finally leveled off at about $1.60 per gallon. The economic ills experienced during the first energy crisis in 1973, such as near-historic unemployment rates, inflation, and high interest rates, dogged the early 1980s.

Class problem: After reading the chapter, return to this problem. Consider this example in light of what you have read. Could such a scenario, perhaps one much worse, happen again? Why? Explain these

---

* This chapter is based on an article by C. H. Eccleston, "Climbing Hubbert's Peak: The Looming world oil crisis," *Environmental Quality Management*, Spring 2008.

impacts in terms of issues such as war, famine, and economic suffering. Is technology the "magical bullet" for preventing such a scenario? What policies should be adopted that would help mitigate such a crisis?

Learning objectives

A well-developed energy policy could minimize the effects of the pending *peak oil* crisis. After you have read the chapter, you should understand the following:

- Why oil is a finite resource.
- World petroleum production will almost certainly peak.
- We may be peaking now, or in the very near future.
- The socioeconomic impacts that would result from peaking could be catastrophic.

11.1   Introduction

While the United States claims a little over 2% of the world's petroleum reserves, it consumes about 30% of annual production. It now imports over 50% of the oil it consumes, and the rate is increasing. Transportation accounts for two-thirds of the oil consumed in the United States. Many European nations do not fare much better.

Although the United States has less than 5% of the world's population, about 35% of the world's vehicles travel down American roads and rack up as many miles each year as the rest of the world combined. This scenario is clearly unsustainable. So, why doesn't the West wean itself off the dependency on imported oil? The principal reason is an abundance of cheap, underpriced oil. Moreover, the cost of pollution is largely not factored into the price of fuel, making it appear a bargain of a deal.

There are other factors as well that account for this energy dilemma. Case in point: By the 1920s, 20 million Americans used trolleys and streetcars. But this was not to last. In the 1940s, General Motors, Standard Oil of California, Firestone, and Phillips Petroleum formed a holding company known as National City Lines. This company purchased privately owned streetcar systems in some 100 major American cities. These older systems were slowly dismantled and replaced by buses and cars that were promoted by the holding company. Then the bus systems were allowed to fail, creating an increased demand for automobiles. These companies were taken to court. The companies were found guilty of conspiracy to eliminate 90% of the American light-rail system. The corporate executive officers were each fined $1, and the companies were each fined $5000. In the end, General Motors had profited $25 million on this racket, not including how much it made on increased bus and car sales.[1]

The American economy is overly dependent on motor vehicles. It has been estimated that one-sixth of every dollar and one-sixth of every non-farm job is related to the motor vehicles sector. Motor vehicles account for one-fourth of the American national trade deficit.[2] But as you will see in this chapter, the true impacts could be much greater than simply pollution or contribution to the national trade deficit.

11.2   Hubbert's prediction

When the American Petroleum Institute met for a conference in March 1956, most of the participants probably expected to hear upbeat news. After all, back in the 1950s, the U.S. petroleum industry was humming and vibrant. The petroleum companies were producing more U.S.-generated crude oil than ever before—in excess of 7 million barrels per day.[3] The sky was the limit, or so it appeared. Then, M. K. Hubbert delivered his paper. His audience did not greet his predictions with enthusiasm.

Dr. Hubbert, then age 52, was a geologist with Shell Oil Company. Officials at his company had received advance word about Hubbert's speech, and they were worried. It is said that the president of the company personally pleaded with him to downplay some of the paper's more controversial claims.[4] Hubbert refused and went ahead with his presentation, entitled "Nuclear Energy and the Fossil Fuels."[5] His theme was grim, and it disturbed many of his distinguished guests: he boldly predicted that U.S. oil production would peak somewhere around 1970 and then begin declining.

11.2.1   Peak oil theory

Hubbert argued that petroleum production would follow a typical statistical "bell-shaped curve." He noted that the quantity of oil available for production in any given region must necessarily be finite and, therefore, subject to depletion at some point.

Whenever a new oil field is discovered, the petroleum yield from that location tends to increase rapidly for a period of years, as drilling infrastructure is put in place and extraction activities are ramped up. Once half the oil field's reserves are pumped out, however, the oil source reaches its peak rate of production. Then, decline sets in, with the rate of production decrease ultimately approximating a "mirror image" of the production-increase rate seen in the oil field's early years.

The actual date when oil production reaches its peak depends on the number of barrels extracted from the oil field. Therefore, plugging different figures into the variables of Hubbert's equations can yield a range of "peak date" estimates. The end result is the same, however: oil production will eventually crest and then begin a decline, sometimes a rapid descent.

11.2.1.1   Trajectory of U.S. oil production

In the years after Hubbert's speech, U.S. oil production continued its steady upward progression. The year 1970 came and went. Many commentators dismissed Hubbert, but trends are not always immediately apparent; it can take a few years of data collection before trends become clear. Soon enough, the statistics were in, and they were indisputable: Hubbert's predictions had been dead-on. U.S. oil production had in fact peaked in 1970—the exact year he had predicted based on his "high oil inventory" estimate.[6]

11.2.2   Repeating the peak

Hubbert's prediction went much further. He argued that his theory applied not only to the U.S. petroleum industry but also to global oil production. Throughout the 1960s and 1970s, Hubbert continued to refine his theory. He eventually predicted that a peak, also referred to as "peak oil," in world oil production would occur around the year 2000.[7] While he appears to have been less accurate on this forecast, many Peak Oil analysts believe that we are very close to witnessing a global Peak Oil scenario.

The concept of peak oil is critical to our modern technological society. Once the peak is reached, the world will be at the highest oil production level it will ever witness. Oil production will decline from that point forward. This decline will occur despite the fact that world demand is increasing and will have grave implications for the world economy, agricultural production, travel, and security. Recall from the Introduction that a temporary gap of only 5% between supply and demand in 1973 caused the price of gasoline to more than double. What will happen following the future global peak, as the gap begins to increase to 10%, 15%, or 20%? The graph in Figure 11.1 shows

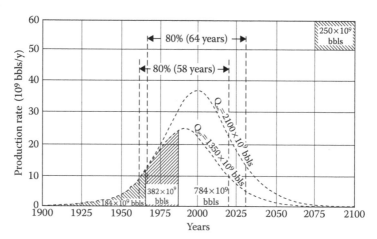

Figure 11.1 Bell-shaped curve of global oil production, known as the Hubbert peak, was first published by Hubbert in 1969.

a forecast that Hubbert made in the late 1960s, based on estimated data regarding world oil supplies. Assuming the production rates shown in this graph, the world would deplete about 80% of its available oil in a period of under 65 years.

11.2.2.1 Developments in world oil production

Within a few years after Hubbert made his prediction about global peak oil, several political events took place that affected his estimates. In particular, the 1973 oil embargo imposed by the Organization of Petroleum Exporting Countries (OPEC) resulted in an "energy crisis" that changed the dynamics of Hubbert's equation. By reducing oil consumption and encouraging greater energy efficiency, the embargo and its aftermath probably delayed the occurrence of the worldwide Hubbert's peak.[8]

11.2.2.2 Heading into the future

A global peak in oil production is still inevitable, however, assuming Hubbert's theory is correct. Some peak oil authorities believe that world production is already beginning to peak. Every oil field has its own unique Hubbert curve. Once its peak is reached, decline is inevitable. To see what this means in practice, consider oil production in Pennsylvania, where the U.S. petroleum industry began in the mid-nineteenth century.

Pennsylvania's production has been in decline for decades. There are approximately 20,000 oil-producing wells in this state. These wells each yield an average of only around one-fourth a barrel per day. Once a mighty producer of crude, today all of the oil wells in Pennsylvania combined produce less than half as much as a single high-producing well in Saudi Arabia.[9] Thousands of oil fields around the world are quickly heading toward the same fate as those in Pennsylvania. In addition, even in regions where production is not yet clearly in decline, petroleum may not be as plentiful as it appears, or as we might hope.

11.2.3 Reserve estimates: Are they trustworthy?

Assumptions about the amount of oil available for pumping are based largely on estimates of oil reserves announced by petroleum producers, but many observers suspect that producer estimates are inflated.[10] These observers note that between 1980 and 1990, the "proven reserves" of many of the largest oil-producing nations suddenly and mysteriously spiked in a dramatic fashion. For example, Abu Dhabi's declared reserves nearly tripled in 1 year, from approximately 31 billion barrels in 1987 to 92 billion barrels in 1988. The same year, Venezuela's declared reserves more than doubled, from 25 billion to over 56 billion barrels.

What could account for such sharp and massive increases in reserves? The answer may have little to do with the actual amount of oil in the ground, but a lot to do with OPEC politics. Since 1985, OPEC has tied member countries' production quotas to their oil reserves. Under this policy, members are allowed to pump only the quantity of oil specified in their individual quotas, and these quotas are based on the estimated reserves declared by each OPEC member. Therefore, the higher a country's reserve estimates, the more oil it is allowed to pump—and the more revenue it can earn from petroleum sales. This, of course, creates a strong incentive for OPEC states to overstate their reserves in order to gain larger production quotas. After all, the bigger the production quota, the more income the country can generate.

Kuwait was the first OPEC nation to suddenly increase its reserve estimate, which soared dramatically by over 40% in a single year (from 1984 to 1985). Within a few years, several other nations (including oil giant Saudi Arabia) responded by dramatically increasing their officially stated reserves. These developments strongly suggest that the estimated reserves of many major oil-producing nations have been purposely and greatly exaggerated.

Moreover, many OPEC members' reserve estimates are not declining appreciably from one year to the next. This means OPEC members are in effect claiming to discover "new" fields—year after year—that coincidently almost exactly match (and replace) the quantities of oil they are pumping out.[11] The overwhelming consensus of the geological community is that oil consumed today was formed millions of years ago through biological and geochemical processes. The Earth's petroleum reserves are therefore finite and declining.

If OPEC producers are in fact overstating their reserves, then the amount of oil yet to be extracted may actually be much smaller than what official projections estimate. This means that the world oil "day of reckoning" may be much closer than previously thought. In fact, some peak oil critics believe that oil production by many OPEC members is already in the process of peaking. Now, here is the problem. Hubbert's equations, like his predictions, are predicated on the volume of oil that remains to be pumped out. If the estimates are exaggerated, the date that world crude production peaks will be sooner.

11.3   Growing oil consumption and lowering of exports by producers

The problem is made worse by growing rates of oil consumption in many countries around the world—including the petroleum-producing nations themselves. A recent article in the *New York Times* noted, "The economies of many big oil exporting countries are growing so fast that their need for energy within their borders is crimping how much they can sell abroad, adding new strains to the global oil market."[12] The article went on to add[13]:

Experts say the sharp growth, if it continues, means several of the world's most important suppliers may need to start importing oil within a decade to power all the new cars, houses and businesses they are buying and creating with their oil wealth.

Indonesia has already made this flip. By some projections, the same thing could happen within five years to Mexico, the no. 2 source of foreign oil for the United States, and soon after that to Iran, the world's fourth-largest exporter.

11.3.1 *"Solution" to peak oil?*

We are unlikely to explore or drill our way out of this pending crisis. As peak oil geologist Colin Campbell states,

> The whole world has now been seismically searched and picked over. Geological knowledge has improved enormously in the past 30 years and it is almost inconceivable now that major fields remain to be found.

Many who dismiss peak oil predictions—particularly some economists—argue that improved technologies and higher oil prices will solve any oil supply problems that may arise. But many petroleum geologists view the situation differently. They point out that traditional economic incentives may not work when the underlying problem is a natural resource shortage. Higher prices cannot make oil appear if it does not exist. Moreover, feverish oil drilling in areas of mediocre potential and other last-minute efforts are unlikely to provide a "lifeboat" once production peaks.

11.3.2 *Peak oil assessment of the U.S. Department of Energy*

A study of the peak oil question commissioned by the U.S. Department of Energy (DOE) paints a sobering picture of the problem—and the level of effort needed to address it. The study resulted in a report (the Hirsch report) entitled "Peaking of World Oil Production: Impacts, Mitigation, and Risk Management."[14]

The report's executive summary opens with an ominous sentence: "The peaking of world oil production presents the U.S. and the world with an unprecedented risk management problem."[15] Table 11.1 summarizes some of the key conclusions made in the Hirsch report.

The authors of the report note, "Without mitigation, the peaking of world oil production will almost certainly cause major economic upheaval. However, given enough lead-time, the problems are solvable with existing

Table 11.1 Key Conclusions from the Hirsch Report

World oil production will peak, although experts differ on exactly when the peak will occur.

Peak oil will have a severe impact on the U.S. economy.

Peak oil is a "unique challenge," something the world has never before faced. The authors note, "Previous energy transitions (wood to coal and coal to oil) were gradual and evolutionary; oil peaking will be abrupt and revolutionary."

The main problems created by peak oil will be concentrated in the transportation sector, which relies primarily on petroleum-derived liquid fuels for which there are no readily available substitutes.

The mitigation efforts needed to avert severe impacts of worldwide peak oil could take decades to become effective. These efforts will involve replacing "vast numbers of liquid-fuel-consuming vehicles" with gas-fueled or electric powered systems and building "a substantial number of substitute fuel production facilities." The authors state, "There will be no quick fixes. Even crash programs will require more than a decade to yield substantial relief."

Demand for oil can be reduced by setting higher oil prices and tighter efficiency requirements. But such reductions will not be sufficient to meet the peak oil challenge. Large quantities of substitute fuels will also have to be produced.

Peak oil "presents a classic risk management problem" that necessitates careful planning and well-timed mitigation efforts.

Government intervention will be needed.

Mitigation efforts are crucial to averting major economic difficulties.

---

technologies."[16] The authors go on to note that the "obvious conclusion" from their overall analysis[17]:

> is that with adequate, timely mitigation, the costs of peaking can be minimized. If mitigation were to be too little, too late, world supply/demand balance will be achieved through massive demand destruction (shortages), which would translate to significant economic hardship.

11.4   Peak oil policy implications

The Hirsch report notes that as the day of peaking approaches, excess production capacity will dwindle, such that even minor supply disruptions will aggravate price volatility as traders, who stimulate speculation, and market participants react to changes in supply versus demand. These effects may prove to be unpredictable. The report concludes that "the world has never faced a problem like this. Without massive mitigation more than a decade before the fact, the problem will be pervasive and will not be temporary. Previous energy transitions were gradual and evolutionary. Oil peaking will be abrupt and revolutionary." A number of major oil company executives are already beginning to publicly acknowledge this fact.

Given the wide range of peak oil forecasts and opinions, it comes as no surprise that battle lines are being sharply drawn. Only a few policymakers appear to appreciate or vocalize the implications of this pending crisis, and fewer still are advocating the need to take immediate action in order to avoid calamity.

A few policymakers understand that peaking has the potential to affect the entire global food chain, which is currently dependent on cheap fossil fuels, and that following the peak there could be ominous worldwide food shortages; some go so far as to warn that as oil production declines, so must the human population. The Western way of life will be in particular danger because we have built our entire infrastructure and society around relatively abundant oil and stable, low oil prices.

Meanwhile, those on the other side suggest that such forecasts are propagated by alarmists and that their warnings seek to simply frighten people. Many on this side of the debate believe that a yet-to-be invented technological solution will save the day. Others argue that market forces will provide answers to the problem—as oil becomes more expensive, fewer people will buy it, and this will fuel the development of new policies and alternatives for coping with the peak. As debate rages over the timing, severity, and implications of peak oil, it is becoming increasingly apparent that the era of "cheap" oil is coming to an end.

As it stands now, petroleum prices and policy are being influenced, if not steered, by OPEC. Any comprehensive national energy policy must consider and respond to future decisions that OPEC will make in setting petroleum supplies. As of today, the United States and many other Western nations lack a comprehensive, scientifically based policy. Politicians need to begin seriously considering policy options, such as the Pickens Plan.

11.4.1 Energy source policy alternatives

The "old guard" of large-scale energy technologies—coal, oil, gas, and nuclear—is unsuitable for solving the peaking problem, which will largely manifest itself in terms of liquid fuels crisis. Over the long term, peak oil solutions must be based on sustainable energy policies. Petroleum has been a unique energy source, and we are unlikely to see alternate energy sources developed in the near future that will provide the same advantages as petroleum:

- Relatively abundant (in the past) and available to most nations with the funds to purchase it

- The presence of a massive and expensive infrastructure that has been developed to produce, process, store, and distribute it
- Very high energy density
- Its use throughout most of the transportation industry
- High versatility

Some of the principal energy policy alternatives will be discussed in Sections 11.4.1.1 through 11.4.1.5. As explained in these sections, the key viable energy supply alternatives are limited.

11.4.1.1 Nonconventional petroleum

The world contains large reserves of nonconventional petroleum (oil shale, tar sands, and so on), which are as yet largely untapped. Nonconventional petroleum resources may have significant potential for mitigating the effects of peak oil; however, economic, geopolitical, technological, and environmental factors may severely hamper the timing and ultimate production rates of these resources.

11.4.1.2 Coal

It is likely that politics and inertia will dictate policies that call for the expanded use of coal in the United States and in other countries that have plentiful coal reserves, regardless of the significant environmental impacts. It may well be feasible to produce transportation fuels through coal gasification, or coal might be used to produce hydrogen fuel for transportation vehicles. However, this use also has its drawbacks (see Section 11.4.1.3).

11.4.1.3 Hydrogen

Hydrogen is frequently touted as a replacement for petroleum. What many fail to realize is that hydrogen is not so much an energy source as it is an energy carrier. Large amounts of energy must be expended to create hydrogen. More energy must be expended on transporting this hydrogen for use by vehicles. Once created, hydrogen may provide a relatively clean form of transportation fuel; yet this is not necessarily the case for the process that is used to create hydrogen. Significant technical challenges must be solved before a "hydrogen economy" can come to fruition. Such a strategy will require the development of new and bold national energy policies.

According to the U.S. DOE, the transition to a hydrogen economy could take many decades. Its chief drawbacks are that it has a low energy density, it is difficult to store and transport, hydrogen fuel cells today are still quite expensive, and society lacks the enormous infrastructure that would be needed to generate and efficiently support a hydrogen economy. Today, 95% of all hydrogen is produced from fossil fuels, which pose the very problem that hydrogen is supposed to solve. Moreover, the inherent inefficiency of using electricity to power an electrolysis process (an electrochemical process for producing hydrogen) makes it worthwhile only when there is an excess of electrical power generation. As a result of such challenges, it is not possible to count on hydrogen in the foreseeable future to mitigate the effects of peak oil in time.[18]

11.4.1.4 Nuclear

In terms of mitigating the coming peak, nuclear energy will play only a limited role. The reason is that only 3% of the U.S. electricity is generated from petroleum.[19] Thus, replacing oil with nuclear energy would lessen current petroleum usage by only an equivalent percentage. Although nuclear could play a major role in furnishing future energy, it will have little effect in terms of reducing dependency on petroleum.

However, it could play a major role in the production of hydrogen through electrolysis, which would significantly reduce the consumption of and dependency on petroleum. The largest drawback of such a policy option is that it would take many years to construct sufficient nuclear power plants that could produce the required quantities of hydrogen.

The Obama administration appears to be increasingly supportive of nuclear energy, yet detractors maintain that nuclear power is unsafe. They also argue that nuclear power generates a long-term nuclear-waste problem. In contrast, advocates of nuclear energy counter that the world has amassed a more than 50-year track record with only two significant accidents and that the next generation of nuclear power plants will be even safer; they also contend that the potential harm of a large nuclear accident pales in comparison to the worldwide effects of global warming. With respect to nuclear waste, many advocates argue that the effects are small compared to those associated with fossil fuels.

11.4.1.5 Wind, tidal, solar, and geothermal

One of the most promising sustainable energy transportation technologies may involve wind, solar, and wave or tidal energy that power a fleet of plug-in hybrids. Wind, tidal, and solar energy technologies have significant long-term electricity-generating potential that could be used to produce hydrogen. However, for many economic and technological reasons, these technologies are unlikely to fill the gap in the near term. Geothermal power also presents a large potential energy resource that some experts believe is more viable than wind, tidal, or solar power. Policymakers should develop plans for promoting these technologies.

11.4.2   Developing a comprehensive energy policy

Significantly more research is needed immediately to support the preparation of any comprehensive national energy policy. Funding for research and development needs to be given top priority. The authors of the Hirsch report dismiss the power of the markets to solve any oil peak. They call for rapid governmental intervention. But in speeding up this decision making, they also point to a need to limit public debate and environmental analysis. Although this move may be necessary, it also raises significant issues that are basic to our democratic way of life.

Any viable policy needs to seriously consider increased tax breaks and incentives in order to assist homeowners and businesses in installing solar hot water heaters as well as solar-heating and geothermal-cooling systems. If this were done, it would place a lesser burden on electricity and natural gas that could otherwise be used to fuel the transportation sector.

Surviving the peak may require developing innovative new policies that encourage widespread relocalization of agriculture and manufacturing centers in order to lessen the burden of transporting such products over long distances. Likewise, many workers typically commute 50–100 miles everyday to work. Policies may be needed to encourage workers to live closer to their workplaces or work in centers closer to their homes. Policymakers may also need to consider significant rethinking of and expansion of mass transportation systems.

Peak oil will require public policy choices and many complex and difficult trade-offs. Consider carbon capture and sequestration (CCS), which remains an infant technology. At present, there is only one demonstrated, commercially successful CCS project in the world (the 1300-MW coal-fired Mountaineer Power Plant in West Virginia), and it is capturing only 1.5% of its emissions. If CCS is ever deployed on a large scale, the technology will consume approximately 30% of the energy it produces to "clean" the coal emissions it produces, and it may add up to 80% in additional production cost. Emphasizing the reduction of CO_2 emissions in a peak oil policy may result in several undesirable outcomes that run counter to the goals of developing an energy mitigation policy. For example, it raises costs substantially while reducing overall power output; to make up for the power lost in the CCS process, more coal is burned, which runs counter to the original purpose of the CCS process. This is not to say that we should not aggressively pursue carbon emission reductions, only that the mutual goals of reducing coal emissions and efficiently generating energy may oppose one another, further complicating the overall problem and trade-offs that will have to be made.

11.4.2.1   Geopolitical repercussions

Much of the world's remaining oil reserves lie in countries that are unfriendly to the West. As oil supply dwindles, these countries will have increasing influence globally. Wealth will flow into these nations as it flows out of oil-consuming nations. Countries like Iran, the former Soviet Union, and Venezuela, having dubious foreign policy motives, will gain in wealth, power, and influence as they hold significant leverage over oil-consuming nations. It is not difficult to envision a future in which some of these nations or perhaps OPEC itself "blackmails" or begins to dictate foreign policy demands on the West. Armed conflict could easily erupt as various groups fight over the control of land, water, food, and energy supplies. We could also witness unprecedented levels of migration and diaspora as people migrate to or invade countries with more resources. Such migrations would probably be accompanied by violent resistance from those already living in the more desirable areas.[20] This, of course, would significantly raise world hostility and the risk of regional or global conflict. Such concerns need to be seriously considered in any viable national energy policy.

11.4.2.2   Preparing for the crisis

Surprisingly, during the course of the 2008 U.S. presidential campaigns, not a single major candidate raised the issue of what may soon become one of greatest crises faced in the last hundred years. As we write this chapter, the authors can point to only one U.S. politician, Representative Roscoe Bartlett, Republication of Maryland, who has seriously and consistently warned of an escalating risk of a peak oil calamity in the future.

Arguably the best way to prepare for peak oil is to consider it in the context of preparing for a natural disaster. The consequences of failing to prepare for Hurricane Katrina were aired live over the television network. In the final analysis, this disaster was the result of failing to prepare for a Category 5 hurricane (which many scientists had warned was a distinct possibility). Katrina resulted from a broad failure to prepare for a disaster across a spectrum of local, state, and federal policymaking levels. No comprehensive plan or policy had even been prepared to face such a disaster. The result was chaos, confusion, uncoordinated responses, and much suffering.

The consequences of Katrina will pale in comparison to the impacts of peak oil. In terms of peak oil, purchasing vehicles with higher fuel efficiency, committing more funding to mass transportation projects, increasing national emergency storage supplies, and car pooling are just a few of the ways that individuals, governments, and policymakers can plan to mitigate the peak.

11.4.2.2.1   A crash program   Some policy platforms (including the Hirsch report) advocate a *crash program* based on how the United States waged World War II following the Great Depression. One of the problems with such an approach is that it does not provide a comparable model. For example, to wage World War II, roughly a quarter of U.S. resources (people and manufacturing capacity), which were idle following the Depression, were mobilized. This mobilization stimulated the U.S. economy and moved the nation out of the Great Depression. In contrast, any crash program to prepare for peak oil will operate within a more functioning economy. Allocating resources toward peak oil will mean diverting them from other productive areas of the economy, which is sure to cause direct or indirect economic repercussions elsewhere.

If not properly conceived, a crash program can easily "crash." Consider some of the U.S. energy program failures of the 1970s—rationing and price controls (which many economists considered to be a failure) and decades of support for fusion-powered electrical energy, with little to show for it. Consider just two of the many options with significant unknowns and limitations:

1. A rapid shift of the vehicle fleet to hybrids requires developing the ability to fuel them on the highway and service them in local repair shops (which must have trained technicians and proper equipment). This cannot simply be done overnight.
2. Building nuclear power plants costs billions of dollars and requires years for permitting and construction It also requires creating university engineering programs to train the engineers (many programs have been shut down over the last several decades). A shortage of trained engineers may represent one of the largest limiting factors to a goal of significantly expanding nuclear power. Moreover, experienced nuclear talent does not grow on trees; it requires years of professional experience after one has graduated from a university program. Finally, capacity to create specialized nuclear components is limited. There are also serious questions to be addressed regarding the supply of uranium fuels.

Added to the complexity of these policy options is a significant body of unknowns. For example, does converting corn to ethanol generate more energy than it consumes? Do we have sufficient coal resources to support a massive clean coal program? Even if we do, how clean and cheap will this energy source really be? Spending money before analyzing alternatives, as we have done with the corn-to-ethanol program, can waste significant time and money.

11.5 Concluding thoughts

The implications of a peak oil scenario in our modern industrial society are almost inconceivable. Consider for a moment, one single segment—agriculture. In his book *The End of Food*, Paul Roberts reports that malnutrition was common throughout the nineteenth century. It was not until the twentieth century that cheap fossil fuels allowed agricultural output sufficient to avert famine. It has been widely argued that an exponential increase in energy supply is the principal reason our food supply has grown exponentially in parallel with the increase in human population. Thus, we have avoided wide-scale famine largely because fossil-fuel supplies expanded geometrically.

David Pimentel and John Steinhart quantified the energy independence of modern agriculture and showed that technological development is almost always correlated to an increased use in fossil fuels.[21] The economic expansion of most countries has been shown to be nearly linearly correlated with energy consumption. If this energy increase is constrained, it tends to hamper economic growth. Fossil fuels are used throughout the agricultural cycle (e.g., in farm machinery, transportation, pesticides, fertilizers, and for drying). Today, it takes approximately 10 calories of petroleum to generate 1 calorie of food. Nearly 20% of all energy used in the United States is funneled into our food system. If this energy supply begins to tighten and prices escalate, the world could face a nightmare scenario.

As the DOE study makes clear, it is time to seriously and honestly acknowledge the severity of the global peak oil emergency. An international program is needed to develop global energy alternatives—something along the lines of the United States' crash program to put a man on the moon in the 1960s. Numerous alternative technologies are already available for implementation. All we need are the requisite funds and the determination to act.

With respect to new technology and governmental action, the evidence of potential impacts has often generally preceded public acceptance and policy actions by several decades. The multifaceted failure of a substantial portion of modern industrial civilization is so completely misunderstood by political leaders that we are virtually completely unprepared to deal with the outcome. The failure to inject the reality and potential impacts of peak oil into the mainstream of public policy forum is a grave threat to modern society.

The important question is not whether peak oil will occur, but a set of questions: When it will do so? What will the shape of the curve be? And how steep will the slope of the curve be? The longer we wait, the worse the impact of peak oil will be. We can only imagine how many people will soon be asking, "Why didn't anyone do something about this crisis when we still could?"

Discussions, problems, and exercises

1. Who was M. King Hubbert? Why is he important?
2. When did U.S. oil production peak?
3. What assumptions could affect the accuracy of Hubbert's predictions?

Class project

As a class project, develop a proposed national energy policy for addressing and mitigating the impacts of peak oil. What direction would you emphasize: development of more fossil-fuel technologies such as coal gasification and CO_2 sequesterization, development of alternative energy, or conservation? What compromise do you believe would have to be made to make the proposal acceptable to business and other lobbies? How would you "sell" such an expensive or controversial policy to a public who is unlikely to want to accept more sacrifices?

Notes

1. L. P. Pojman, *Global Environmental Ethics* (Mountain View, CA: Mayfield Publishing Company, 2000), 284.
2. G. T. Miller, *Living in the Environment*, 10th ed. (Belmont, CA: Wadsworth, 1998), 293.
3. U.S. Energy Information Administration, "U.S. Crude Oil Field Production (Thousand Barrels Per Day), 1859–2006," http://tonto.eia.doe.gov/dnav/pet/hist/mcrfpus2a.htm (accessed July 12, 2010).
4. A tribute to M. King Hubbert, February 28, 2006, "Hubbert's Speech," http://www.mkinghubbert.com/speech/prediction?PHPSESSID=421e8d6f06a74e7afbb37b79f52908ad (accessed July 12, 2010).
5. M. King Hubbert, "Nuclear Energy and the Fossil Fuels," http://www.energybulletin.net/13630.html (accessed July 12, 2010).
6. See note 1 above.
7. N. Grove, "Oil, the Dwindling Treasure," *National Geographic* 145, no. 6 (1974): 792–825. This article quoted Dr. K. Hubbert as stating, "The End of the Oil Age Is in Sight."
8. C. J. Campbell, *The Essence of Oil and Gas Depletion* (Essex, UK: Multi-Science Publishing Company, 2003). This book includes estimates of how much oil production can be obtained from conventional sources, as well as from "unconventional" sources such as heavy oil, tar sands, natural gas liquids, and oil shale.
9. F. de Winter, "Reading Material on the Bleak Future We Can Expect in Petroleum and Natural Gas as Sources of Energy," http://www.energycrisis.co.uk/Dewinter/ (accessed July 12, 2010).
10. C. Simkins, "Open Letter to Daniel Yergin on Optimism and Addressing Peak Oil Seriously," *Energy Bulletin* (September 28, 2005), http://www.energybulletin.net/9335.html (accessed July 12, 2010).

11. Wikipedia, "Oil Reserves," (updated January 11, 2008), http://en.wikipedia.org/wiki/Oil_reserves (accessed July 12, 2010).
12. C. Krauss, "Oil-Rich Nations Use More Energy, Cutting Exports," *New York Times*, December 9, 2007, http://www.nytimes.com/2007/12/09/business/worldbusiness/09oil.html?_r=2&hp&oref=slogin&oref=slogin (accessed July 12, 2010).
13. Ibid.
14. R. L. Hirsch, R. Bezdek and R. Wendling, "Peaking of World Oil Production: Impacts, Mitigation, and Risk Management," February, 2005, http://www.netl.doe.gov/publications/others/pdf/Oil_Peaking_NETL.pdf (accessed July 12, 2010).
15. Ibid., 4.
16. Ibid., 64–6.
17. Ibid., 59.
18. National Hydrogen Energy Roadmap, from the results of the National Hydrogen Energy Roadmap Workshop (Washington, DC: United States Department of Energy, April 2–3, 2002). Paper available online at http://www.hydrogen.energy.gov/pdfs/national_h2_roadmap.pdf (accessed August 13, 2010).
19. Energy Information Administration, 2005, "Net Generation by Energy Source, by Type of Producer," http://www.eia.doe.gov/cneaf/electricity/epa/epat1p1.html (accessed May, 2006).
20. "Resource Wars: An Interview with Michael Klare," *AlterNet*, May 1, 2001, http://www.alternet.org/story/10797/ (accessed May, 2006).
21. Charles A. S. Hall and John W. Day, Jr., "Revisiting the limits to growth after peak oil." *American Scientist* 97 (2009): 230–237.

chapter twelve

Peak food or peak everything: An era of bounty or begging?*

> The sum of intelligence on the planet is a constant; the population is growing.
>
> **Cole's Axiom**

Case study: Dirty Thirties

For aeons, rocks and soils have eroded off the Rocky Mountains and deposited as a fine dust onto the Great Plains of the United States. The Great Plains experienced countless droughts over the geological millennia. However, over time, drought-resistant prairie grass established itself, securing the soil and preventing it from being swept into the air during droughts.

In the 1920s, the U.S. government established policies that encouraged settlement and farming within the Great Plains. Encouraged by the Homestead Act and westward expansion, new farmers came by the trainload to receive their free farmland and launch a homestead. This was an unusually wet period for this semi-arid area. They cleared the land of its native prairie grass, tilled the soil, and planted their crops. Modern agricultural practices such as crop rotation and other techniques for preventing erosion were unknown. This was a boom time, as farmers produced bumper crops. But along with the 1929 stock-market crash came a collapse in the price of grain. Many farmers attempted to clear even more land to make up for their economic losses.

The last nail in the farmers' coffin came in the form of a prolonged drought which struck the Midwest beginning around 1933–1934. Without the natural prairie grass, which had anchored the soil in place, the soil dried, turned to a fine dust, and began to blow eastward and southward in huge dark clouds. These dust clouds were mammoth, covering many states, and sometimes blackened the sky as far east as Washington DC and New York City. Visibility could be reduced to a few feet.

---

* This chapter is based on an article by C. H. Eccleston, "Peak Food? Do Recent Shortages Portend a Leaner Future?" *Environmental Quality Management*, Spring 2009.

The great Dust Bowl that followed is one of the largest ecological and human disasters in history. Dust accumulated in thick deposits within houses. Cattle began to drop dead from lungs that were literally asphyxiated by the dust. Hundreds of thousands of people, particularly children, began to experience debilitating lung problems, which by some estimates caused the death of thousands of people. Hundreds of thousands of people were forced to abandon their farms and homes. The exodus was the largest migration in American history within such a short period of time.

The dust storms were so large that they actually began to alter the climate, reducing rainfall and increasing summer temperatures, which further intensified the dust storms. The drought wiped out natural predators of locusts and insects, which led to infestations of these pests. With the death of natural predators, the jackrabbit population exploded. The natural ecological balance was disrupted, and the resulting population of insects and jackrabbits finished off what little of the farmers' crops had survived the drought.

One lone scientist, Hugh Hammond Bennett, had been trained in soil science and understood the underlying cause of this disaster. Bennett began to campaign for a reformation in farming practices and founded the U.S. Soil Conservation Service. It was not until the Soil Conservation Service began promoting ecologically sustainable soil conservation programs that the Dust Bowl slowly began to abate. Farmers were trained in modern sustainable soil conservation practices. By 1938, the massive conservation effort had reduced the amount of blowing soil by two-thirds. After nearly a decade of dirt and dust, rain finally came in the fall of 1939.

This was not the first, nor the last, time that such a disaster has been witnessed. For example, in 2002, World Press reported that a "devastating famine, provoked by drought, is steadily moving north from Southern Africa, where it has affected more than 13 million lives. Two years of alternating droughts and floods, mismanagement of land and food supplies, political instability, and regional conflicts are being blamed."[1] Today, scientists continue to warn that parts of Asia and China are primed for a repeat of the Dirty Thirties, in which the Dust Bowl occurred.

Interestingly, Bennett was one of the first to understand that the Dust Bowl was the product of many combined factors and that an ecosystem approach was required to restore the region back to its former productivity. He was instrumental in popularizing the concept of ecosystem management.

Class project: After reading this chapter, return to this problem. Consider this example in light of what you have read. This problem may be pursued either as an individual assignment or as a class project. Consider this example also in the context of a hypothetical modern nation. Could such a catastrophe happen again, or was it just a fluke period of drought, unlikely to be witnessed again? What agricultural policy legislation and measures would you consider in ensuring that such a disaster would not happen to your hypothetical nation. How would such measures affect the nation's economy and food supply over the short term versus the long term?

Learning objectives

With the exception of transient localized shortages, recent history has generally been characterized by an ample, if not abundant, food supply. As described in this chapter, this period of relative abundance may be coming to an end. After you have read the chapter, you should understand:

- Policies that have shaped modern food supply
- The factors that control the food supply
- The Green Revolution
- Arguments for and against the Malthusian theory of population and starvation
- Factors that may limit food production in the future
- The argument for a peak food scenario

12.1   Introduction

Throughout human history, the vast majority of people lived in rural settings, working primarily in agriculture. Beginning principally in the nineteenth century, this traditional rural lifestyle underwent an irreversible and revolutionary transformation—one that would have unprecedented implications in terms of environmental and agricultural policy. For instance, in the nineteenth century only about 2% of the world's population lived in urban areas. At the dawn of the twenty-first century, about 50% lived in urban areas. Now, consider the transformation from an agrarian lifestyle. In 1820, 72% of Americans worked on farms. Today, a mere 4% of the largest farms produce more than 50% of the nation's food supply.[2]

During 2008, food riots were witnessed, spreading across one nation after another.[3] Although not always reported in the international press, the riots became increasingly violent. In countries ranging from Indonesia to Mexico, there were protests and riots over the increasing cost of rice. Farmers in Thailand had to hire guards to protect their rice crops from theft. To make things worse, Thailand, once a leading rice exporter, began shifting more than 90% of its rice production to the country's domestic market, leaving little surplus for sale to other countries.[4] The Philippines passed a law imposing life sentences for rice hoarding.[5] Over 80 trucks carrying grain in Sudan were hijacked before reaching the Darfur relief camps. In Pakistan, armed soldiers escorted grain trucks.

Over recent history, increases in food prices were generally short-lived; prices typically returned to normal with the next harvest or so. The recent surge in world grain prices is trend-driven, making it increasingly unlikely to reverse the trends themselves. As of mid-2008, global rice stocks stood at 4% below the previous year's level—the lowest in nearly 25 years. In 6 out of the past 9 years, global grain production has fallen short of consumption. This has resulted in a steady drawdown in grain stocks. At the start of the 2008 grain harvest, global carryover in grain stocks (amount in storage bins as the new harvest begins) was sufficient to last the world for 62 days—a near record low.[6]

There is an increasing need for international cooperation and development of global policies to mitigate the impacts of a future disaster. For most of us, the warnings that modern civilization could disintegrate probably seems preposterous. As food demand increasingly outstrips supplies, the resulting food-price inflation places severe stresses on masses of poor people who may riot, thereby posing a threat to nations and governments already teetering on the edge of chaos. In the twentieth century, the principal threat to international security was superpower conflict. Many observers argue that in the twenty-first century the principal threat will be *failing states*. It is not a concentration of power, but the absence of concentrated power, that presents a real threat to peace.

12.2   Global grocery store

The European Union's stockpile of wheat was only 1 million tons in 2008, a steep drop from 14 million tons the prior year. Worldwide, in 2008 wheat inventories dived to the lowest level seen in 30 years.[7] Droughts and poor weather patterns in the wheat-growing region of Australia slashed output. American wheat reserves sank to a low not seen in 60 years. It was also reported that the world's stockpile of corn amounted to a 5-day surplus, the smallest reserve recorded in recent times. At one point last year, worldwide stockpiles of grain held enough reserves for only 1.7 months of consumption—an alarming 50% decrease in the supply since 2000.[8] The Secretary General of the United Nations (UN), Ban Ki-moon, warned that millions of people worldwide could be in danger of starvation.[9] Moreover, the problem was not limited to the world's poor, as has generally been the case with food shortages in recent decades. The Western middle class was also being squeezed by sharply higher food prices.

Many commentators minimized the significance of the 2008 food shortages, arguing that they resulted from poor planning and the adoption of "just-in-time" production and storage techniques. But a closer examination of the facts reveals that the problem runs deeper than simply scheduling and logistics.

The world's grain-versus-demand ratio balances on a razor-thin edge. As a result, even relatively small upticks in demand or downturns in production can have serious consequences. The increase in the cost of rice in 2008 was principally blamed on increasing demand from large population centers, especially China and India. Both countries were forced to begin importing rice to meet the needs of their growing (and increasingly affluent) populations.

Although widespread famine did not materialize in 2008, we should not assume that humanity's good luck will last forever. With such small surpluses in place, there is little margin for error. To understand the situation we face today, it is useful to have some historical background.

12.2.1   Malthus's admonition

Warnings about resource shortages are not new. Perhaps the most famous was articulated by Thomas Malthus, an English clergyman and economist, in his treatise *An Essay on the Principle of Population*, written in 1798. As described in Chapter 2, Malthus proposed the idea that population, if unchecked, increases at a geometric rate, whereas the food supply grows at a much slower linear or arithmetic rate. Malthus argued that in the absence of natural or human-created constraints (such as wars, disease, famine, or contraception), population growth will tend to increase geometrically without bounds. Because food supply increases only arithmetically, a point will eventually be reached where demand from a rising population will outstrip the food available. The end result will be widespread starvation, which will restabilize population at a lower level, but not before untold human misery has weeded down the population.

Such was the power of Malthus's concept that he was listed as number 80 in Michael H. Hart's book *The 100: A Ranking of the Most Influential Persons in History*.[10] Commenting on the Malthusian threat in early 2008, an article in the *Wall Street Journal* noted that "history is littered with examples of societies believed to have suffered Malthusian crises: the Mayans of Central America, the Anasazi of the U.S. Southwest, and the people of Easter Island."[11]

Despite the many instances of catastrophe, Malthus's concept has been one of the most widely criticized ideas in science. Critics have argued, among other things, that the model is flawed because it fails to take into account scientific and technological advances, which can greatly increase the food supply.

12.2.1.1   Modern-day Malthusianists

Since Malthus, other influential voices have come forward from time to time, warning that human activity would eventually overwhelm the Earth's resources. By the late 1960s, a new wave of commentators was arguing that an ever-increasing global population would soon strain the planet's resources to the breaking point. At the time, famine was beginning to rear its head, presenting a serious threat to Asia.

In his best-selling book *The Population Bomb*, biologist Paul R. Ehrlich predicted, "In the 1970s and 1980s hundreds of millions of people will starve to death in spite of any crash programs embarked upon now."[12] Then, there was the famed Club of Rome. During the oil crisis of the early 1970s, it published *The Limits to Growth*, containing dire forecasts about soaring population levels and future resource limits.[13]

The failure of predictions such as Ehrlich's has led many to assume that our modern-day Cassandra's will always be proved wrong. Each time warning sirens have sounded, they note, new technologies or resources have been discovered that mitigate concerns about resource scarcity.

12.3   Green Revolution

During the 1960s and 1970s, researchers developed higher-yielding seeds and improved agricultural practices, dramatically increasing global food production, especially in developing countries. The process came to be known as the *Green Revolution*, and it resulted in decades of abundance and bumper crops.

The impact of the Green Revolution was nothing short of phenomenal. Within one 4-year period during the 1960s (1964–68), farmers in India increased wheat production by an amount "greater than that achieved during the preceding 4000 years."[14] In 1950, the world produced approximately 692 million tons of grain on 1.7 billion acres of cropland. By 1992, a global grain output of 1.9 billion tons was produced on 1.73 billion acres of cropland. This represented a 170% increase in grain production with only a 1% increase in acreage used.[15]

Wide-scale food shortages were virtually eliminated by the middle of the twentieth century, particularly in the developed world. Although the global population grew rapidly during this time, the Green Revolution's advanced science and technology came to the rescue, averting a period of potentially catastrophic famine.

12.3.1   Has the Green Revolution run its course?

Recent evidence suggests that the Green Revolution may have reached a point of diminishing returns. Between 1950 and 1990, global grain production nearly tripled. Since that time, however, grain production has grown more slowly, creeping upward at a rate of less than 3% during the 1990s.[16] Since the beginning of the Green Revolution half a century ago, we have come to rely on the development of new crop strains to produce continually increasing yields, and have been growing them with ever-increasing amounts of fertilizer.[17]

The stalling of the Green Revolution can be blamed in part on complacency. Beginning in the 1980s, advances in agriculture produced years of bumper harvests. These were dumped on the world market, driving down prices. The overabundance of food led governments to reduce their funding for agricultural R&D. In less developed countries, public agricultural investment fell by 50% in the last two decades of the twentieth century.[18]

The increase in food availability also had another unintended consequence. It allowed world population to increase rapidly, largely unchecked by agricultural supply constraints. Warnings about a "population explosion" fell on deaf ears. As a result, we now are confronted with a new reality: the number of people in the world is much larger than in the 1960s, and population continues to grow (see Chapter 14). Meanwhile, the remarkable production increases of the Green Revolution are slowing. This could create a deadly set of circumstances. Eccleston has coined the term "peak food" to refer to this dilemma.[19]

12.3.2 Understanding the "peak food" concept

To understand *peak food*, it is useful to know some history about the model from which it derives: *peak oil* is a concept pioneered by King Hubbert (see Chapter 11). Increasingly, scientists around the world are coming to the conclusion that Hubbert was correct—it appears to be merely a matter of time before global oil production "maxes out." But Hubbert's model has implications beyond oil production. In fact, his theory can be modified and applied to many key resources.

Of course, there is an important distinction between oil and food: while oil supplies cannot be replenished on anything less than a geologic time scale, food is largely a renewable, human-developed resource. In the case of oil, once half of the recoverable supply is exhausted, production inevitably goes into an irreversible decline. By contrast, food production does not necessarily follow a bell-shaped curve. It may simply increase to a maximum growth rate and then level off, or perhaps continue to grow, though at a slower pace (i.e., approximating a "step" function).

While the total agricultural output may not peak, the available per capita food supply may peak. Thus, the concept of peak food is different, but closely related to that of peak oil. Despite these differences, the authors believe it is appropriate to use the term "peak food," especially in light of the food shortages experienced in 2008. Even if food production does not actually decline, famine could well result if population (and economic affluence) continues to rise faster than the increase in agricultural output.

12.4   "Peak everything"?

The food crisis that emerged in 2008 has characteristics that appear fundamentally different from other modern episodes of scarcity. It was neither localized nor primarily related to the usual transient problems. Instead, the shortages appear to have been more systemic and generalized in nature, perhaps the initial phase of a long-term trend. Moreover, food scarcity is not the only resource issue we face. We have begun to witness escalating price increases in many commodities, ranging from oil and wheat to copper and gold. Author Richard Heinberg warns that we may be facing an era characterized by shortages in many basic resources and commodities, including oil, water, metals, and food.[20] As he notes in the introduction to his recent book—appropriately entitled *Peak Everything*—"it is difficult to avoid the conclusion that we have caught ourselves on the horns of the Universal Ecological Dilemma, consisting of the interlinked elements of population pressure, resource depletion, and habitat destruction—on a scale unprecedented in history."

12.4.1   Reaching our agricultural limits?

On the demand side of the equation, current trends contributing to food scarcity include the ongoing addition of somewhere between 70 and 100 million people a year. There are expected to be approximately 9 billion people on Earth by 2050.[21] Demand pressure is also increased as a growing number of people desire to move up the food chain, consuming high grain-intensive livestock products; moreover, there is a diversion of significant quantities of grain to the ethanol-fuel industry.

Low-income countries, such as India, where grain supplies perhaps 60% of total calories, directly consume a bit more than a pound of grain per person per day. In affluent countries, grain consumption per person is nearly four times this amount, although as much as 90% of this grain is consumed indirectly as meat, milk, and eggs from grain-fed animals.

Food shortages inevitably drive up agricultural commodity prices. As these prices rise, farmers will of course respond by increasing production. Food shortages and rising food prices in the 1960s and 1970s induced farmers to dramatically increase their investment while bringing more land under tillage. It also fueled rapid technological advances. Today, however, resource constraints may make such responses less effective—or even impossible. Our failure to address environmental declines that are degrading the global food supply—most important, eroding soils, falling water tables, and potentially rising temperatures—may result in a collapse of the global social order. Moreover, dependence on "magic bullets" such as fertilizers and pesticides, may have much less effect than in the past. Adding to this problem, the greenhouse effect threatens to make the world's climate warmer than at any time since agriculture first began. Already, intense heat and droughts, perhaps the result of global warming, have resulted in major crop failures.

12.4.1.1   Global warming

Arguably, the most pervasive environmental threat to our food security is a rising surface temperature that can affect crop yields and precipitation patterns worldwide. In many nations, crops are grown at or near their thermal optimum; a minor rise in temperature can substantially shrink the annual harvest. A study by the U.S. National Academy of Sciences concluded that for every rise of 1°C (1.8°F) above normal, wheat, rice, and corn yields decline by 10%. (Also see Chapter 13.)

It may be a race against time. Can we offset carbon emissions fast enough to save the mountain glaciers of Asia? During the dry season, their melt waters sustain the major rivers of India and China. These rivers support vast consumer and irrigation programs. The fate of hundreds of millions of people may hang on the outcome of such questions.

12.4.1.2   Water woes

Of all these environmental trends, water scarcity may pose the most immediate threat (see Chapter 10). Irrigation consumes 70% of the world's freshwater. Irrigation wells in nations around the globe are pumping water out of underground aquifers faster than the water can be replenished (recharge rate).[22] *Fossil aquifers*, which store ancient water and have negligible recharge rates, include the vast Ogallala Aquifer that underlies the U.S. Great Plains, the Saudi aquifer, and the deep aquifer under the North China Plain. At current pumping rates, depletion of many of these aquifers will spell the end of pumping, and by default an end to irrigation.

Lester Brown, president of the Earth Policy Institute, notes,[23]

> Seventy percent of world water use, including all the water diverted from rivers and pumped from underground, is used for irrigation, 20 percent is used by industry, and 10 percent goes to residences. Thus if the world is facing a water shortage, it is also facing a food shortage. Water deficits, which are already spurring heavy grain imports in numerous

smaller countries, may soon do the same in larger countries, such as China or India.[24]

12.4.1.3   Ethanol dilemma

One-fourth of the 2008 U.S. grain harvest, enough to feed 125 million Americans or half a billion Indians at current consumption levels, will go to fuel cars. The grain required to fill a typical 25-gallon sport utility vehicle tank with ethanol is sufficient to feed one person for a year. Moreover, the ethanol fuel program will not make a large change in U.S. petroleum consumption. For example, even if the *entire* U.S. grain harvest were diverted to ethanol fuel production, it would meet less than 20% of U.S. automotive fuel needs.[25] The ethanol fuel industry is leading to an epic competition between cars and hungry mouths.

12.4.1.4   Loss of topsoil and cropland

Topsoil on perhaps a third of the world's farmland is eroding faster than new soil forms. Its destruction from wind and water erosion and poor agricultural practices has doomed countless civilizations preceding ours. An article in *The Economist* made similar disheartening observations with respect to food resource constraints[26]:

> The quickest way to increase your crop is to plant more. But in the short run there is only a limited amount of fallow land easily available. … For some crops—notably rice in East Asia—the amount of good, productive land is actually falling, buried under the concrete of expanding cities.

12.4.1.5   Collapsing fisheries

Other sources of food, particularly fisheries, are also declining at an alarming rate. From 1950 through the 1990s, the global seafood catch more than quadrupled. Since then, however, increases have stalled.[27] Ocean fisheries may be reaching their production limit, or what ecologists refer to as their "carrying capacity." A 2006 study outlined how dire the situation could become[28]:

> An international group of ecologists and economists warned. … that the world will run out of seafood by 2048 if steep declines in marine species continue at current rates, based on a 4-year study of catch data and the effects of fisheries collapses.

12.4.1.6 Lagging yield growth

With arable land diminishing and fisheries being depleted, we will be forced to rely on ever-higher yields from crops. However, this creates its own set of problems[29]:

> Yields cannot be switched on and off like a tap. Spreading extra fertilizer or buying new machinery helps. But higher yields also need better irrigation and fancier seeds. The time lag between dreaming up a new sezed and growing it commercially in the field is ten to 15 years. … Even if a farmer wanted to plant something more productive this year, and could afford to, he could not—unless research work had been going on for years.

12.4.1.7 Phosphorus and fertilizer famine

The price of fertilizers is rising rapidly, which could significantly reduce agricultural yields, particularly in poor Third World nations. Almost all fertilizers contain nitrogen, phosphorus, and potassium. Moreover, global phosphorus supplies may start running out by the end of this century and maybe much sooner. Just four countries—the United States, China, Morocco, and South Africa—contain 83% of the world's reserves. Already the Unites States has become a net exporter of this precious element. China has large reserves, but restricts its export to rest of the world. In the near future, much of the globe, including the Unites States, may depend on a single country for this critical asset.

From an ecosystem perspective, phosphorus-contaminated fertilizer runoff contributes to *eutrophication*, uncontrolled blooms of cyanobacteria (also known as blue-green algae) in lakes and oceans. Under the right conditions, it spurs massive, uncontrolled blooms of cyanobacteria and algae. Once these blue-green algae die and filter down to the bottom, their decay starves other organisms of oxygen, creating dead zones and contributing to the depletion of fisheries.

12.5 Bottom billion

The typical household in India spends approximately 32% of its income on food. Households in Indonesia and the Philippines spend 43 and 36%, respectively. For households in sub-Saharan Africa, the figure can reach 70%.[30] With current trends, the percentage of household income devoted to food can only be expected to rise over time, particularly among the world's poor.

While Western consumers are grumbling about paying more at the supermarket, many people in developing countries are pulling their children out of school to save money or eliminating meat and vegetables from their diet so they can afford a couple of kilos of rice for the week. While members of the American middle class may have to forgo buying a new car next year, the world's poor may have to forgo one or even two meals a day.

Over 1 billion people survive on less than one dollar per day. This segment of the world's population, the poorest of the poor, has been termed the "bottom billion." Another 1.5 billion subsist on less than $2 a day.[31] If food prices rise, these people are in serious peril because they can quickly be priced out of the food market. Given that grains make up a large percentage of people's food intake in much of the world (about 63% in Asia, 50% in sub-Saharan Africa, and 43% in Latin America), increased grain prices can be especially devastating.[32] For the poorest of the bottom billion, increased food costs can be a death sentence.[33] Moreover, soaring food prices can breed greed and corruption. In some poorly governed countries, for example, profits from the nation's harvest enrich corrupt dictatorships, whereas most of the population goes without.

12.5.1 Food for thought: Transitioning into an era of heightened risk

The Green Revolution held the promise of ending starvation. And it basically did—at least for several decades. Now the era of cheap, abundant food could well be coming to a finale. Tens of millions already are living only a step away from starvation. A couple of bad years of weather or drought is all it will take to tip a bad situation into a global catastrophe unwitnessed in modern times. The years ahead will be fraught with risk. Major wars have been fought over issues far less significant than famine.

Some may assume that governments can intervene to avert calamitous outcomes, but poor decisions by governments in the past are at least partly responsible for the agricultural crisis that now looms. Large-scale government investment in scientific research and technology development could reap a bonanza, but most of these benefits are decades away. Other responses, such as adding more land to cultivation, may prove increasingly difficult as cropland becomes more scarce. Capital-intensive programs, such as irrigation projects, are beyond the scope of poor local farming communities and may be impossible where water is in short supply.

Of course, fear of hunger will undoubtedly drive scientific innovation, and may lead to breakthroughs we cannot yet imagine. Some predict that genetic engineering, in particular, could offer a new "magic bullet" equivalent to that provided by the Green Revolution in the mid-twentieth century. Paradoxically, genetic engineering also contains the potential to unleash a Pandora's box of environmental risks or even catastrophes.

Gambling on genetic engineering to save the day could truly be the global equivalent of "betting the family farm."

12.5.2 China: The global joker card

If China turns to the world market for a massive infusion of grain, it will have little choice but to seek such supplies from the United States. U.S. consumers would wake up some morning to find that they were competing for the U.S. grain harvest with 1.5 billion Chinese consumers with fast-rising incomes. For Americans, not to mention the rest of the world, this is a near nightmare scenario. It would be tempting for the United States to restrict exports, as it did in the 1970s when domestic prices on some grains soared. Unfortunately, with respect to China, this may no longer be a viable option. China holds over a trillion U.S. dollars and also finances much of the U.S. fiscal deficit. The Chinese have the capability to make economic life very painful for the United States and, by default, much of the rest of the world. American consumers may have little option but to share their grain with Chinese consumers, regardless of how high food prices rise.

12.6 Food policy perspectives

While the term "peak food policy" is not used in the UN's Millennium Development Goals, its very first goal is "End extreme poverty and hunger." (See Chapter 14, Table 14.2.) This goal has the following three targets to address economic and social equity:

1. Halve, between 1990 and 2015, the proportion of people whose income is less than $1 a day.
2. Achieve full and productive employment and decent work for all, including women and young people.
3. Halve, between 1990 and 2015, the proportion of people who suffer from hunger.

The seventh Millennium Development Goal is "Ensure environment sustainability." The first two of its four targets address sustainable resources issues:

1. Integrate the principles of sustainable development into country policies and programs and reverse the loss of environmental resources.
2. Reduce biodiversity loss, achieving, by 2010, a significant reduction in the rate of loss.

The UN's Food and Agriculture Organization and International Fund for Agriculture[34] are among several key UN organizations that support the above goals and targets.

12.6.1   *International research policy on ways to end hunger*

The International Food Policy Research Institute (IFPRI) receives over $500 million annually from several UN organizations and many individual nations. In its 2008 annual report it recognizes the looming crisis in global food supply:

> In 2008, a year in which the global population—particularly the world's poor—was confronted by both the financial and food-price crises, agricultural systems faced changes that led to market disruptions, reduced growth, mass protests, and a string of political efforts to reshape the design and governance of food systems.
>
> At a time when hopes had been high that reductions in poverty and hunger could be achieved in pursuit of the Millennium Development Goals, developing countries began to suffer severe setbacks. IFPRI sounded an early warning about the magnitude of these crises and quickly adjusted its programs to respond to developments. The Institute also contributed to the search for solutions with additional research-based reports and media outreach; the demand for IFPRI's research findings increased tremendously.[35]
>
> The world is facing protracted and especially difficult food and economic crises, with climate change presenting a growing challenge. Hunger is on the rise and the lives and health of millions of people are being compromised. A global response to the challenges facing the poor remains essential. Governments, development partners, the UN, the private sector, and civil-society organizations have responded at applaudable scale and speed. Still, more action targeting agricultural investment, food security, and nutrition is needed, and it must be sustained for years to come.[36]
>
> One of the emerging risks for the poor—human-induced climate change—has moved up on the global agenda. As the 2009 Copenhagen Conference of Parties to the UN Framework Convention on Climate Change is approaching, it is important that agriculture be included in the negotiations. It is of central importance to both the mitigation of and

adaptation to climate change. IFPRI has scaled up
its research on climate change and has led the sharing of insights on desired negotiating outcomes.[37]

12.6.2   Food industry and its technologies

The following sections briefly discuss the food industry and technologies
and how they affect peak food policy.

12.6.2.1   Agriculture

Research and development since the end of World War II has enabled the
agribusiness industry to achieve major improvements in food production,
both crops and animals. Market distribution technologies have greatly
increased world's food supply while reducing the prices to many millions
of people that would have otherwise done without. The agribusiness
industry is vast and has been a major factor influencing agriculture and
food policies in many countries.

However, these benefits have not been achieved without serious
threats to sustainable agriculture, harmful chemicals in the food supply,
episodes of mass food poisoning, and various adverse effects of genetically engineered crop species. Moreover, there have been serious environmental impacts of agricultural practices that highly depend on large-scale
mechanization along with monocropping and the extensive use of chemical fertilizers, herbicides, and pesticides that pollute waterways and
impact numerous species.

Without an adequate regulatory framework, these practices are neither ecologically nor economically sustainable. The so-called organic or
natural food movements represent just one of several expressions of public concern with these trends. The major challenge to global food policy is
to transform the current food infrastructure to one that is sustainably less
threatening to public health, safety, and the global environment.

12.6.2.2   Seafood

Seafood harvesting, processing, and distribution technologies have
likewise improved seafood availability and lowered market prices.
Unfortunately, this has led to the problem of collapsing fisheries. One
of the major impacts has been species depletion. International policy
responses have included treaties that establish quotas on seafood harvesting in attempts to restore sustainable yields. Another problem is the
contamination of seafood with mercury, PCBs, and other substances that
adversely affect consumers. Fish farming attempts to create sustainable
yields. However, the use of antibiotics and processed feeds can affect the
quality of the food produced. These issues require policies that focus on
the global sources of contaminants to the air and freshwater environments

that wind up in the oceans, as well as direct ocean discharges, all of which enter the global food change.

12.7   Concluding thoughts

The problems in many developing nations stem from a failure to slow their growing populations. For a long period, it has seemed that the views of Malthusianists were simply wrong. Many casual observers, for instance, assume that the predictions made in *The Limits to Growth* have been debunked. In fact, however, those forecasts may have been relatively accurate over the long term. According to a recent study by Graham Turner of Australia's Commonwealth Scientific and Industrial Research Organization (CSIRO), actual observed data for the years 1970 through 2000 match up "favorably" with output from the book's "standard run" model—a scenario that suggests "global collapse before the middle of this [the twenty-first] century."[38]

Any transition to a new global food-supply equilibrium is apt to be prolonged and painful. The Green Revolution may have only forestalled, rather than prevented, a day of reckoning. Modern methods such as just-in-time manufacturing may have been a technological breakthrough for the automobile industry; but when applied to the world's food supply, it and other poorly thought-out policies could paint a bleak picture of tomorrow's world. Rigorous and well thought-out policies and planning (including risk assessment and consideration of worst-case scenarios) could certainly help us make the transition more smoothly, but the resource constraints we face will impose some daunting choices in the years ahead. The current dilemma raises a troubling and perplexing question. Are we entering the era of peak food? Or, are we entering a period of peak everything?

Discussions, problems, and exercises

1. Why did Thomas Malthus believe that the world would eventually face a period of mass starvation?
2. What factors were responsible for the Green Revolution?
3. Why are many policymakers concerned that the Green Revolution may have run its course?
4. What is meant by the term "bottom billion"?
5. Some critics state that the problems faced by poor nations are largely their own fault—corruption, theft by dictatorial leaders, lack of a free enterprise system, failure to provide for internal infrastructure, and an adequate educational system. Assume much of this criticism is correct. If food supplies continue to tighten, imperiling millions with starvation in such countries, do we have a moral obligation to help them? Justify your position. What if feeding these people only

promotes a larger population and even further food scarcity in the future?

6. You are advising national policymakers and need to determine a policy for alleviating food scarcity. One way is to promote vegetarianism, which will release grain currently being fed to animals to feed your population and export to poorer nations. The second is to develop a policy that promotes more efficient water use, which in turn will support greater irrigation and increased crop yields. The first is cheaper but unpopular with the citizenry. Which policy do you promote and why? How will you sell your policy to the citizenry (i.e., changes in lifestyle or increased taxes and a lower standard of living).

7. Global warming is a major issue, but so far it has only received lukewarm acceptance by many nations. It is often explained in terms of its environmental impacts to ecosystems and socioeconomic impacts from changes such as rising sea level. Only lip service has been devoted to explaining the potential impact to the global agricultural system to the public. Do you think that restating the problem in terms of future food scarcity would provide a more effective argument? Defend your answer. If yes, explain how such an argument should be presented to the public. If no, explain the most effective means for educating the public about the risk of food scarcity.

Notes

1. World Press, "Drought Spreads Across Continent," http://www.worldpress.org/Africa/719.cfm (accessed July 12, 2010).
2. L. Pojman, *Global Environmental Ethics* (Mountain View, CA: Mayfield Publishing Company, 2000), iii.
3. CNN.com, "Riots, Instability Spread as Food Prices Skyrocket," April 14, 2008, http://www.cnn.com/2008/WORLD/americas/04/14/world.food.crisis/ (accessed July 12, 2010).
4. C. Nelder, "Grain's Gains: Profits or Pains?" *Energy & Capital*, April 9, 2008, http://www.energyandcapital.com/articles/grain-food+prices-commodities/663 (accessed August 19, 2010).
5. "The New Face of Hunger," *The Economist*. April 17, 2008, http://www.depeco.econo.unlp.edu.ar/catedras/internac/pdfs/food_prices2.pdf.
6. L. R. Brown, "Could Food Shortages Bring Down Civilization?", *Scientific American Magazine*, May 2009, 50–7.
7. See note 1 above.
8. See note 1 above.
9. "Ban Ki-moon Urges Immediate and Long-Term Steps to Fight Escalating Food Crisis," *UN News Centre*, April 14, 2008, http://www.un.org/apps/news/story.asp?NewsID=26310&Cr=food&Cr1=prices.
10. Michael H. Hart, *The 100: A Ranking of the Most Influential Persons in History* (New York: Citadel Press, 2000).

11. J. Lahart, P. Barta, and A. Batson, "New Limits to Growth Revive Malthusian Fears," *Wall Street Journal*, March 24, 2008.

12. Paul R. Ehrlich, *The Population Bomb* (New York: Ballentine, 1971).

13. D. H. Meadows and others, *The Limits to Growth: A Report for the Club of Rome's Project on the Predicament of Mankind*, 2nd ed. (New York: Universe Books, 1972).

14. M. Hennigan, "Global Food Crisis: Malthus, Food Price Surge, Climate Change and a 42% Rise in World Population by 2050," *FinFacts*, April 20, 2008, http://www.finfacts.ie/irishfinancenews/article_1013358.shtml (accessed July 12, 2010).

15. Ibid.

16. G. Lean, "The Era of Scarcity Is upon Us," http://www.unep.org/ourplanet/imgversn/84/brown.html (accessed July 12, 2010).

17. Ibid.

18. See note 2 above.

19. C. H. Eccleston, "Peak Food? Do Recent Shortages Portend a Leaner Future?" *Environmental Quality Management*, Spring 2009

20. R. Heinberg, *Peak Everything: Waking Up to the Century of Declines* (Gabriola Island, BC: New Society Publishers, 2007).

21. *Science* 377, no. 5967 (2010): 797–834

22. See note 6 above.

23. L. R. Brown, "Water Deficits Growing in Many Countries: Water Shortages May Cause Food Shortages," Earth Policy Institute, 2002, http://www.earth-policy.org/Updates/Update15.htm (accessed July 12, 2010).

24. Ibid.

25. See note 6 above.

26. See note 2 above.

27. See note 13 above.

28. J. Eilperin, "World's Fish Supply Running Out, Researchers Warn." *Washington Post*, November 3, 2006.

29. See note 2 above.

30. See note 11 above.

31. See note 2 above.

32. See note 10 above.

33. See note 2 above.

34. Food and Agriculture Organization of the United Nations, http://www.fao.org/; International Fund for Agricultural Development, http://www.ifad.org/

35. International Food Policy Research Institute, "2008-2009 Annual Report," http://www.ifpri.org/annualreport0809/overview (accessed July 12, 2010).

36. International Food Policy Research Institute, "Introduction from the Director General," http://www.ifpri.org/annualreport0809/dg (accessed July 12, 2010).

37. International Food Policy Research Institute, "Introduction from the Chair of the Board of Trustees," http://www.ifpri.org/annualreport0809/chair (accessed July 12, 2010).

38. G. Turner, "A Comparison of the Limits to Growth with Thirty Years of Reality," (working paper series 2008–09, CSIRO, 2008), http://www.csiro.au/files/files/plje.pdf (accessed July 12, 2010).

chapter thirteen

Global climate change

> Congress has a plan to fight global warming. It's
> passed a law that we can lower the temperature
> dramatically by simply switching from Fahrenheit
> to Celsius.
>
> **Anonymous**

Case study

Snowball Earth: Would you like your Earth served frozen or fried sunny-side up? Let us see what the geological record suggests. You are being transported back to a time so long ago that all the continents were still huddled together into a single supercontinent near the Earth's equator. But, you would hardly recognize the place, for the entire planet is entombed in ice. The Earth is gripped in an unimaginable Ice Age so bitterly cold that even the tropics are frozen over. A cap of ice over a kilometer (½ mile) thick covers the oceans.

The initial trigger for this runaway deep freeze remains a mystery. Perhaps geological processes that naturally remove carbon dioxide (CO_2; depositing it as carbonate rock) dramatically increased, forcing the Earth's temperature to plummet. For perhaps 10 million years, the Earth was a desolate, lifeless "cosmic snowball" orbiting quietly around the sun. Ancestral life was all but snuffed out. How could the Earth ever awaken from such a frigid slumber?

While the Earth might at first appear to be silent and dead, looking closer you would discover that volcanoes are working overtime belching out an antidote. Carbon dioxide and other gases are being ejected from deep sea volcanoes and volcanoes high on mountaintops. The frozen Earth has killed off most life forms and shut down geochemical processes that normally keep CO_2 levels in balance. With these processes shutdown, CO_2 belches out of volcanoes unchecked. Ever so slowly, the CO_2 concentration began creeping back to levels associated with normal life forms. As the CO_2 level soars, it triggers a gigantic greenhouse effect. Slowly the Earth begins to thaw. Glaciers begin melting away. Life revives and reclaims its rightful place.

Just as the good old Earth appears to be retuning to normalcy, however, a new emergency arises. An unmerciful and runaway greenhouse effect is now rearing its ugly head. Any life that managed to flourish in the epic days following snowball Earth now had to battle a runaway greenhouse effect. Surface temperatures soared to a sweltering 50°C (122°F), but this too would come to an end, as our friend Earth tries to *right* itself. Slowly, torrential rains scrub CO_2 out of the atmosphere. Geological processes "re-ignite," removing more gas from the biosphere and safely depositing it as carbonate rock on the seafloor. Eventually, the greenhouse effect moderates as the Earth's climate stabilizes.

It is a "miracle" that any life survived this climatic duel. The geological evidence suggests that this cycle may have repeated itself as many as four times over a stretch of several millions of years.[1] The lesson is that our planet has the capacity to respond in drastic and unpredictable ways to even relatively minor environmental perturbations.

Problem: After reading this chapter, return to this case study. Could it be argued that a fluctuating greenhouse effect is simply a natural phenomenon and hence we should not be concerned about altering it? Does this case study have any relevance to the current global climate change debate?

Learning objectives

After you have read the chapter, you should understand:

- The nature and scope of the global warming problem
- The causal factors
- The range of potential environmental consequences that further global warming could cause
- The uncertainties about global warming and its environment impacts that create controversy that affect the policy community
- The range of technology options that have been proposed to reduce the effects of greenhouse gases in the atmosphere
- The history and potential future of global warming policy by the United Nations (UN) Framework Convention on Climate Change
- How to sensibly discuss and critique policy options

13.1   Introduction: Nature of the problem

It began 4.5 billion years ago as a molten core, shrouded by a thick layer of hot gases. As the Earth began to cool, its temperature eventually began to stabilize. But over the aeons, the temperature of the Earth's atmosphere would experience cycles of warming in response to both solar and earthly factors. Volcanic activity periodically released massive plumes of greenhouse gases that sent the Earth's temperature soaring. Likewise, collisions with killer asteroids sometimes turned day into night, resulting in freezing temperatures. Such events caused mass extinctions and altered the environment in ways that forced surviving creatures to adapt. The Earth's atmosphere has experienced both long-term warming and cooling cycles, interspersed with episodic spikes of extreme climate change that have profoundly affected life. The National Oceanic and Atmospheric Administration's global warming Web site reports the following[2]:

> From the paleoclimate perspective, climate change is normal and part of the Earth's natural variability related to interactions among the atmosphere, ocean, and land, as well as changes in the amount of solar radiation reaching the Earth. The geologic record includes a plethora of evidence for large-scale climate changes. Massive terrestrial ice sheets throughout the Northern Hemisphere indicate cold conditions during the last glacial maximum (21,000 years ago). Warm climate vegetation, dinosaurs, and corals living at high latitudes during the mid-Cretaceous (120–90 million years ago) indicate globally warm conditions. More recently during the Little Ice Age (roughly 1450–1890 CE) historic and instrumental records, predominantly around the North Atlantic, indicate colder than modern temperatures.

Perhaps it should come as no surprise that the Earth may again be experiencing a change in climate. Whereas past changes have led to brutal extinctions, this change might also have equally profound implications. The major potential impacts of a modern global warming episode are gaining public attention: shrinking polar ice and glaciers, sea level increases that may threaten continental shorelines and island nations, and a documented trend of increasing average global temperature along with modified weather patterns in many areas. The causes and potential remedies to our global warming problem require at least an elementary understanding of the Earth as a dynamic interactive physical-biological system—in effect a global ecosystem or biosphere.

13.1.1 Global warming uncertainty and controversy

Perhaps no other environmental issue is as controversial as that of global climate change. There are advocates and opponents on both sides of the issue. Both sides of this issue can cite an armada of scientific evidence to support their position and claims. While the evidence is moving toward the consensus of anthropogenic (human-induced) global warming, this phenomenon is by no means proven or settled. Given the Earth's variable history of climate change, it could turn out that global warming may be the combined product of either human-induced and natural climatic variations or other natural causes.

Much hangs on the scientific validity and verification of this theory. The cost of reducing greenhouse emissions to a level that will have a meaningful impact on global climate change will extract a very heavy economic burden on societies around the world; in hindsight, it will have been a grave error indeed, if efforts are taken to significantly reduce emissions, and it later turns out that greenhouse gases are having a negligible impact on our climate. Conversely, if greenhouse gases are profoundly affecting the Earth's biosphere, and nothing is done to mitigate these emissions, the results could be calamitous. Indeed, this is one of the defining issues of the twenty-first century.

13.1.1.1   Debate

The 2007 report of the International Panel on Climate Change (IPCC) concluded that anthropogenic climate change is *very likely* to produce a significant warming in global temperatures within our lifetimes[3]; Here, the term "very likely" is defined as having a likelihood of occurrence of greater than 90%.[4] Many critics in the scientific community argue that the IPCC has overstated its case. Conversely, others such as James Hansen, a climate scientist from NASA, argue that climate change will be much more extreme than the IPCC predicts.[5]

Emotions run high on both sides of this debate. Unfortunately, advocates on both sides of the debate deny any and all evidence that runs contrary to their own opinions. This is particularly true of some global warming dissenters who go so far as to suggest that a conspiracy exists among the scientific community to prejudice the debate toward climate change. But such denials are not a phenomena strictly limited to those who oppose the theory of global warming. Some scientists, who have shown a healthy skepticism toward global warming, claim that they have been denied research funding, ostracized from the scientific community, and denied employment of promotions, and in some cases terminated from their positions. This is hardly the way that the scientific method is supposed to work.

Prominent examples of dissenting literature include climatologist Patrick J. Michaels's *Meltdown: The Predictable Distortion of Global Warming by Scientists, Politicians, and the Media* published by the Cato Institute in 2005, and attorney Christopher C. Warner's *Red Hot Lies: How Global Warming Alarmists Use Threats, Fraud and Deception to Keep You Misinformed*. One reputable critic is Don Aitken, a political scientist and former vice chancellor at Australia's University of Canberra.[6] Aitken contends that the proponents of climate change, including the IPCC, are a political creation rather than a genuine and objective scientific body. He asserts that the IPCC's scientific consensus appears to be manufactured, at least in part, by political pressure. Some dissenting IPCC scientists have made similar claims. Aitken also reasonably points out that the evidential basis underpinning the IPCC consensus is dangerously over-reliant on predictions generated by computer models; reality is much more complicated than our simple models allow (see Chapter 6, which discusses some of the serious limitations of computer modelling). He goes on to argue that our current understanding of the global ecosystem fails to account for the many ways in which causal factors relevant to creating the Earth's climate interact, and thus, it is dangerous to presume that we accurately understand such matters.

Many critics assert, perhaps correctly, that attempting to interfere with the global climate process is problematic, and it is difficult to predict the consequences. Global warming advocates counter that the IPCC computer models and scientific consensus are our best collective estimate of the potential effects of anthropogenic global warming; of course, their case was not strengthened when former president Al Gore received the Noble Prize for his documentary, *An Inconvenient Truth*, which critics claim contains nine significant errors.

There are indeed many unanswered scientific questions and uncertainties that lead to different views of appropriate policy responses. Areas under debate, even by among those advocating stringent policies to limit greenhouse gas emissions, include the following:

• Degree, magnitude, and geographic distribution of potential climate change impacts
• Technical feasibility and affordability of many technological innovations proposed as part of the policy
• Degree of potential adverse or beneficial economic impacts of proposed solutions
• Differing political, economic, and even religious ideologies among stakeholders

13.1.1.2 Framing the global warming policy debate

Spencer Weart, a physicist who specializes in science history, argues that when faced with scientists who publish global warming warnings, the public's natural response is to ask for definitive proof and conclusions. When the scientists fail or are unable to respond with absolute assurance and conclusions, politicians usually tell them to go back and do more research. In the case of global warming, waiting for absolute certainty could mean waiting forever; but it is argued that when society is faced with a new disease or an armed invasion, we do not put off decisions until more research is done. We act using the best evidence available.

Despite the fact that the evidence is moving in the direction of supporting anthropogenic climate change, there are indeed concerns for the efficacy of climate change policies. Hence, we can characterize the global warming policy problem using the following questions:

- What should policymakers believe concerning the reality of global warming?
- What course of policy action should be taken?
- Should we take no action because global warming is a hoax?
- Should we further monitor the data and improve our confidence in the nature and scope of global warming before committing to expensive remedies?
- Should we continue monitoring while taking policy action that we believe will ameliorate and ultimately reverse the global warming trend?

Regardless of the policy that is chosen, the problem of what scientists, statesmen, politicians, and the public accept as fact will not completely go away. Science is, or at least should be, an inherently open-ended process, scrupulous and candid in terms of exploring the evidence and performing experiments that validate or refute current hypotheses. Thus, there will always be room for dissenting views, especially by scientists and science-literate people. However, the problem runs far deeper than this. Powerful lobbyists and business interests, when threatened by proposed policy changes, often launch sophisticated public relations and media campaigns that can overwhelm the efforts of scientific experts. Conversely, the scientific community has been guilty of groupthink in the past, which has slowed the acceptance of more accurate theories, which run counter to established scientific dogma (e.g., Thomas Kuhn's book *The Structure of Scientific Revolutions*).

We will describe the issues and evidence for global warming. This does not mean that we endorse the current thrust. However, whatever their validity, understanding the theory of greenhouse warming is essential for those either dissenting or agreeing with the thesis of global warming. We will describe the scientific evidence in supporting global warming, summarize the environmental impacts of taking no action, outline the history of actions that have already taken place, review the technological opportunities for mitigating global warming and greenhouse gas emissions, and describe the current efforts of international policy makers to deal with the problem.

One final note before beginning this examination: control of emissions, effluents, and solid waste released to the environment is already well embedded in the environmental management practices and infra-structure of most Western countries, albeit with varying degrees of effec-tiveness. Any policies aimed at controlling greenhouse gas emissions should therefore consider the existing pollution control measures and seek to leverage them when feasible. To the extent possible, investments in greenhouse mitigation controls should seek to piggyback on proposals that also simultaneously produce economic benefits.

13.2   Global warming causal factors and research

The Earth is a global ecosystem. All living things collectively constitute the living environment or biosphere. Animals feed on plants and other animals. Plants feed on the remains of dead animals and plants. Some of the carbon emissions are absorbed by plant life in the oceans and on land. The remainder adds adversely to climate change and has other environmental impacts. Climate change stimulates additional environ-mental impacts that cause the global community to consider reduction of emissions and adaption that reduces the effects on the socioeconomic system.

13.2.1   Gaia as a metaphor for the global ecosystem

As described in Chapter 2, James Lovelock proposed the *Gaia hypothesis* as a metaphor for the Earth as a holistic living self-regulating ecosystem. Lovelock argued that any creature adversely altering this balance triggers Earth's self-regulating function. Lovelock's metaphor helps us understand that in terms of a global ecosystem, the behavior of every living creature affects other creatures, all of which cumulatively impact the global envi-ronment. Hence, the Gaia hypothesis is an apt metaphor for the relation-ship between human civilization and the biosphere that it impacts.

13.2.2 Greenhouse gas factor

Greenhouse gases occur naturally in the Earth's atmosphere, but they are also being produced by human activities. The principle source of human-generated greenhouse gases is that of burning of fossil fuels, such as coal, oil, and natural gas, but other activities, such as deforestation and cement production, also have a major influence on increasing CO_2 levels. Proponents and critics largely agree there has been an increase in green-house gas concentrations. As shown in Figure 13.1, there is little debate that CO_2, the principle greenhouse gas component, has increased signifi-cantly since the mid-nineteenth century. Disagreements largely center on what this change in gas concentrations means and what effects it will have on future climate change.

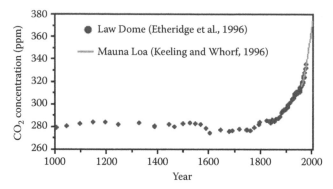

Figure 13.1 History atmospheric carbon dioxide concentration. (From the U.S. National Oceanographic and Atmospheric Administration.)

Due to human activities, atmospheric CO_2 (as measured from ice cores) has increased over the past century from 300 to 386 parts per million (ppm), and the average temperature of the Earth has increased approximately 0.7°C (or about 1.3°F).

Proponents and critics also agree that there is a natural greenhouse effect that warms the Earth. Without this natural warming effect, the Earth would be in a deep freeze. Given what we know about the ability of greenhouse gases in warming the Earth's surface, it is reasonable to expect that as concentrations of greenhouse gases in the atmosphere rise above natural levels, the Earth's surface will likewise warm. Many scientists have now concluded that global warming can be explained as a human-caused enhancement of the greenhouse effect. However, it is on this point that the debate generally ignites.

According to the IPCC Fourth Assessment Report, as of 2004, human activities produced nearly 50 billion tons of greenhouse gas emissions annually (measured in CO_2 equivalency).[7] Ambient concentrations of greenhouse gases do not cause direct adverse health effects (such as respiratory or toxic effects), but public health risks and impacts as a result of elevated atmospheric concentrations of greenhouse gases may occur via climate change.[8]

13.2.2.1 Greenhouse effect: How it warms the Earth

The Earth absorbs radiant energy from the sun and reflects some of the energy back into space. The term "greenhouse effect" describes how CO_2, water vapor, and other atmospheric "greenhouse" gases (methane, etc.) affect the return of energy back to space, and in turn, change the temperature at the Earth's surface (Figure 13.2). Greenhouse gases absorb some of the energy reflected back from the surface, preventing this energy from being lost into space. The lower atmosphere is warmed in the process of absorbing this energy. Nobel Prize-winning chemist Svante Arrhenius

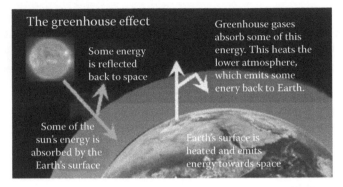

Figure 13.2 The greenhouse effect. (From the U.S. National Oceanographic and Atmospheric Administration.)

first explained this principle in 1896. Life on Earth would be very different without the greenhouse effect. The greenhouse effect keeps the long-term annual average temperature of the Earth's surface approximately 32°C (or about 58°F) higher than it would be otherwise.

13.2.3 Climate change research

As summarized in Sections 13.2.3.1 through 13.2.3.4, there are several ways that climatologists study how the Earth's temperature is changing. Instrumental temperature measurements, which extend the climate record back to the nineteenth century, provide a record indicating that the modern Earth is warming. These measurements indicate that the mean annual surface air temperatures have risen by approximately 0.5°C (0.9°F) since 1860.

13.2.3.1 Paleoclimatology

The study of past climates is known as *paleoclimatology*. Instead of instrumental measurements of weather and climate, paleoclimatologists use natural environmental (or "proxy") records to infer past climate conditions. Paleoclimatic studies use data generated from the study of past indicators such as tree rings, corals, fossils, sediment cores, pollens, ice cores, and cave stalactites to provide an independent confirmation of this recent warming. The paleoclimatic record allows us to look at the variation of climate in the distant past.

Annual records of climate are preserved in tree rings, locked in the skeletons of tropical coral reefs, frozen in glaciers and ice caps, and buried in the sediments of lakes and oceans. These natural recorders of climate are called *proxy climate data*—that is, they substitute for thermometers, rain gauges, and other modern weather instruments.

The instrumental temperature record indicates that the Earth has warmed by 0.5°C (0.9°F) from 1860 to the present. However, we should note that this record is not long enough to conclusively determine if this warming should be expected under a naturally varying climate or if it is unusual and perhaps due to human activities.

13.2.3.2   Research based on proxy data

A review of paleoclimatic data, both from thermometers and paleotemperature proxies, it is clear that the Earth has warmed over the last 140 years. Paleoclimatic studies show that recent years are the warmest, on a global basis, over at least the last 1000 years. While each of the proxy temperature records differs, in part because of the diverse statistical methods, they generally indicate similar patterns of temperature variability over the last 500–2000 years.

Perhaps most striking is the fact that each record reveals a steep increase in the rate or spatial extent of warming since the mid-nineteenth to early twentieth centuries. Compared to the most recent decades of the instrumental record, they indicate that temperatures of the most recent decades are the warmest in the entire record. According to the IPCC, reconstruction of the paleoclimatic record provides relative confidence in the following conclusions:

- Significant warming has occurred since the nineteenth century.
- Recent record warm temperatures recorded over the last 15 years is among the warmest temperatures the Earth has witnessed in at least the last 1000 years, and possibly the last 2000 years.

13.2.3.3   Ice-core evidence

Ancient ice traps bubbles of air, which contains CO_2 in the concentration that it was present in the atmosphere at the time the air bubbles were trapped in the ice. By studying ancient ice cores from Antarctica we can compare atmospheric concentrations of CO_2 that has been present in the atmosphere with temperature variations over the past 400 thousand years (see Figure 13.3).[9] A comparison of these two trends reveals a close correlation between the ways mean temperature has changed with respect to changing CO_2 concentrations. These fluctuations are nearly exactly mirrored across both the temperature and CO_2 plots over the last 400,000 years. Note that in the 1800s, as the Industrial Revolution began to take off, atmospheric CO_2 concentrations also suddenly began an unprecedented upward climb, rising rapidly from 280 parts per million by volume (ppmv) in the early 1800s to a current level of 386 ppmv. This rise in CO_2 concentration is greater than the highest concentrations previously attained in the course of the preceding 400,000 years. The post–Industrial Revolution meant temperatures increased in tandem with the increase in CO_2 levels. Figure 13.4 shows the correlation between rising CO_2 levels and mean temperature increase over the period since the post–Industrial Revolution period.[10]

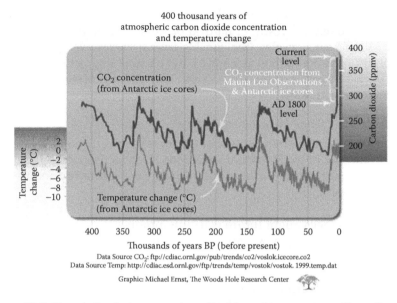

400 thousand years of atmospheric carbon dioxide concentration and temperature change

Figure 13.3 Correlation between carbon dioxide and temperature. (From Barnola et al, 2003; Jouzel/Petit et al; Keeling et al; Neftel/Friedli et al. Data from Oak Ridge National Laboratory. With permission.)

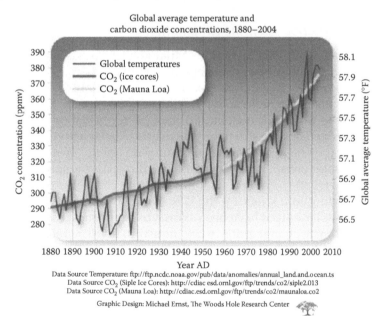

Global average temperature and carbon dioxide concentrations, 1880–2004

Figure 13.4 Carbon dioxide concentration versus mean temperature measured by David Keeling and colleagues at Mauna Loa, Hawaii and from polar ice cores, with average global surface temperature of Earth. (Image from Woods Hole Research Center.)

Most experts do not contest the idea that some of the recent climate changes are likely the result of natural processes, such as volcanic eruptions, Milankovitch cycles (the Earth's orbital cycle), changes in solar luminosity, and variations generated by natural interactions between parts of the climate system (e.g., oceans and the atmosphere). But what about human activities?

13.2.3.4   Some uncertainties

Ice-core data appears to indicate that the increase in CO_2 concentration is well beyond the range of natural variations. This increase correlates well with calculated fossil fuel emissions. Yet, there is uncertainty about some issues, such as:

- What changes might future warming produce?
- Exactly how much warming has occurred due to anthropogenic increases in CO_2 versus other natural phenomena such as sunspot activity?
- How much warming can be expected in the future?

Arguably, the best available estimate is that about 50% of the observed global warming is due to increases in greenhouse gas emissions.

13.2.4   Alternative scientific explanation

The combination of solar variation, Milankovitch cycles, and volcanic effects are all likely to have contributed to climate change, for example during the Maunder minimum. Apart from solar brightness variations, more subtle solar magnetic activity influences on climate from cosmic rays cannot be excluded; neither can they be confirmation as physical models of such effects are still poorly developed.[11]Solar activity measured by satellites and the historical record estimated using "proxy" evidence reveals that solar output varies (over the last three 11-year sunspot cycles) by approximately 0.1%.[12] Interpretations of proxy measures of variations differ. According to a study by the U.S. National Academy of Sciences, variations in solar output have been shown to be the most likely cause of significant climate change prior to the industrial era. The relative impact of solar variability and other factors on climate change during the industrial era is an area of continuing research.[13]

A recent paper by Scafetta and West supports the thesis that solar variability is a major, if not dominant, factor pushing climate change.[14] They have shown an indirect connection between galactic cosmic ray (GCR) intensity and global temperature.[15] This hypothesis suggests that the sun's changing sunspot activity permits more GCR to strike the Earth during high periods of solar activity. The higher intensity of GCR may create more low-level clouds, which cool the Earth. Conversely, when the solar activity is at a lower level, reduced GCR intensity may allow the Earth to warm. Svensmark and Marsh have shown a close statistical correlation between sunspot activity and global cooling and warming over the past 1000 years. The recent rise in global temperature might be at least partially explained by the current lower level in solar activity.

A paper by Benestad and Schmidt contradicts the conclusions of Scafetta and West.[16] Many climatologists believe that the magnitude of recent solar variation is much smaller than the effect due to greenhouse gases.[17] Obviously, more research is needed to determine the effects of solar activity and other natural phenomena on climate change.

13.2.5 Vital statistics on greenhouse gases

While CO_2 dominates in volume greenhouse gas emissions, on a per-molecule basis, methane and nitrous oxide are considerable more potent greenhouse gases. In fact, on a molecule basis, methane as a greenhouse gas is about 30 times more potent than CO_2. Some gases, such as sulfur hexafluoride, a product of electrical insulation and magnesium smelting, are well over three orders of magnitude more potent than CO_2. Nevertheless, CO_2 is the dominant greenhouse driver due to its shear volume. Figure 13.5 shows greenhouse gas contributions from major economic sectors.[18]

A review of global emissions data indicates that China leads the world in total greenhouse gas emissions, whereas the United States leads in per capita emissions of CO_2. On a per capita basis, Canada closely trails the United States, having the second highest emissions on a per capita basis. A paper by Susan Salomon et al. concludes that climate change that takes place as a result of an increase in CO_2 concentration is largely irreversible for a period of 1000 years after the emissions cease.[19]

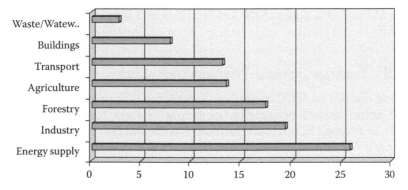

Figure 13.5 Greenhouse gas contributions from major sectors of the economy (%).

13.2.5.1   Deforestation

Trees are natural consumers of CO_2. Destruction of trees not only removes these "carbon sinks," but the burning and decomposition of wood pumps also voluminous amounts of CO_2 into the atmosphere. Some 20% of the world's greenhouse gas emissions are estimated to come from annual deforestation in countries such as Brazil and Indonesia.

Many people know that China has now overtaken the United States as the world's number one emitter of CO_2, but few know that because of its lack of deforestation policies, Indonesia is now the third largest source of CO_2, trailing only the United States and China.[20]

13.2.5.2   Methane

By plugging existing measurements of methane into global climate models, scientists already know that methane is responsible for about a third of the current warming trend.[21] Permafrost is filled with the remains of plant and animal matter that have been locked away in cold storage. Recent evidence suggests that the arctic region is heating up twice as fast as the rest of the globe.

Thawing permafrost is like taking a package of frozen meat from the freezer and thawing it. As the meat warms, microbes oxidize it. On dry land, microbes convert this organic matter principally into CO_2. But in wet, oxygen-starved depths of a lake, like in the polar region, such thawing tends to release methane instead. Methane traps about 30 times as much heat as an equivalent amount of CO_2. Some climatologists are beginning to sound the alarm that melting ice and thawing permafrost might significantly increase methane release, which could significantly accelerate the rate of global warming.

13.3   Current and future impacts of global warming

Table 13.1 presents some of the key findings and projections made by the IPCC.[22]

13.3.1   Findings from the Worldwatch Institute

Relying mainly on IPCC reports and data, the Worldwatch Institute has made some impact projections of future climate change.[23] Warming would be greatest in the high north and in the interiors of the continents. Warming would reduce snow cover, increase thawing permafrost, and decrease extent of sea ice. More frequent heat extremes and heat waves would be experienced, including more intense tropical cyclones, heavier precipitation, and flooding events in many regions. Precipitation can be

Table 13.1 Some Key Findings and Projections Made by the IPCC in 2007

Observed changes in climate and their effects and causes.

Warming of the climate system is unequivocal.

Most of the global average warming over the past 50 years is *very likely* due to anthropogenic greenhouse gas increases and it is *likely* that there is a discernible human-induced warming averaged over each continent (except Antarctica).

Global total annual anthropogenic greenhouse gas emissions (weighted by their 100-year GWP) have grown by 70% between 1970 and 2004.

Anthropogenic warming over the last three decades has *likely* had a discernible influence at the global scale on observed changes in many physical and biological systems.

Drivers and projections of future climate changes and their impacts.

For the next two decades a warming of about 0.2°C per decade is projected.

Continued greenhouse gas emissions at or above current rates would cause further warming and induce many changes in the global climate system during the twenty-first century that would *very likely* be larger than those observed during the twentieth century.

Warming tends to reduce terrestrial ecosystem and ocean uptake of atmospheric CO_2, increasing the anthropogenic concentration that remains in the atmosphere.

Anthropogenic warming and sea level rise would continue for centuries even if greenhouse gas emissions were to be reduced sufficiently.

Impacts are *very likely* to increase due to increased frequencies and intensities of some extreme weather events (heat waves, tropical cyclones, floods, and drought).

---

expected to decrease in most subtropical regions but to increase in high latitudes. By 2050, it is projected that there will be less annual river runoff and water availability in some tropical areas.

Worldwatch cites several scientific studies in their assessment of the potential adverse impacts on global warming.[24] Coupled with a continuing increase in world populations (see Chapter 14), global warming would likely lead to increased human suffering and major loss of life in many areas. The world's socioeconomic systems would likely to undergo radical changes as they adapt to the pace of the looming catastrophe.

13.3.2 U.S. Global Change Research Program

The U.S. Global Change Research Program was established in 1990 by the Global Change Research Act to coordinate U.S. interagency federal research on climate change.[25] Thirteen U.S. federal agencies and organizations participate in the program, with oversight by the White House

Office of Science and Technology Policy, the Office of Management and Budget, and the Council on Environmental Quality. This program encompasses the U.S. Climate Change Science Program, which synthesizes and provides up-to-date results on the science of climate change, including results from the Intergovernmental Panel on Climate Change.

The latest status report by the U.S. Global Change Research Program, titled *Global Climate Change Impacts in the United States*, was issued to Congress in 2009. According to this report, the impacts of a changing climate are already being observed across the United States. The report provides a "state of knowledge" assessment of the science of climate change and climate change-related impacts, now and in the future. It notes, "Observations show that warming of the climate is unequivocal," and " ... is due primarily to human-induced emissions of heat-trapping gases."

13.3.3   Great climate flip-flop

Up until now we have concentrated on global warming. But our recent ancestors had opposite priorities like trying to start a fire to keep from freezing to death. When our hairy ancestors were not busy hunting or dodging wooly mammoths and saber-toothed tigers, they were braving something else—struggling to survive a frigid glaciation.

When one mentions the term "climate change," it generally brings to mind the greenhouse effect with its killer heat waves. But paradoxically, global warming might actually lead to a global deep freeze, which could threaten the very fabric of our civilization.

13.3.3.1   Europe's furnace

The most populated areas of Europe actually lay about 10–15° farther north than the most populous counterparts of North America. The sunny Mediterranean, known for its balmy weather, lies at nearly the same latitude as its chilly cousin Chicago. Europe's climate is abnormally warm compared to other areas of the same latitude in North America and Asia; its climate is moderated by warm water currents flowing north from the tropics. This warm current keeps Europe about 5–10°C (9°F–18°F) warmer in the winter than other countries at comparable latitudes.

If this warm oceanic current gets turned off, not only Europe but also much of the rest of the world would become much colder. Geological evidence indicates that the Earth's temperature can change dramatically in just a few short decades or even less. One of the most striking examples happened about 12,000 years ago as the Earth was thawing from a long cold Ice Age. Something very strange happened. The gradual warming abruptly ceased, and the Earth returned to Ice Age conditions. But, we are left with a puzzle: why?

13.3.3.2   Mighty conveyor belt

New evidence suggests that changes in oceanic circulation may well have been responsible for this rapid and dramatic turn of events.[26] A complex globally interconnected ocean current, collectively known as the *conveyor*, governs much of our planet's climate. Depending on its temperature and salinity, a volume of seawater can "float" or "sink" into the ocean's depths. Water density varies with temperature, being greatest near its freezing point, making that volume of water more prone to sink into a warmer volume. As water warms up it expands, decreasing its density, allowing it to "float" above a denser column. Salt content also increases density, making water more prone to sink.

Today, the oceanic conveyor is driven by cold-salty water from the North Atlantic Ocean. It is nothing short of a very long salt conveyor. Warm surface water flows from southern to northern latitudes. Cold, dry winds blowing eastward from Canada evaporate the surface water as it approaches the northern latitudes. As this water evaporates, it leaves behind a salt load. As the remaining water becomes more concentrated in salt, it also becomes denser and begins to sink. Then as this water column sinks, it literally sucks more warm surface water from southerly latitudes to fill the void.

Like a gigantic conveyor belt, the North Atlantic also has a return loop running deep beneath the ocean surface. The circuit is completed as cold-dense water having sunk, flows back toward the southern latitudes. Thus, a cycle or "conveyor belt" is formed. This process literally pushes water through the Atlantic Ocean.[27]

As far back as 1961, the oceanographer Henry Stommel was beginning to worry that the conveyor system might stop flowing if too much freshwater were added to the surface of the northern seas.[28] If this happens, the conveyor system could come to a screeching halt and would no longer carry heat from the tropics to the polar regions. For example, more rain falling in the northern oceans could do this—exactly what has been predicted to result from global warming. Global warming might also melt Greenland's glaciers, flushing enough fresh water into the ocean surface to stop the conveyor. Is this what happened 12,000 years ago as the continental ice sheets melted? Just perhaps, the melting glaciers might have flushed fresh water into the North Atlantic, shutting down the conveyor.

Between the years 1300 and 1850, the Earth experienced a little Ice Age. Some climatologists suspect that this frigid period resulted from natural climatic cycles that melted Arctic ice, causing the conveyor system to halt. Whatever the cause, the event caused massive suffering throughout Europe: famine, disease, hypothermia, and even the rise of despotic leaders.[29]

While Henry Stommel's conveyor theory has evolved over time, the physics of adding fresh water to denser seawater is pretty clear. In four decades of research, his theory has been improved but not seriously challenged. As Professor Wallace S. Broecker, one of the founders of this theory observes: "The consequences could be devastating."[30] Agriculture on a global scale would be imperiled.[31]

13.4   Reducing greenhouse emissions

Figure 13.6 shows a Worldwatch estimate of the potential energy from non–fossil fuel technologies that can replace fossil fuels.[32] These technologies are all in commercial use today. Various efficiency measures and renewable energy technologies offer similarly large potentials for motored vehicles and the urban transportation sector. Because nuclear power produces no direct greenhouse gas emission, it could also play a vital role in reducing greenhouse emissions, even though its total energy and monetary costs are likely to be quite high.

With the advent of hybrid- and all-electric vehicles, there is the prospect of a national system of recharge and battery exchange that could eventually reduce the need for fossil fuels in road transportation—assuming that renewable vehicles use increases. Agriculture also offers significant opportunities to reduce the methane and nitrous oxide, which are components of greenhouse gases, and reducing deforestation can significantly lower CO_2 emissions.

All this suggests that private investments and innovation, in some cases seeded by governmental incentives, can manufacture new energy devices that will eventually balance our greenhouse gas emissions with the Earth's absorption capacity. However, as long as fossil fuel markets offer lower prices than alternative energy sources, renewable sources cannot compete, while the external costs (see Chapter 9) of greenhouse gas emissions will soar.

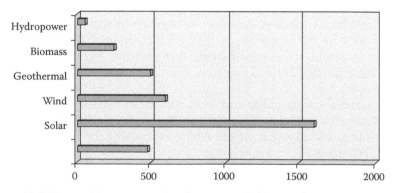

Figure 13.6 Renewable energy technology potential (exojoules per year).

13.4.1 Carbon capture and sequestration option

China's frantic construction of coal-fired power plants, now being built at a rate of one per week, could overwhelm any worldwide attempt to stabilize greenhouse emissions. Although coal is much cheaper, its greenhouse emissions far exceed those of natural gas or even oil. China now uses more coal than the United States, Europe, and Japan combined, making it the world's largest emitter of greenhouse gases.[33] Given the large investment needed to convert to nongreenhouse gas-emitting power sources, there is considerable interest in methods for removing CO_2 from the coal or even the atmosphere itself.

13.4.1.1 Carbon capture

There are three principle ways to capture CO_2 from the burning process of fossil fuel with today's technologies[34]:

1. *Postcombustion capture* involves a demonstrated method for treating flue gases (gases exiting the stack), but has yet to be demonstrated at a commercial power plant. A potential barrier is its high energy consumption.
2. *Precombustion capture* involves gasifying coal to remove CO_2 before it is burned in a coal-powered plant. This technology can capture between 80% and 90% of CO_2. While it is far more efficient than postcombustion capture, it too has not yet been commercially demonstrated.
3. The *oxyfuel process* is largely based on conventional power plant technology in the predemonstration phase. Combustion takes place in 95% pure oxygen enables efficient capture (99.5%) of CO_2.

13.4.1.2 Carbon sequestration

Carbon sequestration involves geological storage of captured CO_2 in ways that prevent it from being released into the atmosphere. After capturing the CO_2, it can be sequestered using one of the following methods:

- Binding CO_2 to silicates, which is in the discussion stage, creates a stable solid that can be buried in the ground.
- Fixing CO_2 with algae to produce biomass, provides animal fodder, bio-diesel, or construction material. This approach is in the discussion stage.
- Pressurizing CO_2 to a quasiliquid and pumping it into deep underground geological formations, deep nonexploitable coal seams, or aquifers. This technology is in the early demonstration phase.

Worldwatch concludes that the cost in energy to capture, transfer, and store CO_2, when added to the costs to acceptably reduce the whole range of environmental impacts of coal production, will not permit large-scale deployment of carbon sequestering methods within the next 10–15 years. If this is the case, then China and other nations that continue to bring coal power plants online pose a serous challenge to any policies aimed at reducing GHG emissions. Other engineering studies present a more optimistic assessment of this technology.

13.4.2   Plan G: The geoengineering option

Chapters 3 and 6 explain why it is so difficult to secure international agreements, such as the Kyoto Protocol, to limit greenhouse gas emissions. But, Plan G, or the geoengineering technology option, has the potential to change the entire policymaking game. Instead of facing a dilemma where one large nation can foil the entire effort to curb global warming, a single nation has the potential to curb the effects of global warming single-handedly. For example, one option such as pumping sulfur dioxide gas into the atmosphere is a lot easier than trying to achieve consensus among 200 countries, each of whom has strong economic incentives not sign onto a protocol or even to cheat on their reported emissions.

Some policymakers view geoengineering as a last-ditch option. Others fear that it could result in unpredictable and potentially calamitous consequences. Regardless of their views, most agree that geoengineering technology is probably powerful enough to reshape the planet's climate and is so cheap and easily implemented that some may be tempted to take actions into their own hands.

Some of the clues for a Plan G contingency come from nature itself, in the form of volcanic eruptions. Consider the 1991 eruption of Mount Pinatubo in the Philippines, which spewed sulfur dioxide and other pollutants into the upper atmosphere. These outgases were sufficient to absorb and reflect sunlight back into space, depressing global temperatures by about half a degree Celsius for several years. Many geoengineering options, such as those outlined below, have been proposed and more can be expected in the years to come.

13.4.2.1   Suspended fire hose method

Under this scheme, "fire hoses" can be suspended from helium balloons or zeppelins hovering 65,000 feet in the air to chemical factories on the ground. These factories would pump sulfur dioxide up through each hose at a rate of perhaps 10 kg/s. At the top of the zeppelin, the hoses would belch the sulfurous pollutant into the upper atmosphere. These sulfurous puffballs have the capacity to shield sunlight, cooling the planet as long as the process continues.

13.4.2.2 Whitening the clouds

One proposal involves "painting" the sky above the oceans white. A fleet of perhaps 1500 ships would use large propellers to spray seawater high enough for the wind to loft it into the clouds. This spray would whiten the clouds, turning them into reflectors that reflect sunlight back into space. By one estimate, the cost of constructing the first 300 ships would be $600 million, plus another $100 million per year to cover operational expenses.[35] This is but a small fraction of maintaining the U.S. Navy for a year.

13.4.2.3 Create forests of carbon-absorbing trees

The above concepts fail to control the underlying and daunting problem of actually reducing CO_2 concentration. Every summer, plants absorb approximately one-tenth of the atmosphere's CO_2 concentration. When these biological processes slow or cease in the winter, they release most of CO_2 back into the atmosphere. Some scientists propose using bioengineering techniques to create immense forests of carbon-absorbing trees to capture CO_2 and to keep it tied up in thick roots that would decay into soil, trapping the carbon.

13.4.2.4 Capturing carbon dioxide in the oceans

It has been observed for some time that natural plankton blooms can ingest large amounts of CO_2. One proposal builds on the observation that plankton blooms only occur when there is a sufficient supply of iron. This proposal involves seeding powdered iron across the ocean to stimulate massive blooms of plankton. But as with other geoengineering proposals, it is plagued with its own set of unknowns. For example, as the dead algae degrade, they might release methane. Methane is a much more potent greenhouse gas than CO_2.

13.4.2.5 Policymaking implications

Such proposals are well within the capability of a single rogue nation taking measures into its own hands. Imagine the plight of Bangladesh, whose population lives in coastal zones that can be flooded if sea level rise. While Bangladesh is among the poorest of the poor, it can easily take on such a program if it finds its existence threatened. And who would have the moral authority to condemn them? Just as a ban on cloning in the United States shifted such research overseas, an outright international agreement to ban or severely restrict geoengineering might motivate some countries to implement potentially dangerous proposals on their own; such bans might simply constrain all but the least responsible nations.

Geoengineering "solutions" also provide governments with an easy excuse for not agreeing to limit their emissions. Or, rather than cutting emissions today, policymakers can simply promise that in a future emergency they will support geoengineering measures to control emissions. Some policymakers argue that we should begin testing this technology now. There is always the possibility that at some time in the future we may have to consider such technology and might need to consider it in a hurry. The least risky option would involve starting with small-scale field experiments and gradually ramping up the scale. With each technology option, comes a host of scientific risks, as well as political, legal, and policymaking implications.

13.4.2.5.1   Cost implications   Geoengineering options are incredibly cheap. Compared to limiting carbon emissions, these proposals are a bargain basement sale, with some ranging from $1 to $100 billion per year. This compares to a cost of capping carbon emissions, which by some estimates might range a thousand times higher (on the order of $1 trillion annually).[36]

They also open a myriad of threatening possibilities, such as the Greenfinger dilemma, in which a rich "lone ranger" is as consumed with saving the planet as James Bond's nemesis Goldfinger was with gold. The world is now host to nearly 40 people worth $10 billion or more. Global geoengineering projects are within the financial wherewithal of any billionaire. Any billionaire could theoretically finance a program to reverse climate change single-handedly.

13.4.2.5.2   Risk factors   This technology has the potential to unleash a Pandora's box of unknown ills. For example, schemes that depend on sulfur dioxide might produce acid rain devastating large swaths of plant and fish habitat. Many concepts might radically shift climate and weather patterns. A change in weather patterns might benefit 4 billion people but might devastate the agricultural system of 2 billion. Or perhaps a scheme cools the Earth but unexpectedly and irreversibly destroys the fragile ozone layer.

13.4.2.5.3   Current status   A 2009 study by a Rutgers University team, published in the journal *Geophysical Research Letters*, compared the pros and cons of the sulfur dioxide approach. The key pros included a cooler planet and reduced or reversed melting of ice sheets and Arctic sea ice. The key cons included more droughts in Africa and Asia, continued oceanic acidification, and fewer blue skies. The team concluded that, given existing technology, the best method of lobbing aerosols into the atmosphere would be with the use of high-altitude military jets.

The view of the Institute of Physics is that geo-engineering should "be considered only as a last resort ... There should be no lessening of attempts to otherwise correct the harmful impacts of human economies on the

Earth's ecology and climate." The institute warned that geoengineering "should be seen as a prudent precautionary measure in case all other attempts to control dangerous climate change fail or are inadequate—for whatever reason." In 2009, the U.S. House Science Committee held its first hearing on the implications of geoengineering.

13.5   Global climate policy

In this section, we review the history of how the global community has attempted to formulate effective policies. We also review policy ideas that are being proposed or are currently in debate. For more information on this topic, the reader is directed to Chapter 3.

The challenges facing the many-nation partners to this undertaking include the following:

- The sheer magnitude and complexity of the policy issues
- Significant scientific uncertainties
- The immense resources and investments required to take effective action
- The virtual certainty that the global human community cannot restore the preindustrial meteorological and ecological status quo in this century
- A host of economic, social, and political impediments to systematic and properly manage preventive and adaptive measures on a global scale

13.5.1   Current status of policy: Cap and trade as a lynchpin

The European Union published a less than enthusiastic assessment of its progress toward a fully functional climate change accord[37]:

> Topping the list of countries that have reduced greenhouse gases are Germany (–2.3%) Finland (–14.6%) and the Netherlands (–2.9%). Bottom of the class in the Kyoto race are Italy, Denmark and Spain. They are ranked bottom because they have increased production in fossil fuel power plants. At present it looks as though these countries will not be able to meet their 2010 targets.

These figures do not stop the *EU* from observing this "progress" with optimism, based on new projections. The predicted *GHG* [greenhouse gas] reduction for 2010 is far better than last year's predictions and the *EEA* believes that even if current projections are maintained, the 15 *EU* member states will exceed the Kyoto Protocol's targets. It is true that 12 of the 15 member states expect to meet their initial objectives through a combination of national measures and European mechanisms. And there's the rub: only by putting in place and making use of additional measures, carbon sinks and Kyoto mechanisms can this ambitious target be met, otherwise *GHG* levels will be reduced by only 4%.

13.5.2   U.S. Waxman-Markey cap-and-trade bill

As this chapter is drafted, the U.S. Congress is seriously debating the most prominent version of a cap-and-trade bill, the Waxman-Markey bill. By any accounts, it will be an expensive proposition that could result in significant economic penalties in terms of economic factors such as unemployment and long-term growth. At the lower end, the U.S. Environmental Protection Agency's own cost projection estimates this bill will cost the average American family an additional $1100 per year. At the higher side, the Heritage Foundation estimates it would cost the typical American family from nearly $3000 annually and up to $4600 by 2035.[38] The nonpartisan Congressional Budget Office reported that most of the cost of meeting a cap on [carbon dioxide] emissions would be borne by consumers, who would face persistently higher prices for products such as electricity and gasoline…[and] poorer households would bear a larger burden relative to their income than wealthier households would.

13.5.2.1   Economy and jobs

Critics respond that a rigid cap-and-trade bill would cost trillions of dollars and result in tens of millions of lost jobs. The Obama administration claims that millions of jobs can be created by expanding the nation's electricity grid to accommodate new wind- and solar-power projects. Few would doubt this claim, however, a recent study by researchers at Spain's King Juan Carlos University questions whether "green jobs" are worth the public investments. Author Gabriel Calzada Alvarez cites Spain's investments as a case study. According to Alvarez, Spain has spent about $752,520 to create each "green" job, including subsidies for wind energy jobs. While jobs were created, other jobs were lost elsewhere in the economy. By Alvarez's estimate, the "U.S. should expect a job loss of at least 2.2 jobs on average, or about 9 jobs lost for every 4 created."[39] Of course, one also has to factor in the economic losses that could result from global warming if GHG levels are not stabilized.

13.5.2.1.1 Free rider problem   More so than the cost, many critics charge that so many allowances have been given away to special interests, to gain support for its passage, that the bill would be both expensive and only marginally effective. As we saw in Chapter 3, unilaterally cutting American and/or European emissions amounts to giving away the sole bargaining lever for pushing other nations to also curb their own emissions. This is the free rider problem (Chapter 3) in its most blatant form. Is it realistic to believe that nations such China, India, or Indonesia will step up to the plate and sacrifice their own perceived economic self-interest, simply because we did it first?

Problems with cap and trade do not end here. Consider the following example (Section 13.5.2.1.2), of what eventually could prove to be a long list of special favors to large lobbyist and powerful special interests.

13.5.2.1.2   Allocation problems   Perhaps no other power company uses as much coal as American Electric Power (AEP). AEP is one of the biggest electricity producers and users of coal in the United States. It seemed strange indeed that Chairman Michael Morris endorsed the Congressional Waxman-Markey bill, which would cap greenhouse gas emissions. The underlying reason for his support has little to do with environmental health and just about everything to do with profits and self-interest. The bill requires many businesses to purchase greenhouse gas emission carbon credits. However, it would give away 85% of the credits initially. As a diverse company, AEP could conceivably tap three of the bill's free credits: 30% to electricity distributors; 2% to electric utilities; and 5% to merchant coal generators.

If AEP accrues more credits than actual emissions, it can sell the extra credits—translation big profits. As a regulated utility (i.e., AEP faces no competition), the state government sets its rates. Thus, it can simply pass any additional costs onto consumers (it is a guaranteed win, no risk proposition). Now for the entrée. AEP's interest in helping to craft the fine print of the Waxman-Markey bill explains the company's elevenfold increase in lobbying efforts and expenditures. AEP has been spending almost $1 million per month on lobbying for this bill. This illustrates how well minded legislation can be re-engineered by big corporate interests and lobbying. For every AEP that is exposed, one can only wonder how many others are incubating behind the closed doors.[40]

13.5.3   Equity issues in global climate policy

Worldwatch cites the ethical issues that need to be dealt with in terms of controlling greenhouse emissions (see Table 13.2).[41]

13.5.4   Assisting developing countries in reducing greenhouse emissions

Worldwatch has compiled a list of proposals for adaptation, mitigation, and technology transfer likely to be considered by negotiators (see Table 13.3).[42]

Table 13.2 Ethical Issues That Need to be Dealt with in Terms of Controlling Greenhouse Emissions

Can climate treaties be built on strong principles of fairness?

How should rights to emit greenhouse gases be allocated?

Should the *egalitarian principle* be invoked, which states that every person worldwide should have the same emission allowance?

Should the *sovereignty principle* be invoked, which argues that all nations should reduce their emissions by the same percentage amount?

Should the *polluter pays principle* be invoked, which asserts that climate-related economic burden be borne by the nations according to their contribution of greenhouse gases over the years?

Should the *ability to pay principle* be invoked, which argues that the economic burden should be borne by nations according to their level of wealth?

Table 13.3 Proposals for Adaptation, Mitigation, and Technology Transfer Likely to be Considered by Negotiators

China and the G-77 (a United Nations coalition of developing countries that now has 130 members) propose to reduce fragmentation in funding and to create a governing board with equal representation from developing and developed nations to determine how much funding would be allocated for programs on adaptation, mitigation, and technology transfer. They also propose a Multilateral Climate Technology Fund to finance activities in developing countries.

India proposes a new global fund for adaptation. Industrial nations would contribute 0.3%–1.0% of their gross domestic product.

South Africa, representing a coalition of African governments, proposes adoption of a program based on assessing costs from developed countries.

The European Union proposes expanding the global carbon market, leveraging private investment flows, and making financing predictable and tuned to the needs of developing countries.

Brazil proposes that developed countries finance a new Clean Development Fund that would finance the costs of climate adaptation for developing countries.

13.5.5 Concerns for cap-and-trade policy

The Unites States, while not yet a party to the Kyoto Protocol, is seeking to introduce a cap-and-trade policy to reduce U.S. carbon emissions. Lobbying efforts are mounting to gain competitive advantage over the potentially lucrative tradable emissions market that the U.S. Congress was debating in 2009. Small companies are claiming that larger corporations are maneuvering to gain an unfair market advantage. Meanwhile, cement makers, mining companies, chemical manufactures, and food processors are angling for more carbon credits to help defer the costs of complying with cap-and-trade hardships. Smaller power cooperatives have already been awarded more permits by the U.S. House in exchange for a promise that they would not mount a grass-roots effort against the cap-and-trade bill.[43] With such a lucrative market at stake, one can only imagine the infighting that will evolve as such a market actually begins to unfold. This simply illustrates environmental policy cannot be divorced from the day-to-day social, political, and economic factors that influence the policy arena.

13.5.5.1 China: The 800-pound gorilla in the global climate policy debate

Under President Obama, the United States appears to be moving in a direction of instilling "unilateral" reductions in fossil fuel use. Meanwhile, China sees fossil fuels as an essential ingredient for economic growth. The Chinese are frantically scouring and purchasing rights to secure natural resources around the world. While China has surpassed the United States in total CO_2 emissions, it currently shows no inclination to sacrifice economic growth for any emission limitation policy, as the United States appears to be doing. Between 1998 and 2008, China's oil consumption was 91%, its coal consumption 96%, its natural gas consumption 240%, and its overall energy use doubled. If anything, it appears that this pattern may only accelerate.[44]

13.5.5.1.1 China's one-child carbon offset credit policy During the 2009 Copenhagen climate summit, Chinese delegates claimed credit for stifling some 400 million births since the 1979 implementation of their one-child policy. This policy generally limited couples to no more than one offspring and was enforced by harsh government controls including coerced abortion and sterilization. By Beijing's calculations, its autocratic family planning program prevented the release of 18 million tons of additional CO_2 from being released.[45]

Critics charged that it amounted to a trade-off—terminating lives to restrict carbon emissions. Peggy Liu, chairwoman of the Joint U.S.-China Collaboration on Clean Energy, defended China's argument for carbon offset credits; in a recent debate sponsored by the *Economist*, Liu argued that China must be given credit "for what it is not doing," explaining that "China's one-child policy reduces energy demand and is arguably the most effective way the country can mitigate climate change." Echoing this policy, Andrew Revkin, a *New York Times* environmental correspondent, even raised the prospect for baby-avoidance carbon credits. The Optimum Population Trust even announced a new campaign called PopOffsets which would pay poor Third World women not to reproduce; their underlying message is that people are the problem.[46]

Critics retort that this is simply the old specter of population control repackaged in terms of a new eco-faddism wrapped around the threat of global warming. Such critics point to the flamboyant predictions of overpopulation and doom voiced by prominent ecologists in the 1970s and 1980s that never materialized.

13.5.5.1.2   Climate change on the international stage   Climate change legislation under consideration in the United States, of and by itself, would have little effect on greenhouse gas emissions. Even a joint reduction by the United States and Europe may be insufficient to counter the significant increase in global GHG emissions. But what about negotiations at the international level?

Unfortunately, the prospects do not seem promising. Over 80% of the energy used worldwide comes from fossil fuels (petroleum, natural gas, and coal). CO_2 emissions are increasing at a rate of about 1% per year. Halting any further climate change "would require a worldwide 60–80 percent cut in emissions, and it would still take decades for the atmospheric concentration of carbon dioxide to stabilize."[47]

Moreover, developing nations like China, Brazil, and India are exempt from emission reduction requirements under the Kyoto Protocol, the main international agreement aimed at curbing global warming.[48] China is now the world's largest emitter of greenhouse gases. It currently builds new coal-fired power plants at the mind-numbing rate of two per week. Other exempt countries are also growing rapidly.[49] The efforts of the United States and other developed economies can have only a marginal effect on global warming if the developing world does not curb its emissions.

Then, there is the "leakage" problem. Leakage refers to jobs that will migrate to China and other countries that have cheaper energy costs because they are not constrained by the need to limit greenhouse gas emissions. Because many industries will simply move their operations, there may be no net decrease in greenhouse gas emissions, even with stringent limits on emissions in developed countries.

13.5.5.1.3   Potential trade wars   Another problem that some economists point to is the historical parallels with the discredited Smoot Hawley tariff of 1930, which was designed to protect American industry and revive the economy from the throws of the Great Depression. Instead, this act spurred international reprisals that led to an international trade war, exasperating unemployment and widespread misery that ensued.

Steven Chu, President Obama's energy secretary, admitted in March of 2009 that "if other countries don't impose a cost on carbon, then we [the U.S.] will be at a disadvantage."[50] To compensate for such a disadvantage, the argument follows that the United States could impose greenhouse penalties (i.e., tariffs) on products purchased in other countries, notably China and India, that do not institute carbon emission controls. However, these countries would in all likelihood, impose retaliatory tariffs on American exports that, like virtually all tariffs, would ultimately harm businesses, workers, and consumers in the United States.

In the end, the world could run the risk of sparking a twenty-first century global trade war that could severely damage the economies around the world. It is important to heed the lesson of the Smoot Hawley tariff of 1930. Rather than a unilateral carbon emissions ban by the United States and perhaps Europe, many critics counter that a more effective strategy would be to work on a cooperative international agreement by all principal players to a negotiated rollback in carbon emissions. However, countries such as China and India show no interest in reducing their greenhouse emissions. This circular reasoning leads one back to the potential for greenhouse penalties and tariffs.

13.5.5.1.4   Cap-and-trade policy dilemma   From a policy perspective, it has been argued that all nations must reduce greenhouse gas emissions. To this proposition, China and India have argued that the United States and other developed nations industrialized their economy with no limitations on greenhouse emissions; it is likewise unfair to burden them with such restrictions before they have also had an opportunity to develop their own economies.

To this argument, the United States and other developed nations have countered that they largely developed their economies before evidence began to indicate the risk that greenhouse gas emissions pose; now that it is becoming increasingly evident that greenhouse emissions could profoundly affect global climate, such arguments by developing nations are no longer valid. Developing nations will simply have to industrialize using alternative energies. Meanwhile, many climatologists argue that if nations like China and India continue along as they are, without making significant reductions, it could doom the planet. In lieu of binding international agreements, some critics have suggested alternatives, such as removing taxpayer subsidies for fuel use, eliminating regulatory barriers to nuclear power, and subsidizing biotechnology efforts to develop a "clean" fuel.

13.5.6   Uncertainty, the precautionary principle, and climate change

As shown in Chapter 2, the *precautionary principle* is largely a political instrument rather than a strictly scientifically based policy strategy. This principle promotes a cessation or avoidance of activities involving uncertain and potentially high-risk human endeavors. In practice, it is often invoked when an influential interest group advocates an issue involving potentially risky and uncertain outcomes. If the interest group is successful in publicly raising concerns or fears, application of the standard scientific investigative method may be circumvented in favor of adopting immediate and stringent limitations of human actions. Government dictates, sometimes with little scientific basis, are imposed. Those who dissent from the popular and prevailing view maybe vilified or even ostracized. Consider this in terms of Dr. James Hansen.

Hansen, who directs NASA's Goddard Institute for Space Studies, is a leading authority and vocal advocate of global warming. He has stated "Global warming has reached a level such that we can ascribe with a high degree of confidence a cause and effect relationships between the greenhouse effect and the observed warming."[51] His scientific credentials are indisputable. He stated his scientific opinion on a well-grounded review of the scientific evidence. So far so good. But here, particularly in terms of the precautionary principle, is where Hansen's views begin to cause some discomfort among his scientific peers and consternation among those whose views run contrary to his own.

Hansen criticizes the "natural skepticism and debates embedded in the scientific process," which he claims have been exploited by the fossil fuel industry as a means of downplaying the scientific world's confidence that CO_2 is fueling global warming. In *The Guardian*, he claimed that oil company executives have been active in spreading disinformation. Critics say that his views have some elements of a conspiratorial nature, just as those on the opposing side of the issue have been accused of advocating conspiratorial views. In an article titled, "Veteran climate scientist says 'lock up the oil men,'" Hanson was quoted as suggesting that those who promote the ideas of global warming skeptics should be "put on trial for high crimes against humanity." Statements like this only inflame the controversy surrounding global warming.

13.5.6.1 Skepticism and scientific scandal?

In 1912, a paleontologist found bones in the Piltdown quarry. Known as the Piltdown Man, these bones for 40 years were heralded as the missing link between primates and humans. Critics have long charged that some of the research on climate change was questionable and that some scientists were guilty of scientific misconduct and unethical behavior. Thanks to purloined e-mails, a new scandal might overtake the Piltdown Man as the most notorious scientific scandal of the last century.

The 2009 release of e-mails from East Anglia University's Climate Research Unit (CRU) has only fueled allegations and controversy over climate change. These e-mails have been alleged to indicate a long and concerted history of scientific misconduct. It began with Geoff Jenkins, chairman of the IPCC's first scientific group, who admitted in 1996, to a "cunning plan" to feed fake temperature data provided to Nick Nuttall, head of media for the United Nation's environmental program. Phil Jones, the CRU's director, e-mailed instructions to "hide the decline" in a recently observed cooling trend. For his part, Jones boasted that he had used a statistical "trick" to do just that.

Michael Mann, director of Penn State University's Earth System Science Center, and a principal author of the IPCC, developed the "hockey stick" graph purporting to prove that global temperatures were at the hottest in recorded history. But in 2008, statistician Steve McIntrey exposed a flawed methodology. Other climate scientists were incredulous that Mann's work had passed two IPCC peer-reviewed rounds without anyone spotting these inaccuracies. Andrew Revkin, a senior *New York Times* reporter was warned about Mann's discredited analysis; Mann replied in an e-mail to Revkin that "those ... who operate outside ... the system are not to be trusted." Instead of investigating the facts, Revkin supported Mann's position. When the CRU's incriminating e-mails came to light, Revkin retorted that they had been "stolen"—his reply was indeed interesting given that the *New York Times* has a long history of printing leaked national security secrets.

Other e-mails revealed efforts to ostracize skeptical researchers from the scientific community. Research that was inconveniently disagreeable to the CRC's mindset was ignored. Efforts were taken to delete opposing research from IPCC reports. Some administrators had threatened to boycott journals that dared print research papers showing evidence to the contrary. Efforts were even made to withhold data and delete some data that researchers did not want revealed to the public.

Most damaging of all is the fact that the CRU maintained a vital database of information, critical to the IPCC analyses and conclusions. The original data measurements had been "corrected." This is an accepted scientific practice, frequently performed to compensate for errors. However, the original uncorrected dataset was destroyed, purportedly because there was insufficient room to store it. At the very least, this is scientific sloppiness at its worst. But, it implies something of profound importance to the science of climate change—there is now no means of verifying the accuracy of the original uncorrected data.

In yet another embarrassing event, the IPCC stated in its latest report that, "Glaciers in the Himalayas are receding faster than in any other part of the world ... the likelihood of them disappearing by the year 2035 and perhaps sooner is very high." In 2009, four leading glaciologists prepared a letter for publication in the journal *Science* arguing that a complete melt by 2035 is physically impossible. In an interview, Professor Georg Kaser, who worked on the IPCC report, added fuel to the fire when he stated that he had warned that the 2035 figure was wrong in 2006, but his comments were ignored. The IPCC has since acknowledged that its report was in error in stating that the glaciers would melt away by 2035. In the *Daily Mail in London*, IPCC author Murari Lal stated that "We thought that if we can highlight it [the melting Himalayan glaciers], it will impact

policy-makers ... and encourage them to take some concrete action."
In calling for the resignation of the IPCC chairman, Andrew Weaver, a
Canadian climatologist who co-authored three chapters in the IPCC
report, stated "There's been a dangerous crossing of the line between neutral science and political advocacy."[52] A scientific commission eventually
cleared several IPCC investigators of scientific fraud. Critics, however,
retorted that the scope of the inquiry was overly narrow. While the commission cleared some investigators of actual fraud, it failed to address perhaps a more troubling question: Were these investigators conducting their
research in an objective and professional manner?

Improprieties and errors such as these have led some commentators
to claim that such events undermine the credibility of the IPCC if not the
entire field of climate science research. Critics have dubbed this scientific
coup "climategate." Sadly, such scientific misconduct, not to mention overt
errors in the IPCC report, is the modern equivalent of scientific heresy.
It has cast a shadow of suspicion over the entire discipline. How far this
scandal spreads, as of now is anyone's guess. It may take years for ethical
scientists to reestablish the trust that has been lost among the public and
policymakers.

13.5.7   Global policy implications

The issue of global warming is among the most complex and controversial
scientific and environmental policy issues we face today. It is apparent
that greenhouse gas emissions research needs to be funded and elevated
to the highest priority so that uncertainties are reduced or confined to
limits.

While attending the December 2009 global warming summit in
Copenhagen, President Obama claimed that it had set an "unprecedented
breakthrough." This was indeed strange given the fact that the Copenhagen
summit committed no nation to any action, let alone a timetable for meeting such commitments. Leaders of developed countries were attempting
to woo the leaders of undeveloped countries with promises of massive
transfusions of cash, while at the same time considering how they would
sell such explosive policies to the taxpayers back home.

Essentially, the Copenhagen agreement accomplished virtually nothing. While some nations set their own goals on emission controls, stating that they would be internationally "codified," many critics contend
that such "commitments" are next to worthless. Absent something akin to
universal cooperation, there seems to be little hope of controlling greenhouse gas emissions—and there is virtually no hope of achieving universal cooperation. So far, the public may not quite understand what all
this means. A recent poll by the *Associated Press* and Stanford University
shows, for instance, that very nearly twice as many Americans think

action to contain global warming would produce more jobs as those who understand it would reduce jobs.

Meanwhile, U.S. politicians and policymakers puzzle over how the reaction of American taxpayers to the economic hardships that will accompany any cap-and-trade bill passed by Congress, or the U.S. Environmental Protection Agency (EPA), that attempts to set enforceable greenhouse gas restrictions on the public. Recently, the EPA propagated a new rule requiring reporting of greenhouse gas emissions that exceed 25,000 metric tons or more of CO_2.

The seriousness of the occasion was occasionally punctuated by periodic moments of humorous folly such as when the tyrant Hugo Chavez rose to address the summit. Chavez, who has crushed freedom and prosperity for the citizens of his own nation, Venezuela, stood up and denounced capitalism. As he did so, he undermined the largely free enterprise system (ostensibly about the global environment) of the developed nations, who were being asked to foot most of the bill for curbing greenhouse gases. It will be difficult to garnish international consensus among the world's taxpayers when despots like Chavez command newspaper headlines, undermining the credibility of global summits.

Reaching an international consensus on reduction of greenhouse gas emissions increasingly appears to be a lofty but unattainable goal. Perhaps the ultimate policy solution is that of a new technology, financed through a collaboration of partly by private and government interests and reinforced by tax and other economic incentives. In the meantime, there is always the possibility that at some time in the future we may have to quickly consider a geoengineering option (described in Section 13.4.2), the global equivalent of a silver bullet. If such an option is needed immediately, we may not have the luxury of performing carefully controlled experiments. If a silver bullet ever needs to be invoked, the least risky option involves starting with small-scale field experiments now and gradually ramping the scale up to demonstration projects.

Discussions, problems, and exercises

1. What are the principal scientific uncertainties about global warming, and what are their implications for policy?
2. Assuming that renewable energies have the potential to replace fossil fuels, what plausible set of policy actions could potentially cause this to happen, and how would you assess the socioeconomic impacts of actions?
3. What are major issues confronting the negotiators for a global warming climate treaty?
4. What are the ethical issues confronting the negotiators?

5. What is most likely combination of technologies likely contribute to the reduction of greenhouse gases in the next 20 years?
6. What are the arguments raised against global action to reduce the global warming effect?
7. What are the direct and indirect environmental benefits of reducing the global warming trend?
8. What analytical methodologies should be used to analyze and compare the relative merits of the various technology solutions that have been proposed?
9. What are the largest potential barriers to a treaty that establishes an effective system for reducing greenhouse gases?
10. What are the costs of combating global warming? What else could these funds be used for if they were not allocated to reducing greenhouse gas emissions?

Notes

1. P. F. Hoffman and D. P. Schrag, "Snowball Earth," *Scientific American*, January 2000, 68–75.
2. NOAA, "Paleo Perspective on Global Warming—Global Warming Sitemap," http://www.ncdc.noaa.gov/paleo/globalwarming/sitemapgw.html (accessed July 30, 2009).
3. IPCC, *Climate Change 207: The Physical Science Basis* (Cambridge, UK: New York, NY: Cambridge University Press, 2007).
4. Ibid., 22–3.
5. University of Oxford, "Policy, Uncertainty, and Global Warming," http://www.practicalethicsnews.com/practicalethics/2008/04/policy-uncertai.html (accessed July 12, 2010).
6. Ibid.
7. Intergovernmental Panel on Climate Change, "Climate Change, Synthesis Report 2007," http://www.ipcc.ch/pdf/assessment-report/ar4/syr/ar4_syr.pdf (accessed July 12, 2010).
8. 74 Fed. Reg. at 66497–98.
9. Woods Hole Research Center, October 16, 2009, http://www.whrc.org/resources/online_publications/warming_earth/scientific_evidence.htm.
10. Woods Hole Research Center, October 16, 2009, http://oceanworld.tamu.edu/resources/oceanography-book/evidenceforwarming.htm.
11. UCAR, "Changes in Solar Brightness too Weak to Explain Global Warming," press release, September 13, 2006.
12. Spencer Weart, "The Discovery of Global Warming," in ed. Spencer Weart (American Institute of Physics, 2006). Retrieved April 14, 2007; R. C. Willson and H. S. Hudson, "The Sun's Luminosity over a Complete Solar Cycle," *Nature* 351 (Cambridge, MA: Harvard University Press, 1991): 42–4.
13. National Research Council, *Solar Influences on Global Change* (Washington, DC: National Academy Press, 1994), 36.

14. N. Scafetta and B. J. West, "Phenomenological Reconstructions of the Solar Signature in the Northern Hemisphere Surface Temperature Records since 1600," *Journal of Geophysical Research* 112 (2007): 1–10.

15. H. Svensmark, "Cosmic Rays and Earth's Climate," *Space Science Reviews* 93 (2000): 155–66; N. Marsh and H. Svensmark, "GCR and ENSO Trends in ISCCP-D2 Low Cloud Properties," *Journal of Geophysical Research* 108(D6) (2003): 1–11; N. Marsh and H. Svensmark, "Solar Influence on Earth's Climate," *Space Science Reviews* 107 (2003): 317–25.

16. R. E. Benestad and G. A. Schmidt, "Solar Trends and Global Warming," *Journal of Geophysical Research–Atmospheres* 114, D14101 (2009): 18 pages. "The most likely contribution from solar forcing a global warming is 7 ± 1% for the 20th century and is negligible for warming since 1980."

17. J. T. Houghton and others, eds., "6.11 Total Solar Irradiance—Figure 6.6: Global, Annual Mean Radiative Forcing (1750 to present)," *Climate Change 2001: Working Group I: The Scientific Basis*, Intergovernmental Panel on Climate Change, 2001.

18. Reformatted from Worldwatch 190. Original source: IPCC.

19. S. Salomon and others, "Irreversible Climate Change Due to Carbon Dioxide Emissions," *PNAS* 106, no. 6 (2009): 1704–9.

20. Mongabay.com. A study released by the World Bank and the British government in 2007. "Indonesia is 3rd largest greenhouse gas producer due to deforestation," March 26, 2007, http://news.mongabay.com/2007/0326-indonesia.html, (accessed October 16, 2009).

21. S. Simpson, "The Arctic Thaw Could Make Global Warming Worse," *Scientific American* (2009): 30–7.

22. See note 7 above.

23. Worldwatch Institute, *2009 State of the World: Into a Warming World* (Norton, 2009).

24. Worldwatch, 20–1.

25. DOE, "NEPA Lessons Learned," U.S. Climate Science Report no. 60, September 1, 2009.

26. C. Rühlemann and others, "Warming of the Tropical Atlantic Ocean and Slowdown of Thermohaline Circulation during the Last Deglaciation," *Nature* 402 (1999): 511.

27. W. H. Calvin, *Atlantic Monthly*, January 1998, 47–64.

28. H. Stommel (Woods Hole, MA: Woods Hole Oceanographic Institution).

29. B. Lemley, "The New Ice Age," *Discover* (2002): 35–40; B. Fagan, *The Little Ice Age* (New York: Basic Books, 2000), 246.

30. W. S. Broecker, "Newberry Professor of Earth and Environmental Sciences," Columbia University's Lamont-Doherty Earth Observatory.

31. See B. Fagan, note 29 above.

32. Worldwatch 132 (based on data from UNDP), Johansson, et al., IEA.

33. *New York Times*, May 11, 2009.

34. Abstracted from Worldwatch, 99–102.

35. G. Wood, "Moving Heaven and Earth," *The Atlantic*, July/August 2009, 70–6.

36. Ibid.

37. European Union, "Kyoto Protocol—Results and Predictions," http://www.eudebate2009.eu/eng/article/28549/kyoto-protocol-predictions-facts-results-overview.html (accessed July 2009).

38. Heritage Foundation Center for Data Analysis (2009).
39. M. Burnham, "Jobs at Issue as Labor-Enviro Coalitions Stump for Climate Bill," *The New York Times*, April 16, 2009, http://www.nytimes.com/gwire/2009/04/16/16greenwire-jobs-at-issue-as-laborenviro-coalitions-stump-10548.html (accessed November 27, 2009).
40. T. P. Carney, "How a Power Giant Profits from Greenhouse Regs," *The Washington Examiner*, September 23, 2009, p. 15.
41. Abstracted from Worldwatch, 161–72.
42. Ibid., 182–3.
43. Bloomberg News, "Greenhouse gas permits spark $100B lobbying fight," August 5, 2009.
44. "Daniel V. Kish is Vice President for Policy of the Institute for Energy Research," *Washington Examiner*, September 20, 2009, http://www.washingtonexaminer.com/opinion/columns/OpEd-Contributor/U_S__-China-taking-widely-different-energy-paths-8264563.html (accessed July 12, 2010).
45. M. Schulz, "Population Control: An Ugly Solution to Climate Change," *Washington Examiner* December 30, 2009.
46. Ibid.
47. D. G. Victor and others, "The Geoengineering Option," *Foreign Affairs* 88, no. 2 (2009), http://www.foreignaffairs.com/print/64829 (accessed July 12, 2010).
48. *Kyoto Protocol to the United Nations Framework Convention on Climate Change* (New York: United Nations, 1997).
49. M. Inman, "China CO_2 Emissions Growing Faster than Anticipated," *National Geographic News*, March 18, 2008, http://news.nationalgeographic.com/news/2008/03/080318-china-warming.html (accessed July 12, 2010).
50. "Cap-and-trade would trigger a new global trade war," *The Washington Examiner*, August 6, 2009, 2.
51. J. Fay, "Veteran climate scientist says 'Lock Up the Oil Men': NASA's Hansen goes DC," *Environment*, June 23, 2008.
52. "EPA should put carbon regs on hold," *The Examiner*, January 28, 2010, 2.

chapter fourteen

Global population paradox: Population explosion or implosion?

[I just read that] somewhere on this globe, every ten seconds, there is a woman giving birth to a child. She must be found and stopped.

Sam Levinson

Case study: St. Matthew Island

The story of St. Matthew Island's reindeer population is a metaphor for our society. For generations, St. Matthew Island, Alaska, had been a poster child for an ecosystem in balance with nature. Then the U.S. Coast Guard decided to build a small station on the 32-mile-long unoccupied island. While trying to stock a remote island with an emergency food source, the U.S. Coast Guard unwittingly set into motion a classic experiment involving an uncontrolled increase in population versus sustainability. In the summer of 1944, a total of 24 female and 5 male reindeer were released by the U.S. Coast Guard. Lichen mats 4 inches thick carpeted areas of the island.

With the end of World War II approaching, the Coast Guard decided to close its station. The men left before they had the chance to shoot the reindeer. The island had all the ingredients for a population explosion—a small group of healthy reindeer, and no natural predators. It also had a limited food supply that could not sustain a large population of reindeer. By 1963, the herd had exploded from 29 to some 6000. The island's carrying capacity was nearing its saturation point. Then it happened. A catastrophic crash occurred the following winter. Partly as a result of extreme snow accumulation, virtually the entire population of 6000 reindeer died of starvation. Nearly 99% of the herd had been destroyed by an ecosystem that was unable to sustain such a population increase. When researchers returned to the island, they found bleached reindeer skeletons scattered across the tundra and a couple dozen half-starved reindeer.

Class problem: In some ways, this story ties together many of the themes presented in this book: natural ecosystems, increasing populations, carrying capacity, and sustainability. After reading this chapter, return to this problem. Consider this example in light of what you have read. Is the story of St. Matthew Island a metaphor for the human population in the twenty-first century? In which ways does it accurately portray or explain what is happening in today's industrialized society? In which ways is it a poor model for explaining what is currently happening? What factors in today's society make this a poor metaphor?

Learning objectives

After you have read the chapter, you should understand the following:

- How fast the human population has grown in the past
- The concept of demographic transition
- The nature, underlying causes, and scope of the global population problem
- The range of potential environmental consequences that further global population increase may cause
- The scope of global population policy as it has evolved
- Critical controversies that confront population policy
- How to sensibly discuss and critique population policy options
- The environmental dilemma between population control and the aging crisis

14.1 Scientific basis for population growth

Figure 14.1 summarizes the historical record typical of "J-shaped" population growth. Note the lack of significant fluctuations related to catastrophes such as famines and wars; the nature of exponential growth depicted by this J-shaped curve is such that episodic reductions in population as a result of such catastrophes have had little effect on the inexorable increase in population. For example, the black death catastrophe in Europe produced only a small downward spike in the curve. Moreover, the large loss of life that resulted from twentieth-century world wars has had only a small perturbation on the exponential upward trend.

As shown in Figure 14.1, the increase in human population over the last century has been nothing short of phenomenal. In the 40-year period following 1950, the population doubled from 2.5 billion to 5 billion. Before the end of the twentieth century, the world population had surpassed 6 billion. As of 2009, the United Nations (UN) estimates world population to be 6.8 billion. China and India have a combined total of 2.6 billion people, or about 38% of the global total.

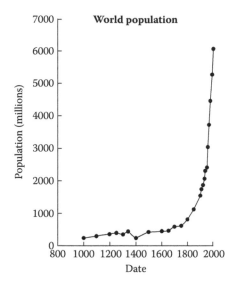

Figure 14.1 J curve showing the growth of human population over time.

Many environmental scientists assume that the world is headed toward a population explosion and that overpopulation is the root of all the world's environmental ills. This may indeed be the case, although a recent report from the UN Population Fund (UNFPA) has revised its earlier projection of future population.[1] The projection for 2050 is now 9 billion, as compared to the earlier 1996 projection of 9.5 billion. The principal reason for the lower projection is that global fertility rates have declined more rapidly than expected, as families have chosen to have fewer children. A global trend toward urbanization also is likewise increasing. The number of world's cities with 10 million or more inhabitants is increasing rapidly. Most of these "megacities" lie in less-developed regions.

Current data evidencing that global population growth has slowed is shown by recent UN projections through the mid-twenty-first century. Figure 14.2 shows the global population growth approaching a plateau.[2] Figure 14.3 shows that the rate of population increase per decade began to decline around 1997 and that this decrease is projected to steepen.[3] Further, the figure clearly shows how the rate of global population growth has been slowing down. Thus, the problem is not as simple as it appears on the surface. The population battleground is controversial and littered with failed prognostications. Moreover, the subject of overpopulation pegs the problems that come with a growing population against the miracles of modern technology. Finally, the chapter concludes with an introduction to the *global aging crisis*, which argues that the largest socioeconomic crisis we face today is actually the result of a population implosion.

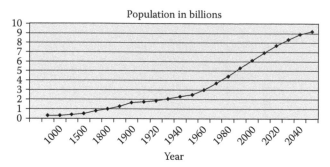

Figure 14.2 Population in billions: year 1000–2050. (From U.S. Census Bureau, "Historical Estimates of World Population up to 1950," http://www.census.gov/ipc/www/worldhis.html (accessed July 13, 2010); United Nations Department of Economic and Social Affairs, "World Population Prospects," http://esa.un.org/unpp/ (accessed July 13, 2010). With permission.)

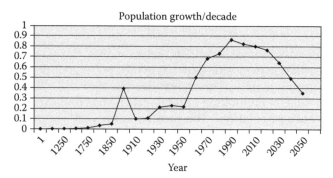

Figure 14.3 Population growth per decade: year 1–2050. (From U.S. Census Bureau, "Historical Estimates of World Population up to 1950," http://www.census.gov/ipc/www/worldhis.html (accessed July 13, 2010); United Nations Department of Economic and Social Affairs, "World Population Prospects," http://esa.un.org/unpp/ (accessed July 13, 2010). With permission.)

14.1.1   Fertility and mortality

Present projections indicate that global population will reach 8–12 billion before the end of the twenty-first century. The human population is expanding at the rate of three births per second. Two principal factors are responsible for the rapid increase in population over the last two centuries: (1) fertility and (2) mortality.

Obviously, the current rate of population growth is driven by fertility. Fertility tends to be controlled by economics and by human aspirations. Population growth and the level of development of a country are clearly linked, with northern developed nations generally having lower fertility rates than the less-developed southern countries. In the developing world, high fertility is partially explained by the large number of bodies needed to perform low-technology agricultural tasks. Families with large numbers of children can often increase their economic well-being. In more technologically advanced societies, parents realize that having more children tends to decrease rather than increase their standard of living.

Similarly, the population death rate, or mortality, also affects population growth. In many developing nations, the death rate has significantly dropped. Improved sanitation in concert with improved diet and medical practices has played a major role in reducing the rate of mortality. A downward trend in mortality has been witnessed in most countries. The combination of decreasing death rate coupled with a decrease in birthrate has led to a profound change in the population growth curve of the developed world. This change is referred to by demographers as the *demographic transition.*

14.1.1.1 Demographic–economic paradox

As explained in Chapter 2, economist Thomas Robert Malthus was the modern equivalent of Cassandra, the Greek mythological figure who prophesied doom. He has been referred to as the world's first professor of political economics. Malthus examined census figures from numerous countries and based his conclusions on a mound of statistics. He was the first to popularize a thesis on the limits to growth in *An Essay on the Principle of Population*, published in 1798. Malthus argued that human population will continue to rise geometrically, while the food supply increases arithmetically; eventually, population will outstrip the food supply and Earth's natural resources.

In opposition to what Thomas Malthus taught, the *demographic–economic paradox* suggests that the rate of reproduction tends to decrease as a consequence of economic progress. This effect may explain the recent trends shown in Figures 14.2 and 14.3. In effect, the demographic–economic paradox states that there is an inverse correlation between wealth and fertility. The higher the degree of education and gross domestic product (GDP) per capita, the fewer children are born in any industrialized country. This is true even though a wealthier population can support more children. At a 1974 UN population conference in Bucharest, a former minister of population in India illustrated this trend by stating, "Development is the best contraceptive."[4]

14.1.2 Demographic transition

Demographic transition refers to the process over the past century that has led to a general stabilization of population growth in developed countries.

This transition is characterized by four separate phases or stages (see Figure 14.4), as follows:

Stage 1 In this first stage of the demographic transition (first observed in Europe), birthrates and death rates are both high. Modern methods had yet to be adopted that lengthen life spans substantially. Birthrates and death rates tended to fluctuate wildly depending on circumstances.

Stage 2 In the second stage, hygiene and modern medical techniques begin to reduce the death rate, leading to a significant increase in population. As much of the economy tends to be based on agriculture, birthrates also remain high. Stages 2 and 3 are indicative of the beginning of the demographic transition.

Stage 3 The population becomes more urbanized, reducing economic incentives for large families. As the cost of supporting a large urban family grows, parents are discouraged from having larger families. The birthrate begins to drop, ultimately approaching the death rate. In Europe, the increased population created social pressures that led to large-scale migration to the United States and elsewhere.

Stage 4 The final demographic transition stage in Europe was characterized by a larger but more stable population size. Birthrates and death rates were relatively low, while the standard of living increased. Much of the developed world is believed to be in this fourth stage.

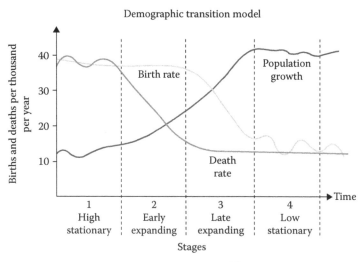

Figure 14.4 Demographic transition consisting of four phases.

Recent data suggests that the demographic–economic paradox holds only up to a point; once a country reaches a certain level of development and economic prosperity, the fertility rate stabilizes and then recovers slightly to *replacement rates.*[5] Much of the developed world is well into this fourth demographic transition stage. China and India now demonstrate evidence of early arrival, as shown in Tables 14.1[6] and 14.2.

Neo-Malthusians maintain that the population problem affecting many undeveloped countries is an inevitable result of high reproductive rates. The theory of demographic transition rejects this view, arguing instead that the population problem is a transitory phenomenon, occurring in the second stage of demographic transition as a result of a rapidly decreasing mortality rate without a corresponding fall in the birthrate. The theory of demographic transition maintains that every country passes through this demographic transition and that this can be proved empirically.

Table 14.1 China Population Data

Year	Population (thousands)	Increase (thousands)	Percentage increase
1950	544,951		
1955	598,226	53,275	9.8
1960	645,927	47,701	8.0
1965	716,270	70,343	10.9
1970	815,951	99,681	13.9
1975	911,167	95,216	11.7
1980	980,929	69,762	7.7
1985	1,053,219	72,290	7.4
1990	1,142,090	88,871	8.4
1995	1,210,969	68,879	6.0
2000	1,266,954	55,985	4.6
2005	1,312,253	45,299	3.6
2010	1,354,146	41,893	3.2
2015	1,395,998	41,852	3.1
2020	1,431,155	35,157	2.5
2025	1,453,140	21,985	1.5
2030	1,462,468	9,328	0.6
2035	1,462,351	−117	0.0
2040	1,455,055	−7,296	−0.5
2045	1,440,289	−14,766	−1.0
2050	1,417,045	−23,244	−1.6

Source: United Nations Department of Economic and Social Affairs, "World Population Prospects," http://esa.un.org/unpp/ (accessed July 13, 2010).

Table 14.2 India Population Data

Year	Population (thousands)	Increase (thousands)	Percentage increase
1950	371,857		
1955	406,661	34,804	9.4
1960	448,314	41,653	10.2
1965	496,934	48,620	10.8
1970	552,964	56,030	11.3
1975	617,432	64,468	11.7
1980	692,637	75,205	12.2
1985	774,775	82,138	11.9
1990	862,162	87,387	11.3
1995	953,148	90,986	10.6
2000	1,042,590	89,442	9.4
2005	1,130,618	88,028	8.4
2010	1,214,464	83,846	7.4
2015	1,294,192	79,728	6.6
2020	1,367,225	73,033	5.6
2025	1,431,272	64,047	4.7
2030	1,484,598	53,326	3.7
2035	1,527,879	43,281	2.9
2040	1,564,763	36,884	2.4
2045	1,593,852	29,089	1.9
2050	1,613,800	19,948	1.3

If these projections are actually achieved by mid-century, China will likely achieve less than zero population growth by about 2055 and India will likely follow suit by about 2070. Most developed countries have already undergone a complete demographic transition and are now experiencing slower population growth rates; a number of them are actually experiencing negative growth rates. However, most developing nations with high population growth rates will constitute an increasingly larger fraction of the world's population. Assuming that other Third World countries will emulate the demographic experience of China and India, many, if not most, demographers believe that a transition similar to this will eventually benefit them as well. But this transition will not occur overnight, as many decades will be needed to achieve the economic and cultural changes that would enable birthrate and death rate to equilibrate. During this period, the population will continue to expand at rates that cannot be sustained because of regional and global resource limitations.

14.2 Nature of the population problem

The UN Population Program describes a projection methodology that takes several causal factors into account.[7] The UN methodology emphasizes models largely based on demographic factors such as age and sex, distribution, birthrates, death rates, and socioeconomic factors such as income, savings, and consumption. These factors support statistical methods of analysis that in turn are the result of more primal causes such as limits to resource availability in relation to a population's demands as Malthus described. Although not directly considered in its population models, the UN acknowledges their importance in a number of other studies.[8] The underlying mathematics of such population forecast models is quite simple. We solve for population at the end of a defined period (P2) using the following two equations:

$$\Delta P = (\text{Births} - \text{deaths}) + (\text{INmigration} - \text{OUTmigration})$$

$$P2 = P1 + \Delta P$$

where ΔP is population increase; P1 is population at the start of a defined period (such as year or decade); and P2 is population at the end a defined period.

The cold calculus involved in these methods conceals the human realities behind the modeled events (births, deaths, and migrations), such as relative wealth and poverty—whether in terms of money, land, water, food and fiber, physical and mental health, and education, or in terms of the many cultural amenities that add to quality of life that lie outside the marketplace.

There is a human price for approaching the limits to growth, and this price is not evenly distributed. As the limits are approached, fewer wealthy people come to own most of the means to ensure their continuous access to life's necessities and amenities, while the rest of the population puts up with scarcity of all kinds and increasing numbers suffer the deprivations of extreme poverty. People living under such extremes experience much higher death rates related to malnutrition, polluted water, tainted foods, polluted air, and diseases that increase mortality in all age groups. Violent crimes are more common among them and add to their toll of premature deaths.

14.2.1 Early calls to action

As described in Chapter 2, in 1968, ecologist Paul Ehrlich wrote a popular but controversial book titled *The Population Bomb*.[9] The book sounded an alarm that the continuing growth in world population would stress the limits of global resources in meeting the most elementary needs of the earth's inhabitants. In 1972, the Club of Rome issued its groundbreaking systems analysis, *Limits to Growth*, which investigated the relationship between the world's exponentially increasing population and its resource limits. The report's principal conclusions are as follows[10]:

> Our world model was built specifically to investigate five major trends of global concern—accelerating industrialization, rapid population growth, widespread malnutrition, depletion of nonrenewable resources, and a deteriorating environment. The model we have constructed is, like every model, imperfect, oversimplified, and unfinished. In spite of the preliminary state of our work, we believe ... that the model described here is already sufficiently developed to be of some use to decision makers. Furthermore, the basic behavior modes we have already observed in this model appear to be so fundamental and general that we do not expect our broad conclusions to be substantially altered by further revisions ... Our conclusions are:
>
> 1. If the present growth trends in world population, industrialization, pollution, food production, and resource depletion continue unchanged, the limits to growth on this planet will be reached sometime within the next one hundred years. The most probable result will be a rather sudden and uncontrollable decline in both population and industrial capacity.
> 2. It is possible to alter these growth trends and to establish a condition of ecological and economic stability that is sustainable far into the future. The state of global equilibrium could be designed so that the basic material needs of each person on earth are satisfied and each person has an equal opportunity to realize his or her individual human potential.

One of its authors, Donella Meadows, a systems analyst, professor, and journalist who directed the Sustainability Institute at Dartmouth, wrote a popular book with same title.[11] Like Ehrlich's book, it was very widely read and influenced the growing environmental movement at the time, but gained little policy traction for its issues. In 2004, Meadows and Jorgan

Randers published *Limits to Growth—The Thirty Year Update*.[12] Although the numbers behind the data and trends, along with global environmental policy thinking and computer modeling capability, had all significantly changed by 2004, their original policy warnings remained unchanged: namely, as the world's population continues to increase, the increasing rate of human exploitation of the Earth's limited resources is unsustainable.

14.2.2 Dissenting view

These earlier forecasts are controversial and the subject of endless debate. As described in Chapter 2, Julian Simon, a senior fellow at the Cato Institute and a prominent professor at the University of Maryland, wrote extensively about population, immigration, and natural resources. He viewed Ehrlich and Meadows as "gloom and doom" scientists. Some have called him a cornucopian who believed that the limits to growth would always be extended by human scientific and engineering ingenuity. Such views were widely held by many mainstream economists who cite revolutions in agricultural production, discovery of new fossil-fuel supplies, and the growing trends in non-fossil-energy technologies as evidence.

Simon reasoned that population control would reduce the likelihood of a future Mozart, Einstein, or Edison being born. He did not view continuous population growth as a problem because the ultimate resource is people—educated, skilled, ambitious people—who will always be there to provide technological solutions to future problems. Simon reasoned that necessity is the mother of invention. He argued that as resources become scarce, in the short run, prices will rise, creating incentives for inventors and entrepreneurs. Although population might grow exponentially, so does science and technology.

It could be that both camps have something of substance to say. To date, the Ehrlich camp has been largely wrong in its original predictions. Simon's camp emphasizes the enormous advances that technology can bring to mitigating the impact that Ehrlich warns against. But over the long term, Simon's optimistic prediction may not keep pace with the extreme demand and burden that an escalating population and increased economic development can place on a limited global resources base.

14.3 Population and the environment

While policymakers debate how the global population can sustain itself using the Earth's limited resources, the tropical rainforests in central African countries and the Amazon basin (the Earth's rainforest) continue to be razed to make way for cattle ranching and farms, and to grow crops for ethanol fuel. In a sobering comparison to Easter Island (see Chapter 2, Case study), the vast deforestation of Haiti, Madagascar, and parts of the Indonesian archipelago have added to the inhabitants' mounting poverty as arable soils are washed away and mudslides bury entire villages during storms. In Chapter 3, on sustainability, we discuss alternative approaches to avoiding such degradation and potential catastrophe.

There are a host of causal links between the environment and the sustainable population that can in principle be modeled within a given region with well-defined population and economic boundaries. However, at the global scale, such models are not sufficiently mature as predictive policy tools, other than at the most macro levels, as general indicators. For example, global community policymakers consider the effects of population increase in their studies of global warming and seek opportunities to reduce its adverse effects (see Chapter 13).

14.3.1   Poverty crisis

According to many scientists and concerned policymakers, poverty is reaching crisis levels and global warming is only likely to make things worse. Much of the world's population survives in poverty with varying kinds and degrees of resource scarcity, both economic and environmental.

It would be simplistic to conclude that excess demands for natural resources alone cause poverty. Major causal factors include economic and social inequities, such as exploitation of indigenous resources by a very small percentage of elites who own most of their nations' wealth—and the policies of international financial institutions that enable inequitable investments (see Chapter 8). The consequences suffered by those whose environments are impacted by unethical behavior include loss of productive land, pollution of air and water, food scarcity, lack of health care, and little opportunity for adequate employment and decent lifestyles. Any one of these factors can have a multiplier effect that reduces productivity and access to resources, while causing disease, shortened life spans, and devastation of families.

14.3.2   Measuring poverty

At the global policy level, there are some difficulties in defining measures of poverty. Citing an August 2008 World Bank report, Anup Shah states the Word Bank's definition of $1.25 per day as the poverty line, while some developing nations define it at $2.00 or $2.50 per day.[13] This may be realistic for relatively self-sufficient, isolated rural communities that live off local land and water resources. But this poverty measure is not likely to apply to urban regions. Shah states that 1.4 billion people live at or below this poverty line, according to the World Bank. Given the continuing large scale of migrations from rural to urban regions, recent global economic stresses, and a more realistic poverty line, the actual numbers living in poverty are undoubtedly much higher. Moreover, the World Bank's figures include China in the global statistics. Yet, Shah reports that from 1981to 2005, China's poverty rate, by its own standard, fell from 85% to 15.9%, representing improved living standards for more than 600 million people. Less dramatic but similar gains have been made by India and a number of other nations in Asia and Latin America. In effect, the so-called Third World has a wide diversity of economic systems that can be divided into two groups—those reducing and those increasing their respective poverty rates.

14.4   Scope of global population policy

14.4.1   United Nations population policy

The primary focus of the UN's global population policy is to improve the standard of living by reducing poverty. Measures designed to reduce population growth may play a part in this policy. However, as social indicators such as earnings and health improve, so does life span, which causes population to increase. This can lead to a vicious cycle. Hence, the goal of this policy is that as poverty is reduced and family planning becomes available, the birthrate will decrease faster than increases in population life spans. The UN's Agenda 21 (see Chapter 3, Table 3.1) promulgates 27 principles pursuant to sustainable development that also apply to its population policy.

The work of the UNFPA is integrated and managed by the UN through a set of key Millennium Development Goals with specific achievement targets (see Table 14.3).[14] It should be clear that the UN's population program focuses on issues that enhance health and establish conditions that enable especially poor people to earn a better living and in other ways to improve their quality of life. An important focus is to enable men and women to control their personal lives by having fewer children, which would also reduce the rate of global population growth.

In effect, the UN's population policy is holistic, in the sense that it encourages conditions of economic, environmental, and social life that are similar to those in wealthier countries where the balance between birthrates and death rates have caused markedly less population increase, and in some cases, a decrease. Nevertheless, some aspects of the program are controversial to some parties because of religious and ethical concerns as well as cultural norms—especially attempts to promote family planning and the use of contraceptives for family planning.

Table 14.3 Summary of Key Millennium Goals and Targets for 2015

Goal	Targets
1. End extreme poverty and hunger	Halve, between 1990 and 2015, the proportion of people whose income is less than $1 a day. Achieve full and productive employment and decent work for all, including women and young people. Halve, between 1990 and 2015, the proportion of people who suffer from hunger.
2. Achieve universal primary education	Ensure that, by 2015, children everywhere, boys and girls alike, will be able to complete a full course of primary schooling.
3. Promote gender equality and empower women	Eliminate gender disparity in primary and secondary education, preferably by 2005, and in all levels of education no later than 2015.
4. Reduce child mortality	Reduce by two-thirds, between 1990 and 2015, the under-five mortality rate.
5. Improve maternal health	Reduce by three quarters the maternal mortality ratio. Achieve universal access to reproductive health.
6. Combat AIDS, malaria, and other diseases	Halt by 2015 and begin to reverse the spread of HIV/AIDS.
7. Ensure environmental sustainability	Integrate the principles of sustainable development into country policies and programs and reverse the loss of environmental resources. Reduce biodiversity loss, achieving, by 2010, a significant reduction in the rate of loss. Halve, by 2015, the proportion of the population without sustainable access to safe drinking water and basic sanitation. By 2020, achieve a significant improvement in the lives of at least 100 million slum dwellers.
8. Develop a global 8 partnership for development	Address the special needs of least-developed countries, landlocked countries, and small island developing states. Develop further an open, rule-based, predictable, nondiscriminatory trading and financial system. Deal comprehensively with developing countries. In cooperation with pharmaceutical companies, provide access to affordable essential drugs in developing countries. In cooperation with the private sector, make available benefits of new technologies, especially information and communications.

Source: United Nations, "Millennium Development Goals 2015," http://www.un.org/millenniumgoals/bkgd.shtml (accessed August 10, 2009).

14.4.2   India's population policy

The UN Environmental and Social Commission for Asia and the Pacific (UNESCAP) briefly describe India's population in a 1999 report about the limited policy successes to date[15]:

> In 1952, India was the first country in the world to launch a national program, emphasizing family planning to the extent necessary for reducing birthrates "to stabilize the population at a level consistent with the requirement of national economy." After 1952, sharp declines in death rates were, however, not accompanied by a similar drop in birthrates. The National Health Policy, 1983 stated that replacement levels of total fertility rate[2] (TFR) should be achieved by the year 2000. On 11 May, 2000 India is projected to have 1 billion people, i.e. 16 percent of the world's population on 2.4 percent of the globe's land area. If current trends continue, India may overtake China in 2045, to become the most populous country in the world. While global population has increased threefold during this century, from 2 billion to 6 billion, the population of India has increased nearly five times from 238 million to 1 billion in the same period. India's current annual increase in population of 15.5 million is large enough to neutralize efforts to conserve the resource endowment and environment.

14.4.3   China's population policy

Excerpts from China's 2001 population law, as displayed on the UNESCAP Web site states the following[16]:

> Article 2: The state adopts a comprehensive policy to control the size and raise the general quality of the population. The state relies on publicity and education, advances in science and technology, multi-purpose services, and the establishment and improvement of the multi-purpose reward and social security systems, in carrying out the population and family planning programs.
>   Article 4: When promoting family planning the people's governments at all levels and their staff members shall perform their administrative duties

> strictly in accordance with the law, and enforce the law in a civil manner, and they may not infringe upon the legitimate rights and interests of citizens.

Article 4 makes it clear that Chinese law has serious enforcement provisions. China, of course, has been cited in many U.S. media reports for its draconian population policy that limits the numbers of children per family. However, one must take cultural conditions into account. In China, the reported birthrate of boys significantly exceeds that of girls. This is a cultural phenomenon in that families are probably aborting and even abandoning female babies because of socioeconomic and cultural factors. Anecdotally, many American families seeking to adopt have found China to be a welcoming source for parents for such children. There appears to be no religiously based cultural prohibitions against abortion in China.

The imbalance between female and male babies has been the subject of both interest and concern among sociologists and political scientists. Some have warned that males, unbound by marital ties, may be much more prone to criminal and antisocial behavior. Moreover, these males may be drawn toward the military, raising further fears that this could lay the foundation for increased propensity toward armed conflict.

14.4.4 Controversies confronting aspects of population policy

China's harsh policy of coercion, including abortion, is perhaps the most controversial population control issue facing the global community. It is the only country to have taken that path and finds little sympathy among Christians, Muslims, and Jews as well as among secular people around the world. Although UN policies do not directly address abortion, a number of nations permit it as a family planning option.

The UN emphasis on empowering women to control their lives is controversial among a number of religious groups in many faiths. Attempts to promote monogamy among polygamous populations are likewise controversial. Such controversies may argue for a fundamental rethinking of contraceptive and abortion initiatives. Perhaps a more effective policy is one that simply attempts to promote socioeconomic advancement, which can lead to female empowerment and smaller families.

14.5 Population paradox: Simmering socioeconomic crisis?

To many, it may seem obvious that population growth is the root of the world's ills and that population control should be a societal goal. Right? Well, ask Peter Peterson, a son of Greek immigrants. He is the quintessential American success story. Very few can rival his economic credentials.

Peterson has led a remarkable career—chairman of the New York Federal Reserve and former U.S. Secretary of Commerce. Today, he lives a lifestyle that most can only dream of, and he will never have to worry about affording retirement. Yet, he is worried; very worried about our future. Here is what troubles Peterson[17]:

> The scenario I see is that one or more developed countries, say Italy, is going to decide that the political cost of reforming their pension systems is just too high. Then they will try running high deficits … in an attempt to finance their way out of the problem. When the financial markets wake up to this news, there will be a broad realization that we have a *global aging crisis* that is going to be unrelenting in its economic consequences.

In Peterson's mind, we are sailing on the *Titanic*. The iceberg Peterson is pointing to could literally sink the world's economy. But will this warning reach the steering tower in time to avert disaster? As a metaphor to the *Titanic*, Peterson has this to say:

> If we seize the wheel of the ship—which is admittedly big and heavy—and turn it now, there is still time to avert the coming collision. If we stand on the deck and debate too long about whether it is or isn't an iceberg out there, valuable time will be lost. Worse yet, … we might … make our way to the deck for a last dance as the ship's orchestra strikes up another jaunty tune. … If we begin now, together, the needed reforms can be gradual, compassionate, and effective. If we wait, if we dither and debate, if we talk endlessly about how to turn the wheel, but never quite get to turning it—then the sharp turn will be sudden and destructive. Maybe even catastrophic.

If you think Peterson is simply a crackpot or another Cassandra yelling, "The sky is falling," consider the following statement by the U.S. congressionally sponsored National Commission on Retirement Policy[18]:

> Looming on the horizon are a number of "economic time bombs," which, if not defused now, threaten the prospects of a secure retirement for many Americans. The imminent retirement of the "Baby Boom" generation soon after the turn of the

> century will strain our nation's fiscal capacity and
> could imperil the standard of living for future gen-
> erations … We are confronted with a rapidly aging
> population, an actuarially unsound Social Security
> system, unsustainable levels of spending for gov-
> ernment programs for senior citizens, under-funded
> private pension plans, and unacceptably low levels of
> private savings. We have [been] promised too much
> and set aside too little. As a result, … retirement is
> increasingly unstable and threatens to collapse.

Here lies one of the greatest environmental policy dilemmas imaginable:
while stabilizing population may conserve environmental resources, the
age demographics could create a shortage of labor of working age people
to meet the pension needs of the retired. This may be a ticking socioeconomic time bomb for societies around the world. This illustrates the complexity of environmental policymaking and the harsh compromises and
trade-offs that must often be made. Perhaps most of all, it illustrates paradoxes that can arise in formulating pragmatic environmental policies.

14.5.1   The birds and the bees

So you might ask, exactly what changed? It is really quite simple. Much
of it has to do with the "birds and the bees," or perhaps the lack thereof.
A sharp decline in the global population growth rate, particularly among
affluent Western societies, has led to a radical shift in age distribution,
which is being propelled by two principal factors:

1. Decline of fertility (sharp fall in birthrates)
2. Increased life expectancy (rising longevity)

First, and foremost, birthrates are declining, albeit at different rates, in
most parts of the world, but particularly in developed nations. Contrary to
the common misperception, the decline in fertility rate, not increasing life
expectancy, is the principal factor responsible for the coming aging crisis.[19]
As noted earlier, the secondary cause is that the average life expectancy
is rising rapidly almost everywhere. Extended life expectancy is among
the most dramatic achievements of the twentieth century. The average life
expectancy across the planet in 1900 was less than 30 years; only 4% of the
American population even reached the age of 65! Today, average life span
expectancy has increased to 76 years. Some scientists are now projecting
that 100 may soon become the new norm.

   Many are surprised to learn that throughout most of history, men have
generally outlived women. It was not until the twentieth century that the

average life expectancy of women increased dramatically, adding more than 30 years to their life expectancy. Men have also added years, but higher rates of smoking, occupational hazards, and other factors have slowed their progress compared to that of women. Notwithstanding, an average male at 67 years of age has fewer health problems and places less burden on the health care system than the average female. Interestingly, the gender gap between men and women is now closing in the United States. In fact, the life expectancy among men is now *rising* at a faster rate than among women.[20]

14.5.2   Making babies

So why are fertility rates so low? In simple language, a change has occurred in human behavior, which is nothing short of a miracle. From America to Zambia, fertility rates are plummeting throughout most of the world. For example, in 1970, children under 5 years old outnumbered Americans aged 85 and by a ratio of 12:1. According to some forecasts, by 2040 the number of elderly will equal the number of preschool children!

14.5.3   Population implosion?

The risks of an exploding population base have been described. Only a few decades ago, most demographers were forecasting overpopulation explosion. Recall that a mere four decades have passed since computer studies sponsored by the Club of Rome showed that population pressures would cause major problems by the year 2000. For example, as recently as the late 1960s, the worldwide average fertility rate stood at about 5 (number of births over an average woman's lifetime). Back in the 1960s, it could easily be computed that with a fertility rate of 5, we were headed for big problems, but the historical trend in birthrates began reversing sharply. The result was an unprecedented and unexpected decline in the global fertility rate to what is now about 2.7. This rate continues to decline and will actually fall below the magical number, 2.1.

So, what is so magical about the number 2.1? Nothing much, except this number just happens to be known as the *replacement rate*—the rate required to merely sustain a constant population or yield zero population growth. That is, for a population to simply remain constant, women must have on average 2.1 children (the extra 0.1 compensates for children who never reach reproductive age). For example, during the last 25 years, the number of children per couple has fallen from 6.6 to 5.1 in Africa, from 5.1 to 2.6 in Asia, and from 5.0 to 2.7 in Latin America and the Caribbean. If the world's fertility rate drops below 2.1, global population will actually begin declining!

To appreciate what we mean, consider the following fact. In 61 countries that represent 44% of the world's population, fertility rates have already fallen to the point where they are now equal to or below replacement levels. In some of the developed world, the average fertility rate has plummeted to an almost unimaginable rate of 1.6. That's right, a negative rate of population growth in 61 countries. It is important to emphasize that *total* world population is expected to continue rising through the middle part of this century, at which point some experts believe the population level might top out around twice its current number. However, the pace at which the global population level is rising is decreasing; the UN predicts that the world's population will continue to increase, albeit at a slower rate.

This is not a theoretical projection or outcome. We know how many people have already been born and they have been accurately counted. Even if the existing fertility rate were to suddenly turn around, it would be too little too late; it would take a generation for a new baby base to get educated, mature, and join the ranks of the working class.

14.5.3.1   World at large

Low fertility is no longer specifically a Western phenomenon, nor is it limited to rich industrial nations. With the exception of principally Muslim countries, mainly located in Northern Africa and the Middle East, it is hard to find any part of the world that is not aging. This should *not* be construed to mean that the total global population is on the verge of falling. A large percentage of the world's population is still of childbearing age. These people will continue to bear children. More people are also living to old age. According to the UN, the world population should increase at an average rate of 1.3% per year over the next 50 years or so. If the fertility rate continues to fall, the world population may actually begin declining somewhere after the middle of this century.[21]

14.5.4   Social insecurity

Pension systems have had great political appeal, particularly in Western developed nations; they promise future recipients a lot and were initially pretty cheap. When the retirement age was set at 65 years in the 1930s, the average American life expectancy was just over 61 years. This was the kind of odds that a Vegas casino would have bet on. The support ratio of retirees to workers has fallen from a ratio of 1:42 in 1940 to 1:3 today; it is headed for a dismal 1:2.5 by 2025![22] Very soon, the costs will begin to greatly exceed annual revenues. Accumulated "reserves" (which exist on accounting sheets but is not safely stashed away in some vault) will be exhausted perhaps before 2030.[23] *Unfunded pension liabilities* are those benefits that have already been earned by today's workers, but for which nothing has actually been stuck away into a "lockbox."

14.5.5   Pay as you go

Now here is something that most Americans and many others Westerners do not realize. The U.S. Social Security system, and virtually every other Western pension system, was initiated as a *pay as you go* system. The term "trust fund" is now often used. Trust fund may suggest a vault safely tucked away somewhere, in which everyone's Social Security taxes are stacked up to the ceiling to be paid out later. This may be the greatest myth ever perpetrated on modern Western society. The money paid into the system does not go into some bank vault, a special reserve account, or even a lockbox. The term "trust fund" is a misnomer. Here is why. Today, workers hand over their hard-earned money to the tax collector. This money is promptly used to pay *today's* retirees. In other words, the government finances today's retirees by taxing today's workers. This is the way virtually all industrial democracies finance their pensions and retirements.

What about so-called annual surpluses, where do they go? When the system is running a surplus, as it recently has, it does not go into some lockbox. Instead, this additional revenue is *lent* by the Social Security Administration to the federal government to pay the cost of other federal programs—everything from jet fighters to building flood-control levies. In return, the Social Security trust fund is credited with special nontaxable debt obligations from the treasury department—in effect, gigantic IOUs. Transferring IOUs from the right to the left pocket does nothing to bridge the coming Social Security and Medicare funding shortfall. It is nothing but an accounting trick to finance one government sector from another. It is estimated that within 30 years, developed countries will need to spend a staggering additional 9%–16% of the GDP simply to meet their retirement obligations.

14.5.6   Swedish experience

Perhaps no country has expended more effort to head off the coming aging crisis by increasing fertility rates than Sweden. In fact, Sweden spends 10 times as much on family support programs as does Italy and 3 times that of the United States.[24] Decades ago, because of a diminishing birthrate, Sweden developed a policy to support larger families. It formulated what was arguably the most generous public-sponsored fertility policy in the world (cash payments, tax incentives and job leave, flexible and part-time work schedules). As expected, Sweden led Europe's birthrate in the 1980s. Its experience roundly suggested that countries could have more children—if only they were willing to ante up and pay the piper. At least, Sweden's initial experience led the experts to believe this.[25]

Then the shocker: In the early 1990s, Sweden's baby boom came to a screeching halt. Birthrates plummeted from 2.1 to an unsustainable 1.6. Experts were perplexed. They blamed the sagging economy. This seemed like a completely reasonable explanation. With a confident wink, experts boldly predicted that birthrates would soon rebound once the economy improved. There was only one problem—birthrates did not pick up when the economy improved. By the late 1990s, Sweden's birthrate had tumbled to a dismal 1.42, about the same rate as Japan. The exuberant confidence that experts shared in the early 1990s deflated like a balloon. Many analysts now believe they are witnessing a fundamental shift in Swedish social and sexual behavior.

14.5.7 Global meltdown?

Retirees tend to be well organized and have a great deal of free time, and they are not shy about battling for various causes, particularly those that personally affect them. By 2038, Peter Peterson estimates that people aged 65 years and older will make up 34% of the electorate—up from a mere 16% in 1966.[26] You think Social Security is a sacred cow now? The battle over entitlements may become quite ugly. Another potential loser in this generational tug-of-war may be education. Consider one retirement community near Phoenix, which tried to secede from the local school district in 1997 in order to avoid paying school taxes. James Button, a political science professor at the University of Florida, studied the voting patterns of older people in six Florida counties. In addition to education, he also found a number of other differences in views held by the young and the elderly. More and more, we may find youth and the elderly at odds, not only over how tax money is spent, but also over core values.[27]

Some demographers and political scientists are postulating a virtual calamity by the time the dust settles. As pension systems go bust around the world, tens of millions of retirees could find themselves living on the streets. Like birth contractions, the reality that lies before us will strike slowly at first. Over time, it will fester into a full-blown crisis. The young, overburdened with unbearable tax rates to finance the collapsing pension system, will be pitted against larger and well-organized groups of voting seniors. The world's economic system may be pushed to the brink. Governments, and nations for that matter, may collapse. If this crisis follows the normal course of events, you may expect to find that it progresses through the following seven stages:

Stage 1 When this little secret becomes widely public, politicians will first be in denial.

Stage 2 When denial no longer works, politicians will offer short-term fixes or hocus-pocus trust-fund-accounting "solutions."

Stage 3 Reality begins to set in—we cannot pay the bills!

Stage 4 Each political entity blames everyone else for this predicament.

Stage 5 Panic sets in.

Stage 6 If luck is on our side, maybe we will make it, albeit it will be a painful experience.

Stage 7 Let us hope it does not make it to this stage.

Discussions, problems, and exercises

1. Summarize the causal factors that influence the rate and direction of population change.
2. Develop a detailed plan for achieving the UN's millennium goals and targets for 2015.
3. Design the scope of an environmental impact analysis for developing a population policy.
4. Write a two-page essay on ways the United States should influence the UN's global population policy.
5. Prepare a policy that attempts to balance the competing goal of population control with the goal of preserving the pension and retirement system (aging problem).

Notes

1. United Nations Population Fund, *The State of World Population* (New York: United Nations, 1999).
2. U.S. Census Bureau, "Historical Estimates of World Population up to 1950," http://www.census.gov/ipc/www/worldhis.html (accessed July 13, 2010); United Nations Department of Economic and Social Affairs, "World Population Prospects," http://esa.un.org/unpp/ (accessed July 13, 2010).
3. Ibid.
4. D. N. Weil, *Economic Growth* (New York: Addison-Wesley, 2004), 111.
5. M. Mikko, K. Hans-Peter and F. C. Billari, "Reverse Fertility Declines," *Nature* 460 (August/June 2009): 741–3.
6. United Nations Department of Economic and Social Affairs, "World Population Prospects," http://esa.un.org/unpp/ (accessed July 13, 2010).
7. United Nations Department of Economic and Social Affairs, "Demographic Manuals," http://www.un.org/esa/population/techcoop/PopProj/PopProj.html (accessed August 10, 2009).
8. See United Nations Department of Economic and Social Affairs, "Environmental Factors" (accessed August 10, 2009).
9. P. R. Ehrlich, *The Population Bomb* (New York: Ballentine Books, 1968).
10. D. Meadows, *The Limits to Growth* (New York: Signet, 1972).
11. See note 10, above.
12. J. Randers and D. L. Meadows, *Limits to Growth—The Thirty Year Update* (White River Jct., VT: Chelsea Green, 2004).
13. Global Issues, "Poverty around the World," http://www.globalissues.org/article/4/poverty-around-the-world (accessed July 13, 2010).

14. United Nations, "Millennium Development Goals 2015," http://www. un.org/millenniumgoals/bkgd.shtml (accessed August 10, 2009).
15. United Nations Environmental and Social Commission for Asia and the Pacific, "India National Population Policy," http://www.unescap.org/esid/ psis/population/database/poplaws/law_india/Indiaindex.htm (accessed August 18, 2009).
16. Population and Planning Law of the Peoples Republic of China (2001), http://www.gov.cn/english/laws/2005-10/11/content_75954.htm (accessed September 16, 2010); United Nations Department of Economic and Social Affairs, "World population prospects," http://esa.un.org/unpp/ (accessed July 13, 2010).
17. P. J. Longman, "The World Turns Gray: How Global Aging Will Challenge the World's Economic Well-being," *U.S. News and World Report*, March 1, 1999.
18. Center for Strategic and International Studies (CSIS), National Commission on Retirement Policy, "The 21st Century Retirement Security Plan: Final Report of the National Commission on Retirement Policy" (Washington, DC: CSIS, 1999).
19. H. J. Aaron and B. P. Bosworth, *The Budget Meets the Boomers: America's Most Famous Generation Approaches another Milestone* (Washington DC: The Brookings Institution, 1996).
20. J. R. Brandstrader, "From Baby Boom to Geezer Glut," *Scientific American Presents*, (Summer 2000): 23–9.
21. See note 17 above.
22. The Editors, "Social Insecurity," *Scientific American Presents*, (Summer 2000): 27–9.
23. See note 19 above.
24. M. Specter, "Population Implosion Worries a Graying Europe," *New York Times*, July 10, 1998.
25. Ibid.
26. E. Pan and E. Roberts, "Home of the Gray," *Newsweek*, March 1, 1999.
27. Ibid.

Epilogue

Acting locally: Six degrees of separation

> Always do right. This will gratify some people and
> astonish the rest.
>
> **Mark Twain**

The authors would like to end this text with an inspirational thought. The phrase *six degrees of separation* refers to an observation that if a person is only one step away from each person they know and two steps removed from each person who is known by one of the people they personally know, then everyone is a mere six steps away from any other person on the planet. Karinthy, an early advocate of six degrees of separation, believed that despite great distances between individuals around the globe, the growing density of human connectivity and networks makes the actual social connection much smaller. By postulating that any two individuals around the world could generally be connected by at most five other acquaintances, Karinthy has been credited for originating the controversial theory of the six degrees of separation.[1]

Several empirical studies, such as *Milgram's small world experiment*, appear to confirm Karinthy's theory. Critics counter that Milgram's experiment did not confirm the six degrees and that such claims amount to an "academic urban myth."[2] Professor Duncan Watts attempted to duplicate Milgram's experiment on the Internet, using e-mail message as the "package" to be delivered; he found that the average (though not maximum) number of intermediaries was close to six.

In 2007, Leskovec and Horvitz examined a set of instant messages amounting to 30 billion conversations among 240 million people; they found the average path length among Microsoft Messenger users to be 6.6.[3] A Facebook application called "Six Degrees," consisting of about 4.5 million users, calculates the degrees of separation among different people; on average, the separation for these users is 5.73 degrees, while the maximum degree of separation is 12.

Consider the following examples showing how six degrees of separation might be used to solve problems and accomplish goals. It has been widely estimated that 80% of jobs are unpublished. Six degrees of separation can be instrumental in making contacts and establishing relationships that may aid students in their job hunt.

So, what does this concept have in common with the goal of formulating environmental policies? If people living in diverse places such as remote rain forests could be brought to within six degrees of separation with environmental organizations, information could be disseminated that might have a marked impact on the importance and awareness of protecting an ecosystem. One hoped-for result would be an increased awareness of the need for local leaders and businesses to take an active role in solving environmental problems.

Capstone problems

Upon completing this book, you should be able to apply the appropriate concepts, principles, and tools described throughout this the text. Most importantly, you should now be able to develop pragmatic policy solutions. The following two capstone problems involve consideration of nearly every key principle and concept presented in this book.

Capstone problem 1: Local economic and environmental development problems

You are the chief environmental policy consultant for the secretary of the environment of the small nation Ecoana. One of Ecoana's islands lies off the coast of Estica. It is a sparsely populated island inhabited principally by indigenous inhabitants. The islanders are poor and have only limited experience with societies and cultures beyond their shores. Outsiders, as well as citizens of Ecoana, have been buying land to develop residential, commercial, and tourist sites on the island.

A mountain divides the island in half. Half of the island consists of luscious, undeveloped beaches and a still-undeveloped rainforest. The other half of the island lies in the rain shadow; this side of the island is a desert, consisting of sparse vegetation and animals, but has magnificent views and beaches. The islanders have been increasingly exploiting the island's natural resources (cutting down trees for wood, slashing native areas for small farms, and hunting native animals for their hides). Pigs have also recently been imported, and many are running wild, wreaking havoc on the native vegetation.

The president of Estica needs capital to support his growing society. He wants to develop the island as a means of collecting revenue. The country's business community wants to develop the entire island into an international resort area. Environmentalists warn that this goal will ultimately destroy the very environment that attracts vacationers. Wealthy people are prepared to pay a premium to obtain a permit to build a residence on the island. The islanders do not want outsiders encroaching on their culture and have threatened violence.

If no action is taken to develop the island, Estica will lose a major source of potential revenue. Moreover, the island may still be randomly developed without a master development policy and land-use plan. If the island is developed, Estica will receive revenue, but this may not be a sustainable alternative from the standpoint of long-term tourism; also, the native inhabitants will be unhappy with this course of action. If the island is sparsely developed, the island can be maintained on a long-term basis, but Estica will lose a significant amount of potential revenue.

Carry out the following steps leading to a plan, policy, and appropriate management mechanisms to ensure the island's future. Review the key principles described in Chapter two while formulating your approach. Prepare a preliminary policy report that addresses the following:

1. Draw a scaled map of the island (use your imagination), and delineate key features (beaches, mountains, rain forests, rivers, waterfalls, and desert).
2. Define the elements of at least two scenarios (in addition to "no action") as a list of projects for public and private investments that respond to needs identified in the description (again, exercise creativity).
3. Outline an economic methodology for assessing and comparing the costs and benefits for the alternative development scenarios. Prepare a qualitative cost–benefit analysis that assesses and compares the economics of each scenario.
4. Identify and discuss the ethical factors to be considered and the potential barriers to their implementation.
5. Outline an environmental assessment (EA) to evaluate the key alternatives in view of all the factors. This assessment should include the following:
   a. List the key environmental, economic, and social issues that the assessment should evaluate.
   b. Describe how the assessment will fairly consider the needs and interests of all potential stakeholders.
   c. Formulate a plausible executive summary of the key findings of the assessment, as if you had actually conducted it, on a table that compares the various environmental, economic, and social effects of the three alternatives.
   d. Recommend a set of mitigation measures to address potential adverse environmental, economic, and social effects of the development policy.
   e. Identify adverse effects that cannot be mitigated.

6. In light of your EA, recommend in outline and/or narrative form the elements of an environmental management system to guide the island's future development (if any) in dealing with issues like monitoring, controls, mitigation measures, and conflict resolution.
7. Outline the elements of a proposed policy to guide the island's economic, social, and environmental development in a sustainable manner. Identify the policy documents that you would recommend, such as treaties, agreements, statement of guiding principles, and regulatory and assurance mechanisms (particularly with respect to the indigenous islanders).
8. Describe in the report how you will ensure that views of all stakeholders are properly considered in the final decision.

Capstone problem 2: Global policy problem

Section four of this book explores the following global problems and policy issues:

Chapter ten: Coming water wars
Chapter eleven: Peak oil, alternatives, and energy policy: The looming world oil crisis
Chapter twelve: Peak food or peak everything: An era of bounty or begging?
Chapter thirteen: Global climate change
Chapter fourteen: Global population paradox: Population explosion or implosion?

The global community is to varying degrees involved in these and other global-scale issues, such as protecting the ozone layer and the oceanic commons, and preventing the extinction of endangered species. Assume that you are working for an international environmental think tank, and that the United Nations Environmental Program (UNEP) has asked you to outline a program that would systematically identify ways in which decisions on behalf of any of these issues can have impacts that should be considered in terms of formulating a global policy that frames decision making. You have been specifically asked to do the following:

1. Organize five teams, one for each of the policy areas (described in Chapters ten–fourteen).
2. Using the principles and concepts described throughout this book, each team is to formulate a comprehensive policy recommendation for addressing their particular issue.

3. Each team should
   a. List ways in which decisions (or nondecisions) in the other policy areas are likely to affect your policy area.
   b. List ways in which decisions (or nondecisions) in your policy area are likely to affect the other policy areas.
   (Note: Nondecision means allowing the issue to evolve without any policy, program, or project intervention by the global community.)
4. Have each team consult with each of the other teams on decision issues of mutual interest, and agree on a proposed crosscutting policy agenda. Distribute copies of the respective agendas to all team members.
5. Convene a conference of all teams to present their results and vote on each of the proposed policy recommendations. Document the results and distribute them to all team members.
6. Have each team member write a personal assessment of the process:
   a. Is the final result useful for policy formulation and planning?
   b. What worked and what did not work in the process that you were a part of? Explain the underlying factors responsible for any proposed policy that did not work.
   c. Each group is to write a summary of the lessons learned about global environmental policy and the process of formulating it.

Notes

1. A.-L. Barabási, *Linked: How Everything is Connected to Everything Else and What It Means for Business, Science, and Everyday Life* (New York: Plume, 2003).
2. M. Blastland, More or Less: Connecting with People in Six Steps, *BBC News*, July 13, 2006.
3. J. Jure Leskovec and E. Horvitz, "Planetary-Scale Views on an Instant-Messaging Network," http://arxiv.org/abs/0803.0939v1 (accessed June 10, 2007).

Index

Milton Keynes UK
Ingram Content Group UK Ltd.
UKHW031138141024
449569UK00024B/1235